ADVANCES IN
MOLECULAR AND
CELL BIOLOGY

Volume 29 • 1999

THE BIOLOGY OF THE PANCREATIC β-CELL

ADVANCES IN MOLECULAR AND CELL BIOLOGY

THE BIOLOGY OF THE PANCREATIC β-CELL

Series Editor: E. EDWARD BITTAR
Department of Physiology
University of Wisconsin—Madison
Madison, Wisconsin

Guest Editor: SIMON L. HOWELL
School of Biomedical Sciences
King's College London
London, England

VOLUME 29 • 1999

JAI PRESS INC.
Stamford, Connecticut

100 Prospect Street
Stamford, Connecticut 06901

ISBN: 0-7623-0547-9

Transferred to digital printing 2006

CONTENTS

SECTION III. REGULATION OF INSULIN SECRETION

SECTION IV. CONCLUSION

LIST OF CONTRIBUTORS

Lydia Aguilar-Bryan — Division of Endocrinology
Baylor College of Medicine
Houston, Texas

Bo Ahrén — Department of Medicine
Malmö University Hospital
Malmö, Sweden

Michael R. Christie — Department of Medicine
King's College School of Medicine and Dentistry
London, England

John W. M. Creemers — Laboratory of Molecular Oncology
University of Leuven
Leuven, Belgium

Philippe Detimary — Department de Physiologie et de Pharmacologie
Unité d'Endocrinologie et Métabolisme
Université Catholique de Louvain
Brussels, Belgium

Kevin Docherty — Molecular and Cell Biology Department
Institute of Medical Sciences
University of Aberdeen
Foresterhill, Aberdeen, Scotland

Décio L. Eizirik — Diabetes Research Centre
Vrije Universiteit Brussel
Brussels, Belgium

Patrick Gilon — Department de Physiologie et de Pharmacologie
Unité d'Endocrinologie et Métabolisme
Université Catholique de Louvain
Brussels, Belgium

Tracey E. Harris — School of Biomedical Sciences
King's College London
London, England

Jean-Claude Henquin — Department de Physiologie et de Pharmacologie
Unité d'Endocrinologie et Métabolisme
Université Catholique de Louvain
Brussels, Belgium

Simon L. Howell — School of Biomedical Sciences
King's College London
London, England

Anne Hoorens — Diabetes Research Centre
Vrije Universiteit Brussel
Brussels, Belgium

John C. Hutton — Barbara Davis Center for Childhood Diabetes
University of Colorado Health Sciences Center
Chicago, Illinois

Robert S. Jackson — Department of Clinical Biochemistry
Addenbrooke's Hospital
Cambridge, England

Jean-Christophe Jonas — Department de Physiologie et de Pharmacologie
Unité d'Endocrinologie et Métabolisme
Université Catholique de Louvain
Brussels, Belgium

Peter M. Jones — School of Biomedical Sciences
King's College London
London, England

Stewart A. Metz — Diabetes Laboratories
Pacific Northwest Research Institute
Seattle, Washington

Shanta J. Persaud — School of Biomedical Sciences
King's College London
London, England

Arun S. Rajan
Division of Endocrinology
Baylor College of Medicine
Houston, Texas

Romano Regazzi
Institut de Biologie Cellulaire et de Morphologie
Université de Lausanne
Lausanne, Switzerland

Yoshihiko Sato
Department of Geriatrics
Shinshu University School of Medicine
Matsumoto, Japan

Jonathan M. W. Slack
Developmental Biology Programme
Department of Biology and Biochemistry
University of Bath
Bath, England

Emile Van Schaftingen
Lab de Chimie Physiologique
Faculte de Medecine
Université Catholique de Louvain
Brussels, Belgium

Frans Schuit
Laboratory of Biochemistry
Diabetes Research Center
Vrije Universiteit Brussel
Brussels, Belgium

Christina Wasmeier
Barbara Davis Center for Childhood Diabetes
University of Colorado Health Sciences Center
Denver, Colorado

PREFACE

As a result of the key advances made more than 30 years ago, specifically the ability to isolate islets of Langerhans from the pancreas, the ability to measure insulin accurately by immunoassay, and the development of microchemical techniques for studying cells and their components, many research volumes, symposium reports, and original papers have been produced. This explosion of interest has probably had at least three stimuli:

1. the inherent scientific interest in understanding secretion of the pancreatic β-cell
2. the β-cells relevance to a very common disease
3. the availability of funding from specific sources related to diabetes research, for instance, Juvenile Diabetes Foundation International and the British Diabetic Association

As a result of all this activity, detailed scientific literature including research reviews are readily available.

Surprisingly enough, there are relatively few attempts to summarize this great bulk of knowledge in a way that is accessible to the newcomer to this field and this book is intended to bridge this gap. We hope it will be useful for postgraduate students setting out on their PhD studies, for established scientific investigators entering this field for the first time, and for advanced undergraduate and postgraduate students seeking to undertake special study of this subject. Each contributor has written an objective overview of his or her particular subject in a way that should allow the reader within a short period to obtain a comprehensive and up-to-date picture of the present state of the art. Where appropriate, the information about human β cells and changes that are known to occur in diabetes are highlighted. Each chapter has references pointing the way to more detailed information.

The contributors and editor hope very much that this book will be useful in providing background information to trigger the imagination of scientists of all ages and experience entering this exciting field for the first time.

Simon L. Howell
Guest Editor

SECTION I

β-CELL DEVELOPMENT, REPLICATION, AND DEATH

DEVELOPMENTAL BIOLOGY OF THE β-CELL

Jonathan M.W. Slack

I. INTRODUCTION

Many misconceptions abound on the subject of the origin of the pancreatic β-cells. This is perhaps surprising because the understanding of β-cell biology is potentially an important practical issue in terms of its significance for novel therapies for diabetes. One commonly encountered error is the idea that pancreatic endocrine cells have their origin in the neural crest of the early embryo, now known to be incorrect. A more subtle misconception is found in most textbooks of pathology or histology insofar as they devote separate chapters to the exocrine and the endocrine pancreas. Because they secrete enzymes into the duodenum, the exo-

Advances in Molecular and Cell Biology
Volume 29, pages 3-19.

ISBN: 0-7623-0547-9

crine cells are seen as an integral part of the digestive system. In contrast, the endocrine cells are seen as belonging in the domain of endocrinology and the control of metabolism. While this is understandable in terms of the different medical specialists who deal with these two respective areas of disease, it is not helpful to an understanding of the biology of the organ, since the evidence is now very good that both cell types arise from a common source in the endodermal epithelium. Another idea that is not necessarily incorrect, but rather takes too much for granted, is that of "precursor cells" for the mature β-cells of the pancreas. It is not self-evident that a precursor cell population should necessarily exist. For example, if the final cell division of a pancreatic epithelial cell gave rise to both endocrine and exocrine daughters, then it would not be correct to refer to endocrine precursor cells. If the final division of an endocrine cell gave rise both to a β-cell and to a cell of another endocrine type, then it might be correct to speak of endocrine precursors, but not of β-cell precursors. If β-cell precursors do exist, then we would need to know about where they come from, when they are produced, and what properties they have.

II. EMBRYOLOGY

Pancreatic development has been studied mainly in rodents (Pictet and Rutter, 1972; Slack, 1995). Although much of the earlier work was done on the rat, it is likely that future experimental work will concentrate on the mouse, which is now the most popular mammal for studies in developmental biology. The reasons for this are twofold: first, more genes have been cloned from the mouse than from the rat; second, the genetic technology in terms of making transgenic and knockout strains is well established for the mouse but not for the rat. Although physiologists are desperate to have transgenic rats at their disposal, it seems likely that they will eventually have to reduce the size of their instruments and settle for working on mice. Thus, the timings given here relate to the mouse, although the overall course of pancreatic development is similar for all vertebrates. The main difference between human and mouse is that the pancreas in the adult is compact in the human but diffuse in the mouse, meaning that there are many small nodules of pancreatic tissue scattered across the mesentery in the vicinity of the main pancreatic mass. In this chapter, the terms "anterior" and "posterior" will be used in the zoological sense to denote the head and tail ends of the body, or the gut, rather than in the human anatomical sense in which they are synonymous with ventral and dorsal respectively.

The pancreas originates from a region of the gut endoderm very close to that giving rise to the liver and gall bladder. This tissue is normally considered to lie in the foregut, although it occasionally is described as being in the midgut. The distinction is that the foregut is that part of the gut tube lying anterior (i.e., cranial) to the anterior intestinal portal (Figure 1). Since the opening in the gut tube is ini-

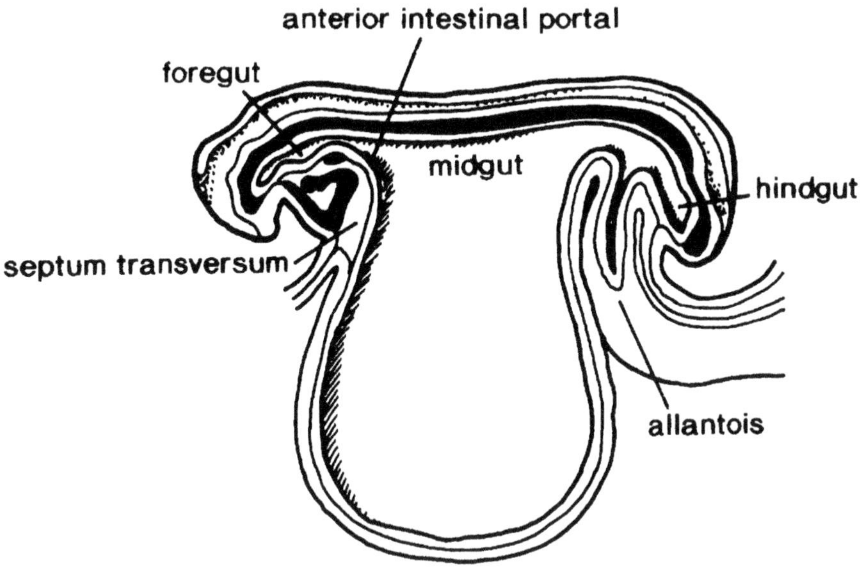

Figure 1. General view of a mammalian embryo at an early stage of gut closure, showing the location of the foregut.

tially enormous and later constricts down to a small orifice at the umbilicus, it is by no means obvious what comes from the foregut and what comes from the midgut. Indeed, it is likely that much of the gut becomes passively transferred from midgut to foregut as the ventral closure of the umbilical opening proceeds. A satisfactory fate map showing which areas in the early endodermal sheet become the pancreatic buds has yet to be produced, and such a fate map is obviously necessary in order to find which genes are active in the pancreatic anlagen prior to outgrowth of the buds.

The first appearance of the pancreas is as two small buds, one on the dorsal and one on the ventral side of the duodenum (Figure 2). The dorsal bud is visible from 10.5 days and the ventral bud from 12.5 days of gestation in the mouse. The two buds are initially separate but quite soon fuse into a single organ. The duct of the ventral pancreas normally becomes the main pancreatic duct, and in the human, the ventral pancreas remains visible postnatally as the uncinate process. Although they look solid, the early pancreatic buds are actually highly convoluted infoldings of the gut epithelium. They undergo repeated branching during growth, with the stems of the branches becoming the ductal system and the acini developing at the tips. The first endocrine cells become visible at 9.5 days by immunostaining with antibodies to glucagon and insulin C-peptide. Over the period from 10.5 to 14.5 days, many of the cells present in the pancreatic buds stain simultaneously with

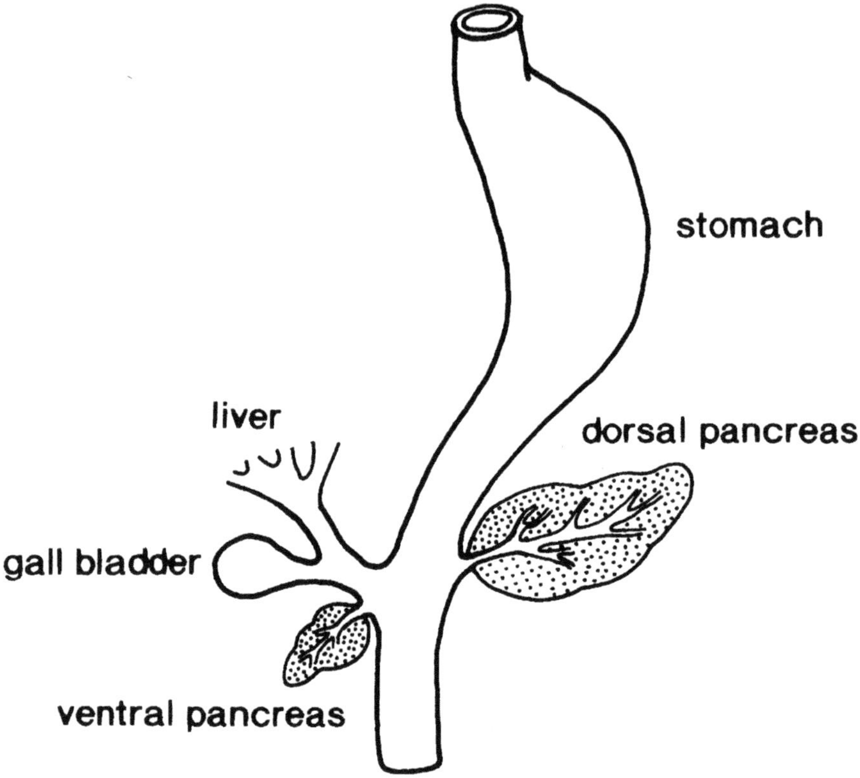

Figure 2. Dorsal and ventral pancreatic buds in the embryo.

antibodies to more than one hormone (e.g., both insulin and glucagon). In the postnatal pancreas the rule is "one cell, one hormone," so there is certainly some difference between these early multihormone-expressing cells and the later mature endocrine cells. It has often been presumed that the multihormone cells are precursors to the mature endocrine cells, but this has not been proved.

There has also been some controversy about the expression of the "PP-fold" peptides in the developing pancreas. There are three peptides—pancreatic polypeptide (PP), neuropeptide Y (NPY) and peptide YY (PYY)—that are very close in primary sequence. Antibodies made against one peptide frequently cross-react with the others. PP is one of the products of a subset of the mature endocrine cells, and may have a physiological function antagonizing insulin secretion. One report suggested that PP was expressed in the early multihormone cell population (Herrera et al., 1991), another that this was actually NPY (Teitelman et al., 1993), and a third that it was actually PYY (Upchurch et al., 1994). The latter view is probably correct, and it seems that most or all of the early endocrine cells express PYY regardless of which of the other hormones they are expressing. Whether PYY-expressing cells are actually the precursors of the mature endocrine cells is, however, not known. Other substances that are expressed in early endocrine cells are the enzyme tyrosine hydroxylase (Teitelman, 1990), and the signalling molecule activin (Yasuda et al., 1993), which has been implicated in a wide variety of developmental processes from mesoderm induction to branching behavior and erythropoietic differentiation.

The pancreas grows very rapidly during the midgestation period. In the mouse, where there are always nodules of pancreas away from the main body of the organ, it is not known whether these arise by migration of cells from the main organ, or by independent small foci of pancreatic development from the gut epithelium. The latter possibility is mentioned because it has been shown to occur quite frequently in humans with the formation of small heterotopic pancreata buried under the muscular layers of the gut (Branch and Gross, 1935). The late embryo pancreas resembles a tree with branches progressively arising from the main trunk, which consists of the two original ducts of the dorsal and ventral buds. The endocrine cells form islets during the last few days of gestation. It is not really known to what extent this occurs by aggregation of preexisting cells or by monoclonal growth from individual precursors. An aggregation chimera study suggested that at least some islets are formed by aggregation of cells (Deltour et al., 1991). In this method, a mouse embryo is formed by fusion of two embryos from different strains whose cells can be distinguished by some biochemical or histological criterion. The cells become well mixed during early development so that all the later organ rudiments contain a mixture of cells of each genotype. If a structure is formed by monoclonal growth at a late stage this means that it will be composed of just one genotype, whereas if it arises from a group of cells it is likely to remain of mixed composition. The integrity of the islet seems to involve cell adhesion. Transgenic mice in which a dominant negative cadherin gene is driven by the insulin promoter show a failure of β-cells to aggregate, although α-cells are unaffected (Dahl et al., 1996).

III. PANCREATIC DETERMINATION

Our understanding of the molecular basis of early animal development is now quite advanced (Slack, 1991). We know that there is a hierarchy of inductive signals, each of which turns on particular transcription factors at different concentrations (Figure 3). The combinations of transcription factors activated in each position determine which body part or cell type will develop there. In classical embryology, the concept of "determination" was formulated to refer to tissues that were irreversibly committed to form a particular structure or cell type but did not at the time show any indication of terminal differentiation. We would now regard

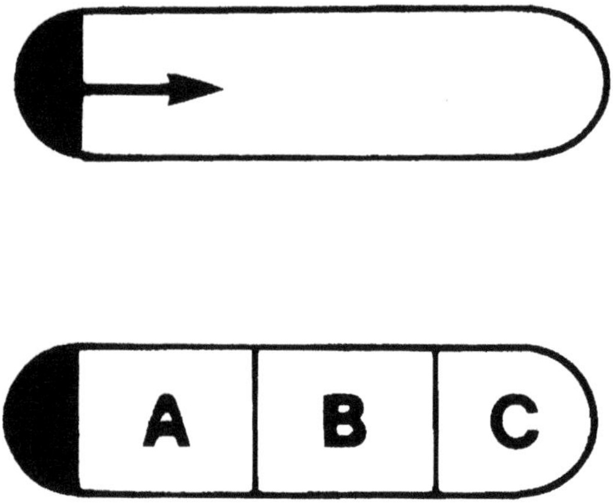

Figure 3. Determination of an embryonic organ rudiment. An inductive signal is emitted from the region on the left and the cells of the rudiment activate several transcription factors, each in response to a different threshold concentrations of the signal. A, B, and C represent regions with different combinations of active transcription factors, which are thereby committed to form different cell types or structures.

the determined state as consisting of the expression of the relevant set of transcription factors.

It is not known exactly when the pancreatic anlage is specified relative to the surrounding parts of the endoderm. A recent fate map shows that the dorsal pancreas arises from midline endoderm at the level of somite 4-7 (Matsushita, 1996), although this region does not display a unique combination of transcription factors at the stage of mapping. It is now known that the proximity of the notochord is required for development of the dorsal pancreas (Kim et al., 1997). The endoderm closes in such a way that its lateral edges meet in the ventral midline, so considerations of symmetry make it likely that there are two further prospective regions at the lateral edges of the embryonic endoderm that come together to form the ventral bud.

After closure of the gut, some transcription factors are expressed in a nested manner, meaning that all of the set are on at the posterior end and different individual genes have different anterior limits of expression. Among these are the HNF3 family of forkhead-related genes, which are expressed in the gut epithelium. HNF3-α and -β have a wide expression, and HNF3-γ has an anterior limit between the stomach and the pancreatic buds (Monaghan et al., 1993). Therefore, this gene, along with others, may be involved in the determination of the pancreas. In the central nervous system and axial mesoderm, it is now well established that the Hox gene cluster specifies anterior–posterior positional value (McGinnis and Krumlauf, 1992; Slack et al., 1993). These genes are not expressed in the endodermal epithelium, but they are expressed in the mesenchyme of the early developing gut (Roberts et al., 1995; Yokouchi et al., 1995). For many years it has been suspected that there are inductive interactions between the mesenchyme and epithelium such that the mesenchyme dictates the positional character of the epithelium (Wessells, 1977). So it is possible that the Hox gene state is transmitted by means of extracellular factors that induce regionally specific non-Hox factors in the epithelium (Roberts et al., 1998).

Certain genes are known to be expressed specifically in the early pancreatic buds. The most attention has been devoted to that which is now called PDX-1, although it has a number of other names including IPF-1, IDX-1, STF-1, IUF-1, XlHbox8 (Wright et al., 1988; Ohlsson et al., 1993; Miller et al., 1994; Offield et al., 1996). It codes for a homeobox containing transcription factor homologous to a family of genes expressed in the leech gut (Lox 3A,B,C) (Wysocka-Diller et al., 1995). As invertebrates have no recognizable pancreas, this suggests that this gene has been associated with gut development for a long evolutionary period and specifically with the pancreas from the time of the origin of vertebrates. PDX-1 is expressed in the pancreatic buds and also in the surrounding duodenum from 8.5 days, and is maximal during the early bud stage of pancreatic development (Guz et al., 1995). At this stage it is not expressed in the early insulin cells, but in later development expression is lost in the surrounding tissues and becomes confined to the β-cells. In the mature β-cell, PDX-1 is among the specific transcription factors regulating insulin expression.

In recent years it has become possible to create mutations of genes to order in the mouse. This is done by introducing a defective version of the gene in place of the normal gene in embryonic stem cells grown in culture. The modified cells are injected into mouse embryos in order to yield chimeric embryos. Germ line chimeras are then selected and bred to obtain homozygous animals lacking the gene in question. The procedure is known as "knocking out" the gene, and the homozygous mutant mice are referred to as "knockouts."

If the PDX-1 gene is knocked out by targeted mutagenesis, then the formation of the pancreas is completely suppressed, although there is some early development of the buds (Jonsson et al., 1994). Interestingly, there are no early insulin cells even though the gene is not normally expressed in these cells (Offield et al., 1996). Experiments in which embryonic epithelium from one mouse line and mesenchyme from another are recombined and grown in vitro show that the defect lies within the epithelium and not the mesenchyme (Ahlgren et al., 1996). Nothing is known of the prior regulation of PDX-1 in mammals. In *Xenopus*, where it has an exactly similar expression pattern, it is much easier to do experiments with early stage embryos. It has been shown that the expression of PDX-1 in endodermal explants is inhibited by fibroblast growth factor (FGF), or by a dominant negative activin receptor (Gamer and Wright, 1995). It is known that FGF is needed for mesoderm formation and so its effect may represent a switch from endoderm to mesoderm. The dominant negative activin receptor is not as specific as its name suggests and is known to inhibit a wide range of transforming growth factor-β (TGF-β) superfamily receptor types. Thus, it shows a requirement for some sort of TGF-β signalling at some stage between the blastula and the regionalization of the gut in the tailbud stage.

Another gene involved in early pancreatic determination is *islet-1*, coding for a lim-homeodomain transcription factor (Ahlgren et al., 1997). This is expressed in all islet cells and in the mesenchyme around the dorsal pancreatic bud. The *islet-1* knockout mice arrest at 9.5 days of development. Explants from the dorsal bud region will not develop in culture unless they are supplied with mesenchyme from a normal embryo, in which case the dorsal bud will form exocrine but not endocrine tissue. This suggests a dual requirement for *islet-1*: firstly, in the mesenchyme to produce a signal required for pancreatic epithelial development, perhaps the mysterious "mesenchymal factor" (see below); and secondly, in the epithelium to enable islet cell differentiation.

Two further genes of importance code for members of the *pax* family of transcription factors. Both *pax-4* and *pax-6* are expressed in scattered cells of the early buds, but the effects of the knockouts are rather different. The *pax-4* knockout lacks β- and δ-cells but has more α-cells than usual (Sosa-Pineda et al., 1997). The *pax-6* knockout has no α-cells (St Onge et al., 1997). The double knockouts, formed by mating heterozygotes of both strains, have no endocrine cells in the pancreas at all. This shows that *pax-4* is needed for β- and δ-cell formation and *pax-6* is needed for α-cell formation.

The pancreatic buds have quite a close anatomical relationship to the liver. It is likely that there is also a closeness in the determined states of liver and pancreas in terms of their respective ensembles of active transcription factors. We know this because it is possible to induce foci of hepatic cells in the pancreas following various insults (Scarpelli and Rao, 1981; Rao et al., 1989; Rao et al., 1990). These insults include treatment with the protein synthesis inhibitor ethionine or removal of copper from the diet for a prolonged period, both treatments causing severe damage to the exocrine tissue. The foci are probably clones and probably represent mutations of one or a few transcription factors occurring during regeneration. The fact that liver is the prodominent phenotype of the transdifferentiated foci suggests that fewer mutational changes are required to convert pancreas to liver than to bring about any other transformation.

IV. CELL LINEAGE STUDIES

It has been supposed at various times that the endocrine cells of the pancreas have a different embryological origin from the exocrine cells and are derived from the neural crest. This idea can be traced to the influence of the "APUD" theory of Pearse, which identified a number of common properties of the neuroendocrine cells found in different tissues and organs (Pearse, 1984). In its original version, this theory predicted that all such cells would be found to have a neural crest origin. Although this turned out to be correct in a few cases, such as the C cells of the thyroid, it has not been supported in others. In the case of the endocrine cells of the gut, the theory was tested experimentally by Andrew using quail to chick embryo grafts (Andrew, 1976). The principle of the method is that the quail has cells with a large mass of DNA associated with the nucleolus. This is not found in the chick and can easily be detected by Feulgen staining (Le Douarin, 1982). If a region of neural tube is grafted from a quail to a chick embryo at an early stage, then the neural crest cells will migrate out in the usual way and colonize more or less their usual destinations. If the specimens are fixed and Feulgen stained at some later time, then it can be determined which structure are of neural crest origin by finding whether they are composed of chick or quail cells. Andrew's results showed clearly that the enteric ganglia are of neural crest origin but that the endocrine cells, both of the gut and of the pancreas, are not.

This experiment does not of course prove that the endocrine cells do come from the endodermal epithelium, although descriptive studies make this the most likely origin. It is easy to obtain differentiation of endocrine cells from pancreatic buds cultured in vitro. This shows that they must arise from some cell type present in the endodermal bud or in the surrounding mesenchyme (Wessells and Cohen, 1967); see Figure 4. The "mesenchyme" is a complex tissue that contains precursors for smooth muscle, fibroblasts, peritoneum, and various cells of the immune system. In recent experiments, my colleagues and I have recombined epithelium from lacZ-expressing mouse embryos with mesenchyme from unlabelled embryos and have shown that the β-cells arising from the cultured recombinants are all lacZ expressing (Slack and Percival, 1999). This is good evidence for an endodermal origin of the β-cells.

Although an endodermal origin seems secure, it is less easy to know whether the endocrine cells arise from all over the pancreatic rudiment, or from some particular part of it. This latter hypothesis is not excluded by current data. In this connection it may be noted that some fish have a separate endocrine organ (Brockmann body [Falkmer, 1985]), and it seems more likely that this organ would arise from a separate rudiment than from a segregation of cells at a late stage of development.

A popular view at present is that all the endocrine cells of the pancreas are formed from the early multihormone-expressing population. This view is depicted in Figure 5, in terms of the expression of PYY, insulin, and glucagon in the various cells. This perspective does represent the sequence of cells seen in the developing pancreas, but as a cell lineage it is a hypothesis rather than a fact. One argument in its favor comes from a cell line called MSL, which was isolated from a transplantable insulinoma (Madsen et al., 1996). In vitro this differentiates in a somewhat heterogeneous way to form a wide variety of endocrine cells secreting various gut or pancreatic hormones. In vivo, it can become either an insulinoma or

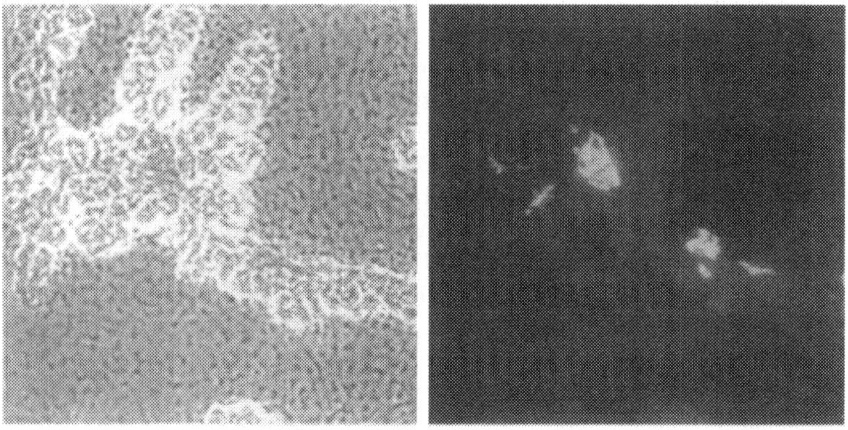

A B

Figure 4. Development of a mouse embryo pancreatic bud in vitro for 6 days. **A.** Phase contrast. **B.** Immunofluorescence for insulin.

a glucagonoma. The parent cell expresses PDX-1, and this is maintained in the insulinomas but switched off in the glucagonomas. So the MSL cell does resemble the embryonic multihormone cells. But its existence does not prove any particular pathway of endocrine cell formation in normal development, as it derives from an X-ray-induced tumor and has been passaged for many years, and therefore has doubtless collected a number of mutations and abnormalities.

On the other hand, there are various lines of evidence against the idea that all endocrine cells are formed in one burst, and cannot be generated at late embryonic or postnatal stages. Studies of the cell kinetics of the endocrine population suggest that there is some recruitment from nonendocrine cells before and after birth (Kaung, 1994). Study of the expression of the glucose transporter type 2 (GLUT-2) protein, restricted to β-cells in the adult, show that, like PDX-1, it is expressed more widely in embryogenesis, and it has been suggested that GLUT-2 positive cells could fuel a second round of endocrine cell formation (Pang et al., 1994). Another pancreatic cell line, the AR42J, are normally amylase-secreting, but treatment with activin A and betacellulin causes some cells to become neuronlike and secrete insulin or pancreatic polypeptide (Mashima et al., 1996). If this cell line corresponds in its behavior with cells in vivo, then it would argue for long-term ability to generate endocrine cells from nonendocrine precursors. Furthermore, there are many descriptive histological studies that indicate the existence of new islet formation in postnatal pancreas. It has been argued repeatedly that the ductal tissue represents, or contains, a stem cell population (Githens, 1988), and remarkably, incubation of postnatal ductal tissue with embryonic mesenchyme seems able to provoke islet formation (Dudek et al., 1991).

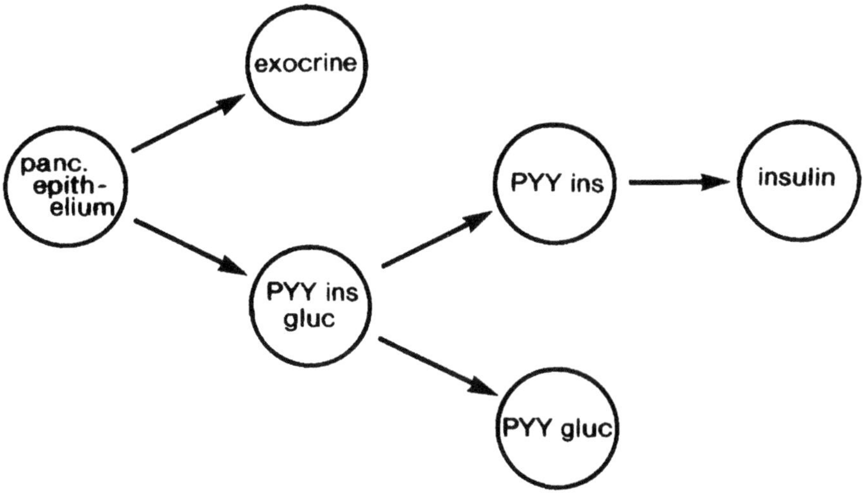

Figure 5. A possible cell lineage for the formation of the pancreatic endocrine cells.

V. GROWTH CONTROL

Despite a great deal of work on the cell cycle and on the mechanism of action of growth factors on tissue culture cells, there is remarkably little understanding of how growth is controlled in a whole animal. Exponential growth is never encountered in vivo, although it may be approximated by the very rapid growth of mammalian embryos around midgestation. Adult tissues are either postmitotic, such as neurons and myotubes; show rapid cellular turnover without growth, like the gut epithelium or the blood; or show very limited cell turnover but with some capacity to respond to external stimuli, like the liver (Leblond, 1964). The pancreas belongs in this last group. Normally there is little mitosis, but it is possible to provoke a remarkable growth of the exocrine cells by treatment with cholecystokinin (Elsasser et al., 1994). This hormone is, of course, also a normal physiological stimulus for secretion.

During the early development of the pancreas, the presence of mesenchyme is required to support the growth and differentiation of the bud. This was investigated at some length in the 1960s and it was shown that at the earliest stage, there was an absolute requirement for pancreatic mesenchyme; that later, other mesenchymes, such as that form the salivary gland would do; and that later still, growth could be supported by a chick embryo extract (Wessells and Cohen, 1967). The significance of this finding is a little unclear since at the earlier stages there is no mesenchyme abutting the dorsal pancreatic rudiment in vivo. This question has been reexamined more recently. The effect of various culture systems was investigated by Gittes et al. (1996). They showed that some development could be obtained in the absence of the mesenchyme if the epithelia were cultured in Matrigel, which is a complex mixture of extracellular matrix components, or if the grafts were grown in vivo underneath the kidney capsule of nude mice. These are both conditions in which a number of poorly characterised factors are supplied. Even so, the differentiation favors ductal and endocrine tissue at the expense of exocrine. Sanvito et al. (1994) treated cultures with a number of growth factors. They reported no effect of fibroblast growth factor (bFGF), hepatocyte growth factor (HGF), insulin-like growth factor II (IGF-II), PDGF, or nerve growth factor (NGF). Epithelial growth factor (EGF) promoted ductal development and TGF-β promoted endocrine development. Miralles and colleagues (1998) showed that follistatin, an antagonist of activin, is made by the mesenchyme and can inhibit endocrine development by the epithelium. These experiments have not so far solved the old problem of the biochemical nature of the mesenchymal factor and remains unclear what effects growth factors and extracellular matrix components have on the growth and differentiation of the embryonic pancreas *in vivo*.

The formation of β-cells postnatally can occur either by growth of preexisting β-cells or by neogenesis. There is an increase in the size of the entire pancreas during growth of the whole animal, and the mass of β-cells rises considerably for several months (Finegood et al., 1995). In the perinatal period, this seems to involve some neogenesis (Kaung, 1994), and subsequently it is probably maintained by intrinsic growth. Finegood et al. (1995) have produced a model from the available cell kinetic data supporting the idea of continuous slow turnover and suggesting an average turnover time of about 20 days for β-cells in the adult rat. An important control on β-cell mass is that operating during pregnancy when there is a considerable increase. This seems to be due to intrinsic growth under the control of placental lactogen (Sorenson and Brelje, 1997).

It remains unclear whether there is any normal role for neogenesis following the postnatal period, but it does seem to occur in various pathological processes, particularly those giving rise to chronic trauma or inflammation of the pancreas. This is found, for example, in hamster pancreas wrapped with cellophane (Rosenberg et al., 1983), or in transgenic mice expressing γ-interferon from an insulin promoter (Gu and Sarvetnick, 1993; Gu et al., 1994). There is also the rare human condition of nesidioblastosis, in which new islets appear to be formed continuously (Goossens et al., 1989) . All of these examples suggest that it is still possible to form both endocrine and exocrine cells from a common precursor cell type during adult life.

The study of regeneration of internal organs has been somewhat conditioned by other examples such as the amphibian limb, where it is possible to remove a large part of the organ and see it regrow over a few weeks. Only the liver and kidney regenerate fast enough to provide convenient experimental models. But for other organs it is of interest to know whether they can regenerate at all. There is a methodological problem here. If regeneration is not very rapid, it is necessary to factor

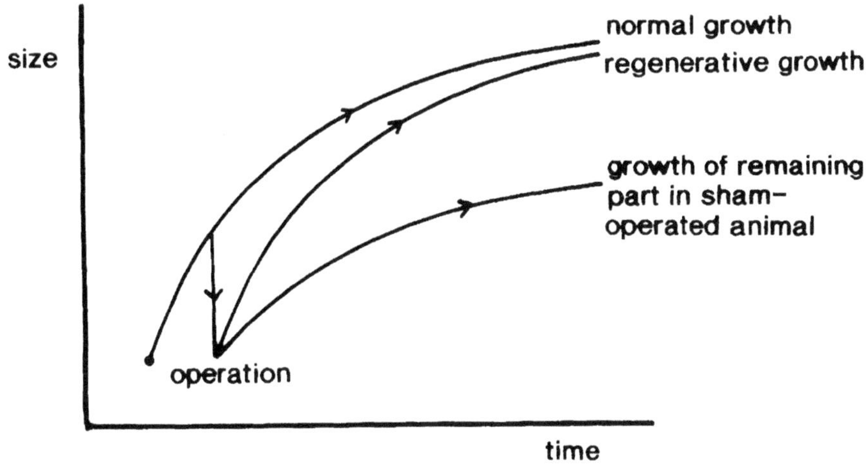

Figure 6. Definition of tissue regeneration in a situation in which the whole animal is growing.

out the normal growth that occurs as the animal gets bigger. For example, with the pancreas it is possible to remove about 90% from a rat and the remaining 10% is sufficient to maintain digestion and reasonable blood glucose control. It is then necessary to estimate the size of the residual 10% in unoperated and in sham-operated animals and to compare the growth of this region to the growth in the experimental cases (Figure 6). The results suggest that the small piece does grow somewhat faster than usual, although it never produces a pancreas as big as normal. Descriptions of regeneration suggest that there is sprouting of new acini and islets from the ducts at the cut surface (Sidorova, 1978). So there is some regenerative capacity in the postnatal pancreas, although there is presumably no mechanism for sensing the total size of the organ, as must exist for example in the liver.

The idea that there is a fixed stock of β-cells in the pancreas and that they cannot recover following ablation is based on two well known facts. First, there is the continued lack of function in severe cases of insulin-dependent (type 1) diabetes mellitus. Second, there is the failure of recovery following experimental induction of diabetes in animals by treatment with streptozotocin. However, these effects do not mean that there is no normal turnover of β-cells. Type 1 diabetes arises from an autoimmune attack on the β-cells and may be initiated by viral infection, or by an abnormal expression of class II major histocompatibility genes. The immune attack responsible for the original destruction of the β-cells will presumably continue to destroy any new ones as well, and if they are formed slowly the balance of destruction over regeneration will ensure the presence at any one time of few or no viable β-cells. As far as streptozotocin treatment is concerned, there is some regeneration following lower doses. Although the drug is often supposed to be specific for β-cells, there are side effects and it cannot be assumed that other cells remain unaffected. If there were an inhibition of a normal slow cell production by pancreatic stem cells, then this could explain the nonrecovery of the β-cells, but any other consequences for pancreatic morphology or function would not be apparent over the duration of an experiment.

VI. CONCLUSIONS

As with many organs of the body, there is still a great deal that we do not know about the development of the pancreas. However, the available evidence points clearly toward an origin for all cell types of the pancreas—exocrine, endocrine, and ducts—from a common rudiment present in the endoderm of the early embryo. Although the earliest endocrine cells differ from the later ones in being multihormone producing, it seems likely that endocrine cells can be produced continuously during development, at least up to the postnatal period. Beyond this, it is likely that the turnover of β-cells is normally due to a limited amount of division balanced by cell death. But there are in addition various pathological circumstances in which β-cells seem to arise from nonendocrine precursors during adult life.

It is certainly important that those interested in diabetes should not ignore the biology of the pancreas. Type 1 diabetes may be an autoimmune disease, but some attention should be given to the target cell as well as to the immune cells. Type 2 (noninsulin-dependent) diabetes mellitus of course involves insulin resistance in peripheral tissues, but there are pancreatic adaptations to insulin resistance and in many cases the failure of adaptation may be just as important to the development of disease as the insulin resistance itself. The lesson for those interested in potential novel therapies for diabetes is that a proper understanding of the controls on pancreatic growth and differentiation are likely to be the only route to the eventual control of β-cell production and regeneration.

REFERENCES

Ahlgren, U., Jonsson, J. and Edlund, H. (1996). The morphogenesis of the pancreatic mesenchyme is uncoupled from that of the pancreatic epithelium in IPF1/PDX1 deficient mice. Development 122, 1409–1416.

Ahlgren, U., Pfaff, S.L., Jessell, T.M., Edlund, T. and Edlund, H. (1997). Independent requirement for isl-1 in formation of pancreatic mesenchyme and islet cells. Nature 385, 257–260.

Andrew, A. (1976). An experimental investigation into the possible neural crest origin of pancreatic APUD (islet) cells. J. Embryol. Exp. Morph. 35, 577–593.

Branch, C.D. and Gross, R.E. (1935). Aberrant pancreatic tissue in the gastrointestinal tract. Arch. Surg. 31, 200–224.

Dahl, U., Sjodin, A. and Semb, H. (1996). Cadherins regulate aggregation of pancreatic beta cells in vivo. Development 122, 2895–2902.

Deltour, L., Leduque, P., Paldi, A., Ripoche, M.A., Dubois, P. and Hami, J. (1991). Polyclonal origin of pancreatic islets in aggregation mouse chimaeras. Development 112, 1115–1121.

Dudek, R.W., Lawrence, I.F., Hill, R.S. and Johnson, R.C. (1991). Induction of islet cytodifferentiation by fetal mesenchyme in adult pancreatic ductal epithelium. Diabetes 40, 1041–1048.

Elsasser, H.P., Biederbick, A. and Kern, H.F. (1994). Growth of rat pancreatic acinar cells quantitated with a monoclonal antibody against the proliferating cell nuclear antigen. Cell Tissue Res. 276, 603–609.

Falkmer, S. (1985). Comparative morphology of pancreatic islets in animals. In: *The Diabetic Pancreas.* (Volk, B.W. and Arquilla, E.R., Eds.), pp. 17–52. Plenum, New York.

Finegood, D.T., Scaglia, L. and Bonner-Weir, S. (1995). Dynamics of beta cell mass in the growing rat pancreas. Estimation with a simple mathematical model. Diabetes 44, 249–256.

Gamer, L.W. and Wright, C.V.E. (1995). Autonomous endodermal determination in Xenopus: Regulation of the expression of the pancreatic gene XlHbox8. Dev. Biol. 171, 240–251.

Githens, S. (1988). The pancreatic duct cell: Proliferative capabilities, specific characteristics, metaplasia, isolation and culture. J. Pediatr. Gastroenterol. Nutr. 7, 486–506.

Gittes, G.K., Galante, P.E., Hanahan, D., Rutter, W.J. and Debas, H.T. (1996). Lineage specific morphogenesis in the developing pancreas: Role of mesenchymal factors. Development 122, 439–447.

Goossens, A., Gepts, W., Saudubray, J.M., Bonnefont, J.P., Nihoulfekete, Heitz, P.U. and Kloppel, G. (1989). Diffuse and focal nesidioblastosis. A clinicopathological study of 24 patients with persistent neonatal hyperinsulinemic hypoglycemia. Am. J. Surg. Path. 13, 766–775.

Gu, D., Lee, M.S., Krahl, T. and Sarvetnick, N. (1994). Transitional cells in the regenerating pancreas. Development 120, 1873–1881.

Gu, D. and Sarvetnick, N. (1993). Epithelial cell proliferation and islet neogenesis in IFNγ transgenic mice. Development 118, 33–46.

Guz, Y., Montminy, M.R., Stein, R., Leonard, J., Gamer, L.W., Wright, C.V.E. and Teitelman, G. (1995). Expression of murine STF-1, a putative insulin gene transcription factor, in cells of pancreas, duodenal epithelium and pancreatic exocrine and endocrine progenitors during ontogeny. Development 121, 11–18.

Herrera, P.L., Huarte, J., Sanvito, F., Meda, P., Orci, L. and Vassalli, J.D. (1991). Embryogenesis of the murine endocrine pancreas; early expression of pancreatic polypeptide gene. Development 113, 1257–1265.

Jonsson, J., Carlsson, L., Edlund, T. and Edlund, H. (1994). Insulin promoter-factor 1 is required for pancreas development in mice. Nature 371, 606–609.

Kaung, H.L.C. (1994). Growth dynamics of pancreatic islet cell populations during fetal and neonatal development of the rat. Dev. Dyn. 200, 163–175.

Kim, S.L., Hebrok, M., and Melton, D.A. (1997). Notochord to endoderm signaling is required for pancreas development. Development 124, 4243–4252.

Le Douarin, N. (1982). *The Neural Crest.* Cambridge University Press., Cambridge, UK.

Leblond, C.P. (1964). Classification of cell populations on the basis of their proliferative behaviour. Natl. Cancer Inst. Monog. 14, 119–150.

Madsen, O.D., Jensen, J., Blume, N., Petersen, H.V., Lund, K., Karlsen, C., Andersen, F.G., Jensen, P.B., Larsson, L.I. and Serup, P. (1996). Pancreatic development and maturation of the islet B cell. Studies of pluripotent islet cultures. Eur. J. Biochem. 242, 435–445.

Mashima, H., Shibata, H., Mine, T. and Kojima, I. (1996). Formation of insulin-producing cells from pancreatic acinar AR42J cells by hepatocyte growth factor. Endocrinology 137, 3969–3976.

Matsushita, S. (1996). Fate mapping study of the endoderm of the 1.5 day old chick embryo. Roux's Arch. Dev. Biol. 205, 225–231.

McGinnis, W. and Krumlauf, R. (1992). Homeobox genes and axial patterning. Cell 68, 283–302.

Miller, C.P., McGehee, R.E. and Habener, J.F. (1994). IDX-1: A new homeodomain transcription factor expressed in rat pancreatic islets and duodenum that transactivates the somatostatin gene. EMBO J. 13, 1145–1156.

Miralles, F., Czernichow, P., and Scharfmann, R. (1998). Follistatin regulates the relative proportions of endocrine versus exocrine tissue during pancreatic development. Development 125, 1017–1024.

Monaghan, A.P., Kaestner, K.H., Grau, E. and Schutz, G. (1993). Postimplantation expression patterns indicate a role for the mouse forkhead/HNF-3 α, β, and γ genes in determination of the definitive endoderm, chordamesdoerm and enurectoderm. Development 119, 567–578.

Offield, M.F., Jetton, T.J., Laborsky, P.A., Ray, M., Stein, R.W., Magnuson, M.A., Hogan, B.L.M. and Wright, C.V.E. (1996). PDX1 is required for pancreatic outgrowth and differentiation of the rostral duodenum. Development 122, 983–995.

Ohlsson, H., Karlsson, K. and Edlund, T. (1993). IPF-1, a homeodomain-containing transactivor of the insulin gene. EMBO J. 12, 4251–4259.

Pang, K., Mukonoweshuro, C. and Wong, G.G. (1994). Beta cells arise from glucose transporter type 2 (Glut2) expressing epithelial cells of the developing rat pancreas. Proc. Natl. Acad. Sci. USA 91, 9559–9563.

Pearse, A.G.E. (1984). Islet development and the APUD concept. In: *Pancreatic Pathology*. (Klippel, G. and Heitz, P.U., Eds.), pp. 125–132. Churchill Livingstone, Edinburgh.

Percival, A.C. and Slack, J.M.W. (1999). Analysis of pancreatic development using a cell lineage label. Exp. Cell Res. 247, 123–132.

Pictet, R. and Rutter, W.J. (1972). Development of the embryonic endocrine pancreas. In: *Handbook of Physiology*, Sec. 7. (Steiner, D.F. and Frenkel, N., Eds.), vol. 1, pp. 25–66. Williams & Wilkins, Washington, DC.

Rao, M.S., Dwivedi, R.S., Yeldandi, A.V., Subbarao, V., Tan, X., Usman, M.I., Thangada, S., Nemali, M.R., Kumar, S., Scarpelli, D.G. and Reddy, J.K. (1989). Role of periductal and ductular epithelial cells of the adult rat pancreas in pancreatic hepatocyte lineage. Am. J. Pathol. 134, 1069–1086.

Rao, M.S., Heldandi, A.V. and Reddy, J.K. (1990). Stem cell potential of ductular and periductular cells in the adult rat pancreas. Cell Diff. Dev. 29, 155–163.

Roberts, D.J., Johnson, R.L., Burke, A.C., Nelson, C.E. and Morgan, B.A. (1995). Sonic hedgehog is an endodermal signal inducing BMP-4 and Hox genes during induction and regionalization of the chick hindgut. Development 121, 3163–3174.

Roberts, D.J., Smith, D.M., Goff, D.J., and Tabin, C.J. (1998). Epithelial–mesenchymal signaling during the regionalization of the chick gut. Development 125, 2791–2801.

Rosenberg, L., Brown, R.A. and Duguid, W.P. (1983). A new approach to the induction of duct epithelial hyperplasia and nesidioblastosis by cellophane wrapping of the hamster pancreas. J. Surg. Res. 35, 63–72.

Sanvito, F., Herrera, P.L., Huarte, J., Nichols, A., Montesano, R., Orci, L. and Vassalli, J.D. (1994). TGF-1 influences the relative development of the exocrine and endocrene pancreas in vitro. Development 120, 3451–3462.

Scarpelli, D.G. and Rao, M.S. (1981). Differentiation of regenerating pancreatic cells into hepatocyte like cells. Proc. Natl. Acad. Sci. USA 78, 2577–2581.

Sidorova, V.F. (1978). *The Postnatal Growth and Restoration of Internal Organs in Vertebrates*. PSG Publishing, Calcutta.

Slack, J.M.W. (1991). *From Egg to Embryo. Regional Specification in Early Development*. 2nd ed. Cambridge University Press, Cambridge.

Slack, J.M.W. (1995). Developmental biology of the pancreas. Development 121, 1569–1580.

Slack, J.M.W., Holland, P.W.H., and Graham, C.F. (1993). The zootype and the phylotypic stage. Nature 361, 490–492.

Sorenson, R. and Brelje, T.C. (1997). Adaptation of islets of Langerhans to pregnancy: Beta cell growth, enhanced insulin secretion and the role of lactogenic hormones. Horm. Metab. Res. 29, 301–307.

Sosa-Pineda, B., Chowdhury, K., Torres, M., Oliver, G. and Gruss, P. (1997). The pax-4 gene is essential for differentiation of insulin-producing beta cells in the mammalian pancreas. Nature 386, 399–402.

St Onge, L., Sosa-Pineda, B., Chowdhury, K., Mansouri, A. and Gruss, P. (1997). Pax-6 is required for differentiation of glucagon-producing alpha cells in mouse pancreas. Nature 387, 406–409.

Teitelman, G. (1990). Insulin cells of the pancreas extending neurites but do not arise from the neurectoderm. Dev. Biol. 142, 368–379.

Teitelman, G., Alpert, S., Polak, J.M., Martinez, A. and Hanahan, D. (1993). Precursor cells of mouse endocrine pancreas coexpress insulin, glucagon, and the neuronal proteins tyrosine hydroxylase and neuropeptide Y but not pancreatic polypeptide. Development 118, 1031–1039.

Upchurch, B.H., Aponte, G.W., and Leiter, A.B. (1994). Expression of peptide YY in all four islet cell types in the developing mouse pancreas suggests a common peptide YY-producing progenitor. Development 120, 245–252.

Wessells, N.K. (1977). *Tissue Interactions and Development.* W.A. Benjamin, Menlo Park, CA.

Wessells, N.K. and Cohen, J.H. (1967). Early pancreas organogenesis: Morphogenesis, tissue interactions and mass effects. Dev. Biol. 15, 237–270.

Wright, C.V., Schnegelsberg, P. and De Robertis, E.M. (1988). XIHbox 8: A novel homeo protein restricted to a narrow band of endoderm. Development 105, 787–794.

Wysocka-Diller, J., Aisemberg, G.O. and Macagno, E.R. (1995). A novel homeobox cluster expressed in repeated structures of the midgut. Dev. Biol. 171, 439–447.

Yasuda, H., Inoue, K., Shibata, H., Takeuchi, T., Eto, Y., Hasegawa, Y., Sekine, N.T, Yotsuka, Y., Mine, T., Ogata, E. and Kojima, I. (1993). Existence of activin A in A and D Cells of rat pancreatic islet. Endocrinology 133, 624–630.

Yokouchi, Y., Sakiyama, J.I. and Kuroiwa, A. (1995). Coordinated expression of AbdB subfamily genes of the HoxA cluster in the developing digestive tract of chick embryo. Dev. Biol. 169, 76–89.

PANCREATIC β-CELL LINES

THEIR ROLES IN β-CELL RESEARCH AND DIABETES THERAPY

Shanta J. Persaud

Advances in Molecular and Cell Biology
Volume 29, pages 21-46.

ISBN: 0-7623-0547-9

I. INTRODUCTION

The islets of Langerhans are well-organized endocrine units found dispersed within the exocrine pancreas and they compose only 1 to 2% of the pancreatic tissue. They are clusters of approximately 3,000 to 5,000 endocrine cells, only 70% to 80% of which are insulin-secreting β-cells. Islet isolation from experimental animals (usually rats and mice) is a time-consuming process with low tissue yields. The islets are usually retrieved by collagenase digestion of the exocrine pancreas and are picked free from the exocrine tissue under a dissecting microscope. Up to 1,000 islets (equivalent to ~3-5 $\times$ 10^6 cells) may be retrieved from each rat pancreas, but a more typical yield is 500 islets, and the process of obtaining a few thousand islets for biochemical studies can take several hours. In contrast, measurement of signaling events that may be important in β-cells can be carried out in more accessible tissues. For example, examination of glucose transport and metabolism in the liver is relatively easy given that hepatocytes can be readily prepared from experimental animals. Likewise, investigation of the signal transduction pathways regulating exocytosis in model systems such as platelets is not hampered by limited availability of tissue. However, although there are similarities between β-cells and hepatocytes in terms of glucose handling, and β-cells may utilize some of the exocytotic mechanisms that have been identified in platelets, it is clear that β-cells or islets must be used for unequivocal identification of signaling events involved in the regulation of insulin secretion. The yield of islets from human pancreata is considerably greater than that from rodents, but studies with human islets are severely restricted by tissue availability. The option of isolating islets and stimulating them to proliferate in culture, as can be achieved with vascular cells such as endothelial and smooth muscle cells, is not feasible since adult islets have a very limited proliferative capacity. It would seem apparent that virtually limitless supplies of pancreatic β-cells would radically ease the burden of islet preparation and would open the way for extensive studies of β-cell stimulus–response coupling mechanisms that might not be possible to carry out using isolated islets because of constraints on tissue availability. In addition, the availability of β-cells of defined characteristics in abundance could be invaluable as starting material for transplantation in the treatment of insulin-dependent diabetes mellitus (Type 1 diabetes). Over the past 20 years or so, many β-cell lines, which can be maintained in continuous cultures, have been developed by a number of methodologies. These cell lines are the subject of this chapter.

II. REQUIREMENTS OF β-CELL LINES

For β-cell lines to be useful as models for investigation of the function of normal β-cells within islets or as the starting material for treatment of Type 1 diabetes, they must be as similar as possible to primary β-cells. Several basic requirements are necessary for a regulated secretory response to the major physiological stimulus, glucose; these are summarized in the discussion that follows.

A. Glucose Transport

It has been known for many years that the ability of β-cells to respond to increases in glucose with increased insulin secretion is dependent on the entry of glucose into the β-cells by a facilitated glucose transporter. The glucose transporter type 2 (GLUT-2), expressed by β-cells, was cloned more than a decade ago (Thorens et al., 1988), and this high K_m (~15 mM) transport system has been considered to be essential for appropriate glucose sensing by β-cells. It has been suggested that the observed decline in both GLUT-2 expression and glucose-induced insulin secretion from transplanted mouse islets in a hyperglycemic environment might reflect the importance of GLUT-2 in insulin secretory responses (Ogawa et al., 1995). However, it is clear that GLUT-2 is not rate-limiting for β-cell responses to glucose since glucose uptake is reported to be up to 100-fold in excess of the rate of glucose usage (Tal et al., 1992). Furthermore, it has recently become apparent that GLUT-2 expression in human islet cells is considerably lower than in rat islet cells, but that reductions in glucose transport do not compromise glucose usage (De Vos et al., 1995). It therefore seems likely that the low K_m (~3 mM) glucose transporters, GLUT-1 and/or GLUT-3, both of which are expressed by human β-cells, are sufficient for glucose uptake (De Vos et al., 1995). These issues are discussed in detail in the chapter "Signal Recognition: Glucose and Primary Stimuli" later in this volume.

B. Glucose Metabolism

Once inside the β-cell, the first stage of glucose metabolism is its phosphorylation by glucokinase, a high K_m (~10 mM), high specificity hexokinase (Matschinsky et al., 1993). Although glucose transport into β-cells is rapid, with equilibration of extracellular and intracellular concentrations within seconds, the rate of glucose phosphorylation is slower. The low rate of glucose metabolism relative to glucose transport has led to the notion that glucokinase activity is rate-limiting for insulin secretion in response to glucose. Human islets are deficient in GLUT-2 expression; however, they express glucokinase at levels similar to those found in rat β-cells (De Vos et al., 1995). The importance of β-cell glucokinase in the maintenance of normoglycemia is highlighted by studies carried out on transgenic mice in which glucokinase expression was reduced or abolished (Grupe et al., 1995). Mice containing only one copy of the glucokinase gene were hyperglycemic and showed reduced insulin secretory responsiveness to glucose. Glucokinase-deficient (–/–) mice died 3 to 5 days postnatally, and were markedly diabetic. However, mice deficient in liver glucokinase expression but containing β-cell glucokinase were able to maintain glycemia (Grupe et al., 1995). These observations are consistent with glucokinase being an important β-cell glucose sensor.

C. Insulin Biosynthesis and Processing

The biosynthesis and packaging of insulin take place in the β-cell endoplasmic reticulum and Golgi network. Insulin synthesis occurs via two intermediate single chains: preproinsulin and proinsulin. The hydrophobic "pre" region is thought to promote transportation of the newly formed polypeptide through the membrane of the endoplasmic reticulum and it is then cleaved. Proinsulin is folded in the endoplasmic reticulum and transported by an energy-dependent process to the trans-Golgi network (Howell, 1972), where it is packaged into secretory granules. The conversion of proinsulin to insulin occurs within the maturing secretory granules at sites marked by two pairs of basic amino acids. Two distinct Ca^{2+}-dependent acidic endoproteases, whose substrate specificities are defined by the primary structures of the dibasic cleavage sites, are involved in the conversion of proinsulin to insulin (Davidson et al., 1988). The endoproteases have been identified as PC2 and PC3, members of the subtilisinlike family. PC3 cleaves at the B chain–C peptide junction (Arg–Arg) and this is followed by cleavage at the C peptide–A chain junction (Lys–Arg) by PC2. Refer to the chapter by J. Hutton later in this volume.

The rate of insulin biosynthesis by the β-cell is under the control of glucose and other sugars and is dependent on sugar metabolism (Ashcroft et al., 1978). There is a sigmoidal relationship between glucose concentration and the rate of proinsulin synthesis, with a threshold value of 2.5 to 3.9 mM (Pipeleers et al., 1973). This is lower than the threshold for glucose-stimulated insulin secretion (4 to 6 mM), possibly to ensure that adequate supplies of insulin are maintained.

D. Regulated Insulin Secretion

In normal β-cells, the majority of insulin is secreted via the regulated pathway, while less than 1% is released via a constitutive route (Rhodes and Halban, 1987). Thus, insulin forms an insoluble crystalline hexameric complex with zinc in the secretory granules and is only released into the circulation in response to an appropriate stimulus, such as an increase in circulating glucose concentrations. Exocytotic release of fully processed insulin from the granules occurs following granule fusion with the β-cell plasma membrane. (Refer to the chapter "Mechanism of Insulin Exocytosis" later in this volume.) This regulated secretion of insulin allows fluctuations in blood glucose levels to be rapidly restored to normal values.

E. Other Requirements

The requirements for glucose transport and metabolism and insulin synthesis and regulated secretion have been highlighted because these are areas in which β-cell lines have been found to be compromised. In particular, the β-cell–derived lines may have deficiencies in glucose transport and/or phosphorylation, whereas the lines engineered from non-β-cell tissue often have restricted abilities to synthesize and package insulin and to release it in a regulated manner. Obviously many other factors should be taken into account when considering what makes a β-cell a β-cell. For example, adenosine triphosphate (ATP)-sensitive K^+ channels are required so that glucose metabolism can be converted into a depolarizing signal. In addition, receptors for modulators of insulin secretion (such as acetylcholine, norepinephrine, glucagon, VIP, PACAP, etc.) should be present, as well as the intracellular effector systems such as adenylate cyclases, phospholipases, protein kinases, and so forth, which are necessary for coupling external stimuli to secretory responses (see Figure 1). Furthermore, the cell line phenotype should be stable such that, with continued life in culture, the expression of key signal transduction elements and secretory responsiveness are maintained. It is not easy for β-cell lines to fulfill these criteria given that they should both possess the differentiated functions of primary β-cells and also proliferate readily in culture. It is scarcely surprising that many of the currently available β-cell lines differ dramatically from primary β-cells in terms of glucose recognition and phosphorylation and also with respect to insulin content and secretory responsiveness to glucose (Table 1).

III. GENERATION OF β-CELL LINES

Several different approaches have been used to establish β-cell lines, and in general these have become rather more sophisticated in recent years as advances in cellular biology have allowed production of transgenic mice in which pancreatic tumors have been developed by targeted expression of transforming genes. Attempts have also been made to engineer non-β-cells to synthesize, process and secrete insulin in a regulated manner, and these cell lines, as well as conventional β-cell lines, are being developed as potentially transplantable material for the treatment of type 1 diabetes.

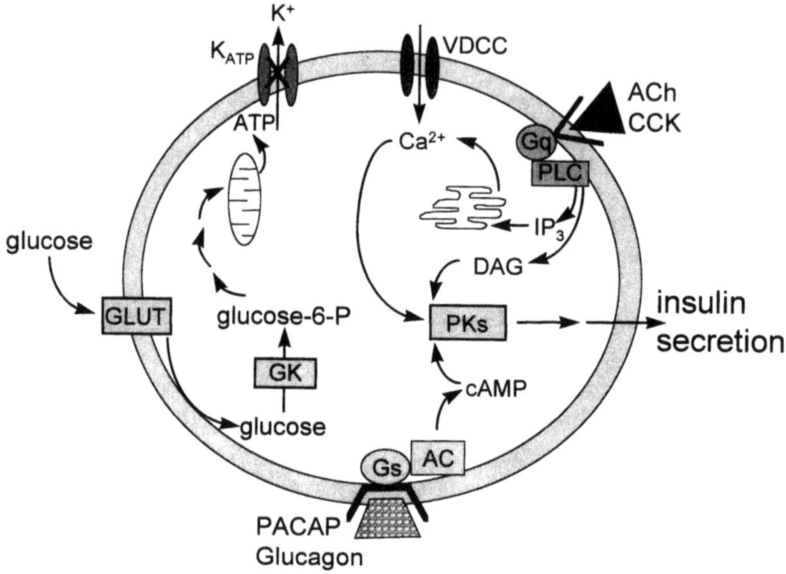

Figure 1. Regulated insulin secretion from pancreatic β-cells. GLUT, a facilitated glucose transporter is responsible for glucose entry into cells. In rodent β-cells this is the high capacity GLUT-2 transporter, while in human β-cells glucose transport is via the low capacity transporters, GLUT-1 or GLUT-3. The rate-limiting step in insulin secretion in response to elevated concentrations of glucose is its phosphorylation to glucose-6-phosphate (glucose-6-P) by the high specificity hexokinase, glucokinase (GK). The increase in intracellular adenosine triphosphate (ATP) concentrations (or ATP:ADP ratio) following glycolytic and oxidative metabolism of glucose results in closure of ATP-sensitive potassium channels (K_{ATP}), decreased K^+ efflux and membrane depolarization. The consequent opening of voltage-dependent Ca^{2+} channels (VDCC) results in Ca^{2+} influx and activation of Ca^{2+}-sensitive enzymes such as protein kinases (PKs). Agonists acting at cell surface receptors can potentiate insulin secretion through Gs and Gq, heterotrimeric guanosine triphosphate (GTP)-binding proteins, which regulate the activities of adenylate cyclase (AC) and phospholipase C (PLC). AC catalyses the generation of cyclic adenosine monophosphate (cAMP) and PLC activation results in the generation of inositol 1,4,5-trisphosphate (IP_3) and diacylglycerol (DAG). These second messengers can regulate insulin secretion through the activation of PKs.

A. Transplantable Insulinomas

The obvious starting point for β-cell lines is the pancreas. The initial studies on development of β-cell lines were carried out in the late 1960s and 1970s when transplantable islet cell tumors were induced in the golden hamster (Sodoyez et

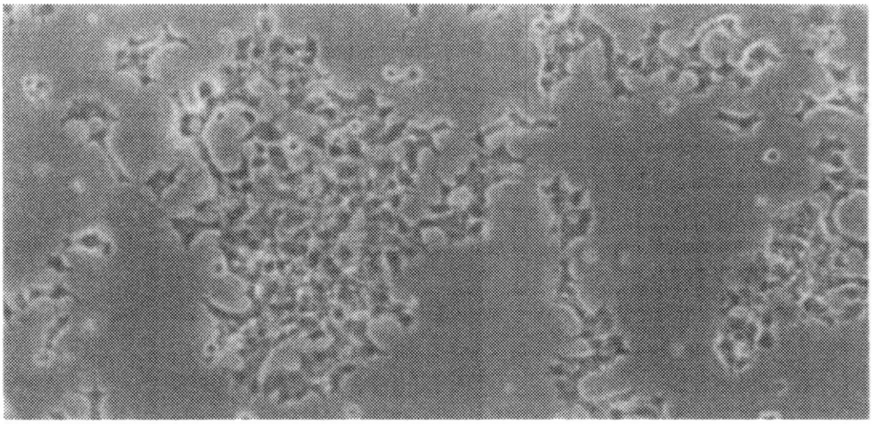

Figure 2. Phase-contrast photomicrograph of RINm5F cells. RIN cells grow as monolayers with cytoplasmic processes extending from cells at the periphery. Magnification: ×100; passage: 65.

al., 1967) and rat (Chick et al., 1977). The hamster insulinoma had extremely low insulin content, and its use for general β-cell studies was limited. However, the rat transplantable insulinoma proved to be very successful for establishing β-cell lines, and the RIN-r and RIN-m lines, which were established from a pancreatic tumor induced in an NEDH rat by X-irradiation, are still much used today. The RIN (*r*at *in*sulinoma) cell lines were initiated from tumors developed in recipient NEDH rats (RIN-r) or nude mice (RIN-m) and were established as continuous cultures after separation of tumor cells from connective tissue and elimination of fibroblasts (Gazdar et al., 1980). RIN-r and RIN-m cell lines were similar in appearance, growing as monolayers with distinct cytoplasmic processes when cultured in tissue culture plastic (Figure 2), and at least some clones of both the rat- and mouse-derived lines secreted somatostatin as well as insulin (Gazdar et al., 1980). More recently, a novel β-cell line, INS-1, has been established from the X-ray induced rat transplantable insulinoma by growth of isolated tumor cells in a mercaptoethanol-supplemented medium (Asfari et al., 1992). The INS cells originally grew as free-floating aggregates which were decanted from adherent fibroblasts, and they required trypsinization before they could be propagated (Asfari et al., 1992).

When considering the requirements for β-cell lines with respect to signaling in normal β-cells, as outlined in the previous section, RIN cells do not perform particularly well. Their major disadvantage is their lack of responsiveness to glucose. INS-1 cells, however, are closer in their characteristics to native β-cells.

The most widely used RIN-m cell line, RINm5F, does not show any increase in insulin secretion in response to glucose over the range of 2.8 to 33.4 mM (Praz et al., 1983). The deficiency in secretory response to glucose is not a consequence of its inability to enter the β-cells, as RINm5F cells have been shown to express a high capacity glucose transport system (Trautmann and Wollheim, 1987), likely to be GLUT-2. The defect appears to be at the level of glucose metabolism, which is far higher in RINm5F cells than in islet β-cells and reflects the predominance of a low K_m (hexokinase) glucose-phosphorylating activity (Halban et al., 1983).

Like RIN-m cells, RIN-r cells also fail to respond appropriately to glucose. Most experiments have been carried out on the RINr1046-38 (RIN-38) clone. These do respond to glucose at low passages, albeit maximally at 600 μM glucose, and even this response is lost at high passages (Clark et al., 1990). It has been reported that the loss of responsiveness to glucose in these cells is associated with decreased expression of GLUT-2 and glucokinase and that glucose-stimulated insulin secretion can be reinstated by stable transfection of the RIN-38 cells with GLUT-2 (Ferber et al., 1994). In cells in which GLUT-2 expression had been restored, there was also an increase in glucokinase activity, despite there having been no introduction of glucokinase cDNA (Ferber et al., 1994). This suggests that there is a close coupling between GLUT-2 expression and that of glucokinase. Despite the elevations in GLUT-2 and glucokinase in the transfected RIN-38 cells, the secretory response to glucose was less than threefold basal and was maximal at only 50 μM. Thus, it would appear that the presence of GLUT-2 and glucokinase per se are insufficient for a normal glucose-induced secretory response, and it is likely that this is because RIN-38 cells also express low K_m hexokinases which shift the dose-response curve leftward. When these were inhibited the maximal responsiveness to glucose shifted by two orders of magnitude to ~5 mM (Ferber et al., 1994).

INS-1 cells, which are related to RIN cells in terms of derivation, have very different characteristics. The major difference is their ability to respond to glucose, albeit with only a maximal doubling in insulin secretion, over the normal physiological range (Asfari et al., 1992). In addition, the insulin content of INS-1 cells is considerably higher than that of RIN cells (see Table 1), and their phenotype is stable for at least 2 years. The ability of INS-1 cells to respond to glucose is most likely related to their relatively high levels of glucokinase expression, although overexpression of glucokinase in INS-1 cells has been shown to cause a left-shift in the concentration-dependence of glucose-stimulated insulin release such that maximal responses were obtained at only 6 mM glucose (Wang and Iynedjian, 1997). It has been suggested that with a fourfold induction of glucokinase over wild-type levels in INS-1 cells, glucose phosphorylation occurs so rapidly that a downstream metabolic event(s) takes over as the rate-limiting step(s) in insulin secretion, and that this is responsible for the modified glucose concentration-dependence profile in these cells (Wang and Iynedjian, 1997).

B. Electrofused β-Cell Lines

Novel clonal β-cell lines, known as BRIN (***b***eta ***r***at ***in***sulinoma) cells have been established from RINm5F cells by their fusion with rat islet cells (McClenaghan et al., 1996a). The fusion of the insulinoma cells with normal islet cells was achieved by exposing them to an alternating current of 1.5 kV/cm followed by three pulses of a direct current of 4 kV/cm. Hybrid fused cells secreting the most insulin were cloned for further investigation.

Three BRIN cell lines (designated BRIN-BG5, BRIN-G7 and BRIN-BD11) have been characterized to date and all show improvements over the parental RIN cells. Thus, unlike RINm5F cells the BRIN cells are glucose-responsive (McClenaghan et al., 1996a, 1996b), but like the INS-1 cells, the maximal secretory response to glucose is at most a doubling over unstimulated levels. The BRIN cells also preferentially express glucokinase over hexokinase and glucose uptake is via the GLUT-2 transporter (McClenaghan et al., 1996a, 1996b). Surprisingly, GLUT-2 immunoreactivity could not be detected in the parental RINm5F cells (McClenaghan et al., 1996a), despite an earlier report that RINm5F cells possessed a high-capacity, low-affinity glucose transporter (Trautmann and Wollheim, 1987), characteristic of GLUT-2. This observation again highlights the changing characteristics of RIN cells with passage. This is also noticeable when the insulin content is considered: early studies with RINm5F cells indicated that they expressed ~0.2 pg of insulin per cell, but the parental RIN cells used for establishing the BRIN cell lines were found to contain 0.03 pg/cell (McClenaghan et al., 1996c) or as low as 0.6 fg/cell (McClenaghan et al., 1996a). In addition, the doubling times of the BRIN cells (~20 to 30 hours) were considerably more rapid than those reported originally for RIN cells (>50 hours) and may again be indicative of the alterations in characteristics of the parental cells with time in culture.

The low insulin content of BRIN cells is mirrored by their low rates of secretion. Maximal stimulation of BRIN-BD11 cells with 16.7 mM glucose resulted in insulin secretion rates of around 3 ng/10^6 cells/20 min, and when cells were incubated with depolarizing concentrations of KCl (at a substimulatory glucose concentration) secretion was ~12 ng/10^6 cells/20 min (McClenaghan et al., 1996a). This is particularly low when compared to insulin secretion from normal rat islets of ~5 to 10 ng/islet/h when stimulated (equivalent to ~1250 to 2500 ng/10^6 cells/h, assuming 4,000 β-cells/islet), and even some of the cell lines that might be considered to be reasonably poor models for normal β-cells, such as the AR42J-B13 cells (see Section III.E), are reported to have stimulated secretion rates of up to 150 ng/10^6 cells/h. Therefore, while generation of clonal cell lines by electrofusion is undoubtedly a useful way of establishing β-cell substitutes, it might be more efficient to use an early passage, high insulin-content, reasonably well-differentiated cell line, such as may be generated from transgenic mice (see Section III.D) as the starting material, and fuse them with normal mouse β-cells.

C. SV40-Transformed β-Cells

The Simian virus 40 (SV40) T antigen is a protein coded for by an oncogene in the SV40 genome. It inhibits growth-suppressing proteins in host cells, resulting in unregulated DNA replication and cell growth. Rat islet cells have been transformed using SV40, but the subcultured cells rapidly lost the ability to synthesize insulin (Niesor et al., 1979). The best known and most widely used continuous transformed β-cell line is the HIT line, which was established from SV40-infected Syrian hamster primary islet cells (Santerre et al., 1981). Exposure of trypsinized islet cell monolayers to SV40 resulted in the development of colonies of transformed cells that could be recloned and propagated continuously. Like RIN cells, HIT cells are epithelioid in appearance, but they tend to grow as multilayered cell clusters (Figure 3), rather than in the monolayers that are characteristic of RIN cells (see Figure 2).

HIT cells, like RIN cells, are not ideal models for normal β-cells. They are improved with respect to secretory responsiveness to glucose in as much as glucose stimulates a dose-dependent increase in insulin secretion. However, the dose-response curve is left-shifted, with maximal responses at 10 mM glucose and a decline in insulin output at 20 mM glucose (Ashcroft et al., 1986). It seems that

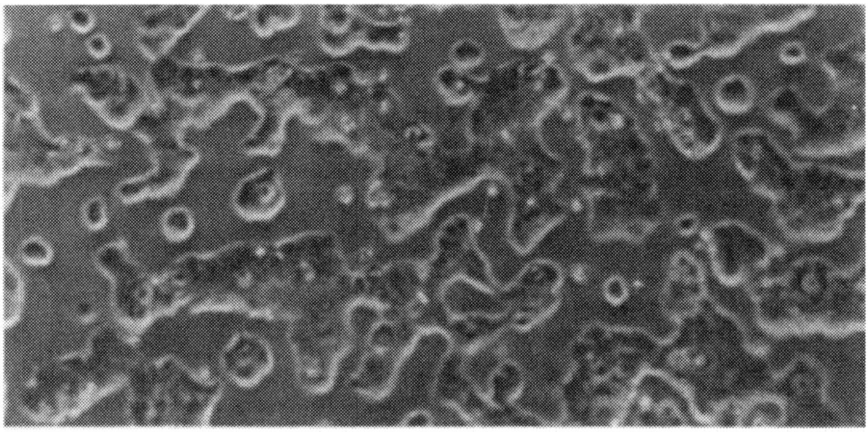

Figure 3. Phase-contrast photomicrograph of HIT-T15 cells. HIT cells grow as multilayered cell clusters in which individual cells cannot be distinguished. Magnification: ×100; passage: 63.

the left-shift in the concentration-dependence of glucose-stimulated insulin release from HIT cells reflects the expression of a low K_m (~4 mM) glucose transporter, which is insufficient for rapid equilibration of glucose between the extracellular and intracellular environments. Thus, the rate-limiting stage in insulin release from HIT cells in response to glucose appears to be the low rate of glucose transport into the β-cells (Ashcroft and Stubbs, 1987). The most commonly used HIT clone, T15, had an insulin content of 300 fg per cell, which was up to 10-fold greater than the other originally established clones and which was reported to represent around 5% of the insulin content of normal hamster islets (Santerre et al., 1981). However, with continued propagation HIT-T15 insulin content declined and by passage 110 it was reported to be only 8 fg per cell (Welsh et al., 1995). The decrease in insulin content is accompanied by a decrease in the rate of insulin release, with a 30-fold reduction over 47 passages (Santerre et al., 1981). In spite of the skewed dose-responsiveness to glucose, the low insulin content, and the instability of insulin release, HIT-T15 cells are useful in at least two respects: they mimic normal β-cells in terms of requirement of sugar metabolism and extracellular Ca^{2+} for insulin secretion; and secretion is biphasic, potentiated by activators of protein kinases A and C and inhibited by epinephrine and somatostatin (Santerre et al., 1981; Ashcroft et al., 1986).

D. Transgenic Mice Insulinomas

Transgenic technology has led to the generation of a number of novel β-cell lines in the past decade, including βTC (*b*eta *t*umor *c*ell) (Efrat et al., 1988), MIN (***m***ouse ***in***sulinoma) (Miyazaki et al., 1990), and βHC9 (*b*eta *h*yperplastic *c*ell) (Radvanyi et al., 1993). These cell lines have been established by development of β-cell tumors, or, in the case of βHC9 cells, preneoplastic foci, after microinjection of recombinant insulin/SV40 genes into fertilized mouse embryos. Targeted expression of the SV40 T antigen to β-cells of the transgenic mice is achieved by ligating sequences from the β-cell–specific insulin promoter to the SV40 T antigen coding region. More recently, the transgenic approach has become rather more sophisticated, with the development of conditional oncogene expression in β-cells of transgenic mice. This has been achieved by the generation of double-transgenic mice bred from mice expressing an insulin promoter-driven fusion protein containing the tetracycline-resistance operon repressor (*tet*R) linked to a viral activating domain, and mice in which the SV40 T antigen was introduced under the control of *tet* operator sequences, but not expressed (Efrat et al., 1995). In the double-transgenes, β-cell tumor development occurred following binding of the *tet*R fusion protein expressed in β-cells to the *tet* operator sequences, resulting in activation of T antigen expression. Cells from these tumors, known as βTC-tet cells, were expanded in culture and, since the expression of T antigen is switched off in the presence of tetracycline, cells maintained in tetracycline-supplemented medium failed to proliferate (Efrat et al., 1995). The ability to generate large quan-

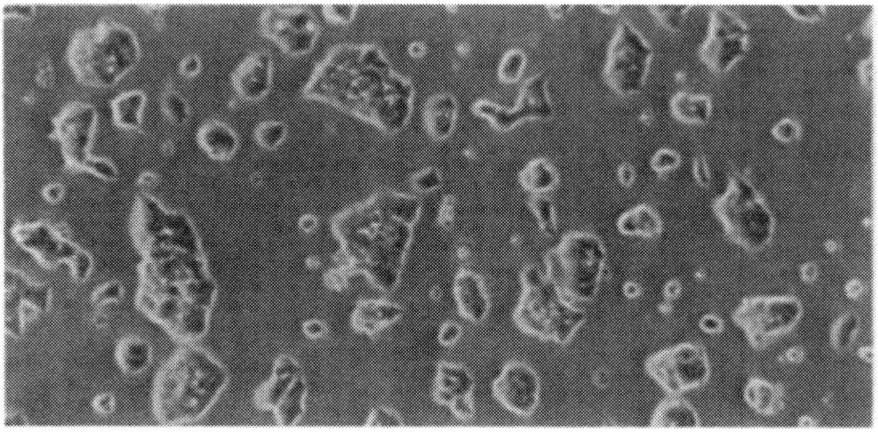

Figure 4. Phase-contrast photomicrograph of βTC3 cells. βTC3 cells form small, well-defined clusters in culture, and grow to a maximal confluency of approximately 60% to 70%. Magnification: ×100; passage: 29.

tities of β-cells and then growth-arrest them is moving very close to the ideal of β-cell lines: virtually limitless supplies of β-cells with maintained differentiated functions.

The first β-cell lines to be established from pancreatic insulinomas excised from transgenic mice harboring insulin-SV40 T antigen chimeric genes were the βTC lines (Efrat et al., 1988). The βTC lines are nonclonal since the cells are poorly proliferative at low confluency (Efrat et al., 1988), and they tend to grow in monolayers as well-defined clusters (Figure 4). Several βTC cell lines have been established and characterized. Some (e.g., βTC1 and βTC3), express GLUT-1 transporters in preference to GLUT-2 and phosphorylate glucose predominantly by the low K_m hexokinases (Whitesell et al., 1991; Efrat et al., 1993). These lines do respond to glucose, but with left-shifted responsiveness (βTC1 maximal at 1.25 mM), probably as a result of the preferential expression of hexokinase over glucokinase. The βTC7 line is reported to be more similar to β-cells from normal islets than either βTC1 or βTC3 cells. βTC7 cells express GLUT-2, have similar glucokinase and hexokinase levels to those of normal islets, and show a 10-fold increase in insulin secretion over the glucose concentration range of 5 to 16.7 mM (Efrat et al., 1993). However, the secretory response of βTC7 cells to glucose was dependent on the inhibition of cyclic adenosine monophosphate (cAMP) phosphodiesterases with isobutylmethyl xanthine (IBMX), and was not stable with continued propagation due to increased hexokinase activity (Efrat et al., 1993). A clonal βTC cell line (βTC6-F7) has been established from βTC6 cells, which do not maintain glucose-responsiveness, by soft-agar cloning, and these clonal cells appear to possess all of the features required of an ideal rodent β-cell line: they have high glucokinase activity, have low hexokinase activity, express GLUT-2, have low levels of constitutive insulin release, and show normal glucose-induced insulin secretion which is stable for at least 55 passages (Knaack et al., 1994). However, the normal secretory response to glucose may reflect the ability of βTC6-F7 cells to synthesize glucagonlike peptide 1 (GLP-1) or a related peptide (N. Fleischer, personal communication). It is possible that such peptides could exert autocrine effects to increase β-cell cyclic AMP, thus potentiating insulin secretion at physiological levels of glucose.

Several other cell lines have been established in the past decade from insulinomas developed by targeted expression of SV40 T antigen in transgenic mice. Two MIN cell lines have been characterized. While the MIN7 cells do not secrete insulin in response to glucose, the MIN6 cells appear to be ideal candidates for normal β-cell substitutes in as much as they transport and phosphorylate glucose via GLUT-2 and glucokinase, have a relatively high insulin content (up to ~20% of that of normal mouse β-cells), and secrete insulin in an appropriate glucose-dependent manner (Miyazaki et al., 1990; Ishihara et al., 1993). Low passage MIN6 cells grow in clusters and are glucose-responsive (Miyazaki et al., 1990), but the MIN6 phenotype is not stable: with prolonged culture the morphology of these cells begins to resemble that of RIN cells (Figure 5), and their secretory responsiveness to glucose diminishes progressively (S. Persaud, unpublished observations).

The βHC lines have been developed with the aim of circumventing the problems associated with tumor cells becoming de-differentiated with time. The βHC9 line, established from hyperplastic islets removed from the pancreata of insulin promoter-SV40 T antigen transgenic mice before insulinoma development, shows a

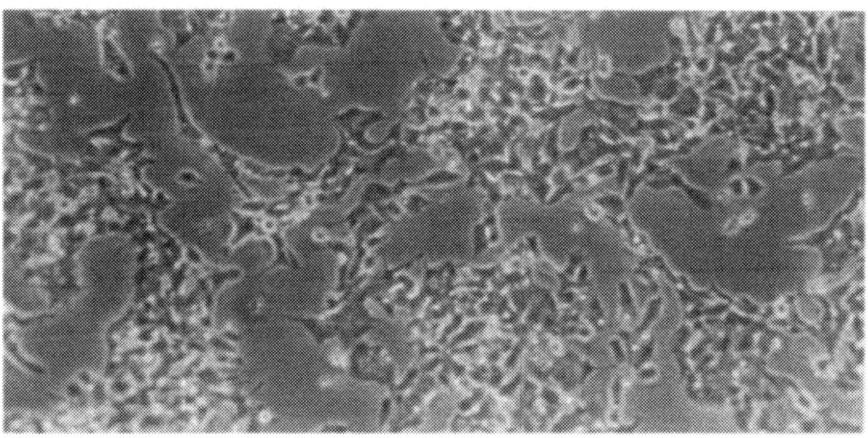

Figure 5. Phase-contrast photomicrograph of MIN6 cells. MIN6 β-cells of relatively late passage grow as monolayers and extend cytoplasmic processes to neighboring cells. Magnification: ×100; passage: 46.

robust (40-fold stimulation at 25 mM glucose) secretory response to glucose (Radvanyi et al., 1993). The ability of βHC9 cells to respond appropriately to glucose was associated with preferential expression of glucokinase over hexokinase, and βHC9 cells also express the high capacity glucose transporter, GLUT-2 (Radvanyi et al., 1993; Liang et al., 1996). Unfortunately, these characteristics, which come close to the ideal requirements for a β-cell line, are unstable. Although glucose-stimulated insulin secretion is maintained in early passage cells, it is lost with continued culture, an effect which is attributable, in part, to elevated basal secretion (Noda et al., 1996).

E. Insulin-Secreting Acinar Cells

Ductal cells within the pancreas are thought to be precursors for the formation of both exocrine acinar cells and islet endocrine cells, but the processes by which this occurs have not been defined in detail (see the first chapter in this volume). An acinar cell line, AR42J, has recently been shown to differentiate to an insulin-secreting cell line after exposure to hepatocyte growth factor (HGF) and activin A (Mashima et al., 1996). HGF caused a concentration-dependent inhibition of amylase content of AR42J cells, and combined treatment of the AR42J-B13 subclone with both HGF and activin A for 5 days resulted in nearly 90% of the cells becoming insulin-positive (Mashima et al., 1996).

Although cells derived from a pancreatic acinar carcinoma may appear to be an unusual starting point for generating an insulin-secreting cell line, AR42J cells have the advantage of already expressing the glucose phosphorylating enzyme glucokinase. Furthermore, when treated with HGF the normally insulin-deficient cells were shown to express insulin mRNA (Mashima et al., 1996). Exposure to HGF also resulted in induction of GLUT-2 mRNA, and this was further enhanced by co-treatment with HGF and activin A. AR42J-B13 cells secreted insulin at relatively high levels (maximally at ~15 ng/10^5 cells/h) in response to depolarizing concentrations of KCl, to the sulphonylurea tolbutamide and to the receptor agonists carbachol and GLP-1, but despite expression of both GLUT-2 and glucokinase they did not respond to glucose (Mashima et al., 1996). The reason for the lack of responsiveness to glucose was not established, but, given that a high hexokinase:glucokinase ratio increases glucose flux at low glucose concentrations, it is possible that high hexokinase activity may have played a role. The AR42J-B13 cells clearly require further characterization before their usefulness as β-cell substitutes can be evaluated more fully.

F. Non-β-Cell Insulin-Secreting Cell Lines

There have been a number of attempts to engineer nonislet cells to synthesize, package, and process insulin, and secrete it in a regulated manner. In general, cells have been transfected with human proinsulin cDNA, usually driven by a viral promoter, and the ability of these cells to secrete insulin in a regulated and glucose-dependent manner has been assessed. When considering the development of surrogate cells for diabetes therapy, non-β-cells are attractive because they should not be susceptible to the autoimmune attack responsible for β-cell destruction in Type 1 Diabetes. Several different cell types have been chosen as recipients for the insulin gene and these can be loosely characterized as somatic cells such as fibroblasts (Moore et al., 1983; Selden et al., 1987); hepatic cells (Lu et al., 1996; Hayashi et al., 1997); and neuroendocrine cells, such as pituitary GH3 (Kaufmann et al., 1997) and AtT-20 (Moore et al., 1983; Hughes et al., 1992) cells. Each of these systems has inherent advantages: somatic cells can be taken from donors, transfected with the insulin gene, and reinserted without immunological intolerance; hepatic cells may also be amenable to ex vivo gene transduction, and they contain the enzymes responsible for glucose recognition; and endocrine cells already contain the machinery required for hormone production and storage.

The success of attempts to engineer non-β-cells to take on the functional characteristics of native β-cells has depended to a large extent on the degree of similarity between the transduction systems operative in the recipient cells and those of normal β-cells. The mouse fibroblast cell line, Ltk^-, has been successfully transfected with proinsulin cDNA, but because these cells lack glucose-recognition elements, secretory granules, proinsulin-processing enzymes, and a regulated secretory pathway, they secrete proinsulin in a constitutive manner, independent of stimulus (Moore et al., 1983; Selden et al., 1987). Intraperitoneal injection of proinsulin-secreting Ltk cells into streptozotocin-induced diabetic mice resulted in initial normalization of blood glucose levels since proinsulin has up to 10% of the biological activity of insulin, but the animals soon died of hypoglycemia, most likely caused by a combination of dysregulated tumor growth and constitutive proinsulin secretion (Selden et al., 1987; Kawakami et al., 1992). This obviously limits their usefulness as β-cell substitutes.

Although hepatocytes are an improvement over fibroblasts as starting material for generation of glucose-responsive insulin-secreting cell lines, in that they express GLUT-2 and glucokinase, they also suffer from the absence of the prohormone convertase enzymes required for proinsulin cleavage. However, the processing sites of human proinsulin can be engineered by site-directed mutagenesis to contain sequences cleavable by the PC-related ubiquitous protease furin, and mature, processed insulin is secreted from cells transfected with this modified proinsulin (Groskreutz et al., 1994). This approach has been applied to H4EII hepatoma cells, which have been transfected with a furin-cleavable proinsulin under the control of the cyclic AMP–regulated promoter phosphoenol pyruvate carboxykinase. The transfected H4EII cells secreted mature insulin, albeit at exceedingly low levels (~5 to 15 pg/10^6 cells/h), and this could be increased up to fivefold with increases in cyclic AMP (Lu et al., 1996). Although this semiregulated release of mature insulin is a useful step forward, the H4EII model is still hampered by the lack of glucose-responsiveness of the secretory response. This

Table 1. Characteristics of Insulin-secreting Cell Lines

Cell Type	*Source*	*Doubling Time*[a]	*Transporter*	*Hexokinase*[b]	*Insulin (pg/cell)*[c]	*Response to Glucose*	*Stable Phenotype*	*Other Islet Hormones*
RIN	Transplantable insulinoma	~50–80 h	High capacity (GLUT-2)	HK>>GK	~0.2[c]	No	No	Glucagon, SST
INS-1	Transplantable insulinoma	~100 h	GLUT-2	GK>>HK	~8	2.8–11.2 mM	Yes? (>2 y)	No
BRIN-BD11	RIN/islet cell electrofusion	~20 h	GLUT-2	GK>HK	~0.08	4.2–16.7 mM	Yes? (>50 passages)	No?
HIT-T15	SV40-infected islet cells	~30 h	Low capacity (GLUT-1)	GK	~0.3	1–10 mM	No	Glucagon
βTC	Transgenic insulinoma	~60 h	GLUT-1[TC1] GLUT-2[TC7] GLUT-2[TC6-F7]	HK>GK[TC1] GK>HK[TC7] GK>>HK[TC6-F7]	~0.2[TC1] ~2[TC7] ~3[TC6-F7]	0.5–1.25 mM[TC1] 2–17 mM[TC7] 5–30 mM[TC6-F7]	No[TC1] No[TC7] Yes?[TC6-F7]	Glucagon[TC1]
MIN	Transgenic insulinoma	~60 h	GLUT-2[MIN6] GLUT-1/2[MIN7]	GK>>HK[MIN6] HK>GK[MIN7]	~6[MIN6]	5–25 mM[MIN6] No[MIN7]	No	No
βHC9	Transgenic hyper-plastic islets	~70–100 h	GLUT-2 (also some GLUT-1)	GK>>HK	~5	7–15 mM	No	PP[βHC8]
AR42J-B13	Acinar carcinoma	?	GLUT-2	GK	~1.5	No	?	PP
AtT-20ins	Pituitary tumor + insulin gene	?	GLUT-1	HK>>GK	~0.001	No	Yes?	No
Islets	Adult rodent pancreas	Nonproliferative	GLUT-2	GK>>HK	~10–20	5–20 mM	Not applicable	Glucagon, SST, PP

Notes: [a] Doubling times vary with culture medium, serum concentration, and culture passage.

[b] β-Cell lines that initially show preferential expression of glucokinase tend to show progressively higher hexokinase expression with passage.

[c] Insulin content tends to decrease in all cell lines with prolonged culture. Although RINm5F cell insulin content was initially described as being ~0.2 pg/cell (Praz et al., 1983), more recent estimates have calculated it to be as low as 0.6 fg/cell (McClenaghan et al., 1996a).

problem has been addressed by studies in which the insulin gene was transfected into a differentiated, nontumorigenic cell line developed from regenerating mouse liver. The transfected cells (43-15) secreted unprocessed insulin, but its rate of secretion did increase with glucose concentration (5.5 to 25 mM), albeit over a prolonged time-scale (24 to 48 h) (Hayashi et al., 1997). One major advantage of the 43-15 cell line is that although it proliferated well in vitro, it did not proliferate when transplanted into the liver of recipient mice; the lack of tumor formation would be a useful attribute for gene therapy in vivo. Despite this, it is clear that even if proinsulin can be engineered to be fully processed by nonendocrine cells, the absence of hormone storage in secretory granules means that regulated second-to-second control of insulin secretion will not be obtained, and at best these cells would provide a slow increase in insulin levels following activation of glucose-dependent gene transcription.

Stable expression of human proinsulin cDNA in neuroendocrine cells has proved to be rather more successful than transfection of somatic and hepatic cells, since endocrine cells naturally store and secrete hormones. Human proinsulin cDNA has been expressed in the pituitary cell line, GH3, but mature insulin was not obtained in the absence of co-expression of either PC2 or PC3, which were required for proinsulin cleavage to insulin (Kaufmann et al., 1997). Another pituitary cell line, AtT-20, contains PC2 and PC3 and when transfected with the human proinsulin gene the cells (AtT-20ins) stored and processed proinsulin and secreted mature insulin from their secretory granules in response to nonnutrient secretagogues, but not in response to glucose (Moore et al., 1983; Hughes et al., 1992). The AtT-20ins cells contain glucokinase; stable transfection with GLUT-2 cDNA resulted in increased insulin content, indicative of increased biosynthetic activity, and glucose stimulated a significant increase in insulin secretion from these cells. However, the increase was maximally threefold of basal and was obtained at only 10 μM glucose (Hughes et al., 1992). This left-shifted responsiveness to glucose probably occurred because AtT-20ins cells express more hexokinase than glucokinase, thus making them maximally responsive to glucose at subphysiological levels. Another problem with AtT20ins cells is that they still express large amounts of their native hormone, adrenocorticotropic hormone (ACTH), which is co-localized with insulin in the secretory granules. Thus, engineering of these cells for possible therapy of diabetes would require both correction of the enhanced sensitivity to glucose and inhibition of ACTH synthesis so that adrenal function in vivo would not be affected.

IV. USES OF CELL LINES FOR β-CELL RESEARCH

One of the main reasons why so much effort has been expended in the generation of insulin-secreting cell lines is that it has long been realized that a ready, abundant supply of β-cells would radically simplify investigations of β-cell function. The cell lines available to date have been used widely as substitutes for primary β-cells, and a large body of information has been obtained from experiments carried out using the cell lines. However, it is also clear that β-cell lines are in themselves distinct groups of cells, none of which is identical to another β-cell line, and all of which differ in some respects from primary β-cells. Therefore, although studies based on cell lines can give some insight into the signal transduction pathways operative in normal β-cells, it is essential that such pathways also be shown to exist in β-cells from experimental animals or human pancreata before the data can be used to build up a fuller picture of the systems normally regulating insulin secretion.

A. Advantages of β-Cell Lines

The cell lines described in the previous section exhibit, to varying extents, the characteristics of primary β-cells within normal islets of Langerhans. They have, therefore, been used in many studies aimed at understanding the stimulus–response coupling mechanisms responsible for the regulated release of insulin. Obviously the lines derived from β-cells are more useful for this purpose than those engineered from non-β-cell material. The vast number of publications using β-cell lines over the past two decades precludes reference to the numerous individual studies, but there are several areas where cell line use has been of great advantage. In particular the β-cell lines have lived up to expectations in providing sufficient material for large-scale experiments, which would have been exceedingly difficult to carry out using islets isolated from experimental animals. Thus, experiments requiring preparation of enriched subcellular fractions have made good use of cell lines as starting material, and many phosphorylation studies have been carried out using β-cell lines in attempts to identify substrates for protein kinases. Furthermore, there are specific areas where β-cell lines have been of use not only in β-cell research, but in a wider biochemical context. Thus, molecular cloning studies using β-cell lines have led to the identification of PKCι, an atypical isoform of protein kinase C (Selbie et al., 1993), and cloning of the sulfonylurea receptor was made possible by identification of a partial amino acid sequence of the receptor from HIT-T15 cell membranes, which was used to generate a cDNA fragment for screening insulinoma libraries (Aguilar-Bryan et al., 1995; see also the later chapter on ion channel signal transduction in this volume).

As well as being a source of large quantities of β-cells, β-cell organelles, and β-cell mRNAs, most β-cell lines are also amenable to stable modification of their genome by transfection with exogenous DNA. Stable expression of foreign DNA by cells is a low-efficiency process, and obtaining sufficient numbers of cells, all of which have been modified, depends on the ability of the recipient cells to divide and on the selection of those cells into which the vector has integrated into the genome. The ability of β-cell lines to express foreign genes is a valuable attribute and one that often makes their use preferable to islet β-cells which, because of their nonproliferative nature, cannot be stably transfected. There have been numerous reports in recent years of the functional consequences of changes in protein expression by β-cell lines (e.g., Blondel et al., 1994; Ferber et al., 1994; Ishihara et al., 1994), and these have given us further insight into the signal transduction mechanisms that may be operative in normal β-cells. In addition, stable transfection of β-cell lines with genes required for expression of human insulin and appropriate glucose-sensing proteins is currently being carried out as a means of engineering rodent β-cells to take on the characteristics of human β-cells as a strategy for treatment of type 1 diabetes (see the following section).

One further advantage of β-cell lines is that their use limits the number of animals sacrificed for experimentation. It is becoming increasingly difficult to justify large-scale use of animals for research, and it makes sense to use β-cell substitutes in the first instance to build up a body of data before repeating the experiments using islets isolated from experimental animals. In this way, the initial screening process can be carried out on cell lines, but, because of variations in signal transduction systems between some cell lines and normal β-cells, pivotal experiments must still be repeated with authentic islet tissue.

B. Limitations of β-Cell Lines

The major limitation of the β-cell lines is that, although they are derived from pancreatic tissue, they are not faithful reproductions of β-cells within islets. There are many general examples of this, relating to glucose transport and phosphorylation, insulin content, secretory responsiveness, and so forth, and there are clearly problems with the stability of phenotype of the cell lines (see Table 1). In addition, there have also been many reports, too numerous to cite, of specific areas of variability between β-cell lines and islets. These reports usually relate to differences in expression of particular proteins or protein isoforms, so functional data obtained using the cells lines differ from those obtained using normal islets and, in these circumstances, cell line studies can hinder rather than aid our understanding of normal β-cell physiology. It is important, therefore, as has been stressed continually throughout this chapter, that key observations made using β-cell lines be confirmed in normal islets.

V. POTENTIAL APPLICATION OF β-CELL LINES FOR DIABETES THERAPY

A. Theoretical Considerations

Although type 1 diabetes can be treated fairly effectively by exogenous insulin therapy, the only cure is through replacement of β-cell tissue so that the body can once again synthesize and secrete insulin in response to appropriate stimuli. Problems with donor rejection of transplanted tissue are being addressed, and improvements are being made in microencapsulation techniques for islets, but it is clear that there will never be sufficient human pancreata available to fulfill the transplant requirements of all insulin-dependent diabetics. The most obvious options available to circumvent the shortage of human islets are to establish human β-cell lines, to use cells from the prospective recipient to create artificial β-cells that can be implanted without problems of rejection, or to engineer rodent β-cell lines to take on the characteristics of human β-cells. The human β-cell line would be the most suitable option, as, ideally, the cells would already be programmed to secrete human insulin in response to appropriate stimuli. The non-β-cell option, although immunologically attractive, would require too much intervention to make it feasible in the short term. Given the difficulties in establishing human β-cell lines (see the discussion that follows), there has been a move toward modifying rodent β-cells to assume the characteristics of human β-cells, with the aim that suitably engineered animal β-cells may be used in the future as islet surrogates in the treatment of type 1 diabetes.

B. Progress to Date

Human β-Cell Lines

There have been several reports of attempts to create stable cell lines from human pancreatic tissue, but invariably these have not succeeded. These studies have generally used either human islet cell tumors (Chick et al., 1973; Thivolet et al., 1988) or islets isolated from pancreata of human cadaver organ donors (Soldevila et al., 1991). In the case of the human islet cell tumors, cells were dissociated from insulinoma fragments, plated, and passaged by trypsinization, while the human islet cells were transfected with SV40 T antigen DNA and cell lines were established from the resultant colonies. The human insulinoma cells were either glucose-unresponsive and became overgrown by fibroblasts (Chick et al., 1973), or showed small (maximum 1.5-fold) responses to glucose, but failed to proliferate (Thivolet et al., 1988). The SV40-transfected cells proliferated well in culture and maintained most of the features of the parental islet cells for the first eight passages. However, the phenotype was unstable, and the cells progressively lost the capacity to synthesize and secrete insulin (Soldevila et al., 1991). Progress may be made with the development of stable human β-cell lines if human fetal pancreas is used as the starting material, since the β-cells of the fetal pancreas have the ability to replicate, but this is progressively lost with age. A recent preliminary report suggests that this is a possibility (Wang et al., 1997) and the possibility of generating appropriately regulated insulin-secreting cell lines following pancreatectomy of children with persistent hyperinsulinemic hypoglycemia of infancy is currently being explored (Chapman et al., 1998). However, we are still some way from having plentiful supplies of well-defined human β-cells available for transplantation purposes.

Artificial β-Cells

Progress in the engineering of insulin-secreting cells from non-β-cell material has been covered earlier in this chapter, and the rationale for developing such cells is clear: treatment of patients with type 1 diabetes would simply be a matter of removing cells from the diabetic patient, preferably by biopsy; adding the genes essential for regulated insulin synthesis and secretion; and reintroducing the surrogate β-cells into the patient, where they would perform the function of the native β-cells that had succumbed to autoimmune attack. At present, work carried out with the AtT-20ins cells has provided the most encouraging data for the artificial β-cell approach, but even this model is far from ideal (Hughes et al., 1992). It is clear that further developments are needed in our understanding of the precise nature of glucose sensing, metabolic coupling factors, and the intracellular mechanisms required for regulated insulin secretion from normal β-cells before non-β-cells can be engineered to respond appropriately to initiators and potentiators of secretion.

Rodent β-Cell Lines

Much energy is currently being applied to the engineering of rodent β-cell lines for diabetes therapy (reviewed by Newgard et al., 1997). The aim of this approach is to modify β-cell lines derived from rodent tissue so that they take on the characteristics of human β-cells. In theory, this is a sensible strategy since all of the β-cell lines established to date have the capacity to synthesize, process, store, and secrete insulin; many of them are glucose-responsive; and they tend to show normal responses to modulators of insulin secretion such as incretins and neurotransmitters. In spite of this, rodent β-cells are not ideal for transplantation into patients with type 1 diabetes. In addition to the problems inherent in xenograft transplantation, rodent β-cells do not secrete human insulin; the glucokinase:hexokinase ratio is often unfavorable, resulting in inappropriate secretory responses to nutrients; their phenotype may change with passage; and they are by nature proliferative cells, which could lead to tumor growth following transplantation. In addition, the expression of the low K_m GLUT-1 transporter in many β-cell lines is often considered to be a disadvantage (Newgard et al., 1997), despite the predominance of GLUT-1 in human β-cells (De Vos et al., 1995).

Of the many rodent β-cell lines established so far, two (RIN-38 and βTC-tet) are considered to have potential as surrogate β-cells for insulin replacement in patients with type 1 diabetes. RIN-38 cells might seem to be an unusual choice given that they have an unstable phenotype, are essentially nonglucose-responsive, and have a high hexokinase:glucokinase ratio. However, RIN-38 cells have been engineered to express glucokinase, GLUT-2 genes, and multiple copies of the human insulin gene, and expression of these genes was found to be stable after transplantation of the engineered cells into nude rats (Clark et al., 1997). RIN-38 cells engineered to express human insulin, glucokinase, and GLUT-2 contained between 100 and 200 ng human insulin/10^6 cells and showed up to eightfold increases in insulin secretion in response to 5 mM glucose (Hohmeier et al., 1997). These promising observations are tempered by the fact that the cells still responded to subthreshold concentrations of glucose, with maximal responses at concentrations as low as 250 μM (Hohmeier et al., 1997). However, inhibition of low K_m hexokinases resulted in a rightward shift in the glucose dose–response curve (Hohmeier et al., 1997), suggesting that genetic manipulation to suppress hexokinase activity in engineered RIN-38 cells may produce more physiologically appropriate models.

The βTC-tet cell line is an intriguing possibility for β-cell replacement therapy in type 1 diabetes because proliferative activity of the cells is controlled by tetracycline. Thus, exposure of βTC-tet cells to tetracycline resulted in reversible growth arrest in vitro, and, more importantly, tetracycline also regulated the growth of βTC-tet cells in vivo after their intraperitoneal injection into mice (Efrat et al., 1995). None of the tetracycline-treated mice developed tumors, and all maintained normoglycemia. This was in marked contrast to those maintained in the absence of tetracycline, which developed tumors and hypoglycemia (Efrat et al., 1995). Also promising are the observations that the tetracycline derivative anhydrotetracycline, a weak antibiotic, was 100-fold more potent in inhibiting βTC-tet cell proliferation, that βTC-tet cells restored normoglycemia when implanted into streptozotocin-diabetic mice (Efrat et al., 1995), and that they showed concentration-dependent increases in insulin secretion in vitro over the range of 1 to 10 mM glucose, albeit in the presence of IBMX (Efrat, 1996). Although the βTC-tet line is far from perfect in that it secretes mouse, rather than human, insulin, requires the continuous presence of tetracycline or a derivative to prevent tumor growth, and is still prone to immune attack by the host system, it will undoubtedly make a useful contribution to the development of genetically engineered β-cells for transplantation therapy in type 1 diabetes. Should a stable human β-cell line be developed, the methodologies applied to the βTC-tet cell line could be refined to provide the much-needed abundant supplies of human β-cell substitutes required for transplantation in type 1 diabetes.

VI. FUTURE PERSPECTIVES

There is no doubt that β-cell lines have provided us with an opportunity to investigate signal transduction mechanisms involved in insulin biosynthesis and secretion, which would not have been possible using primary islets, and they have also paved the way for genetic manipulations that, in the future, may be applied in transplantation therapy for type 1 diabetes. It is also clear that none of the cell lines developed to date and discussed in this chapter are ideal models for normal β-cells and that the development of a well-characterized human β-cell line would be an enormous step forward. Not only would human β-cells be of great use for basic research, they would be invaluable for transplantation purposes and would circumvent the problems associated with xenograft rejection. When we reach the stage of having human β-cells of the desired characteristics available for transplantation, new problems will have to be addressed. Given that secretory responses of dispersed β-cells are considerably less than those of equivalent numbers of β-cells within islets, it will need to be considered whether cells in isolation perform sufficiently well, or if it is necessary to transplant artificial islet cell clusters that show cell–cell communication. The future is promising and the use of cell lines for β-cell replacement therapy in type 1 diabetes is becoming an ever more likely possibility.

REFERENCES

Aguilar-Bryan, L., Nichols, C.G., Wechsler, S.W., Clement J.P. IV, Boyd A.E. III, Gonzalez, G., Herrera-Sosa, H., Nguy, K., Bryan, J. and Nelson, D.A. (1995). Cloning of the β-cell high affinity sulfonylurea receptor: A regulator of insulin secretion. Science 268, 423–426.

Asfari, M., Janjic, D., Meda, P., Li, G., Halban, P.A. and Wollheim, C.B. (1992). Establishment of 2-mercaptoethanol-dependent differentiated insulin-secreting cell lines. Endocrinology 130, 167–178.

Ashcroft, S.J.H., Bunce, J., Lowry, M., Hansen, S.E. and Hedeskov, C.J. (1978). The effect of sugars on (pro)insulin biosynthesis. Biochem. J. 174, 517–526.

Ashcroft, S.J.H., Hammonds, P. and Harrison, D.E. (1986). Insulin secretory responses of a clonal cell line of simian virus 40-transformed B cells. Diabetologia 29, 727–733.

Ashcroft, S.J.H. and Stubbs, M. (1987). The glucose sensor in HIT cells is the glucose transporter. FEBS Lett. 219, 311–315.

Blondel, O., Bell, G.I., Moody, M., Miller, R.J. and Gibbons, S.J. (1994). Creation of an inositol 1,4,5-trisphosphate-sensitive Ca^{2+} store in secretory granules of insulin-producing cells. J. Biol. Chem. 269, 27167–27170.

Chapman, J.C., Macfarlane, W.M., Cosgrove, K.E., Shepherd, R.M., Kamimura, N., James, R.F.L., Aynsley-Green, A., and Dunne, M.J. (1998). NES 2Y: A novel human insulin-secreting cell line. Diabetic Medicine 15, S23.

Chick, W.L., Lauris, V., Soeldner J.S., Tan, M.H. and Grinbergs, M. (1973). Monolayer culture of a human pancreatic beta-cell adenoma. Metabolism 22, 1217–1224.

Chick, W.L., Shields, W., Chute, R.N., Like, A.A., Lauris, V. and Kitchen, K.C. (1977). A transplantable insulinoma in the rat. Proc. Natl. Acad. Sci. USA 74, 628–633.

Clark, S.A., Burnham, B.L. and Chick, W.L. (1990). Modulation of glucose-induced insulin secretion from a rat clonal β-cell line. Endocrinology 127, 2779–2788.

Clark, S.A., Quaade, C., Constady, H., Hansen, P., Halban, P., Ferber, S., Newgard, C.B. and Normington, K. (1997). Novel insulinoma cell lines produced by iterative engineering of GLUT2, glucokinase, and human insulin expression. Diabetes 46, 958–967.

Davidson, H.W., Rhodes, C.J. and Hutton, J.C. (1988). Intraorganellar calcium and pH control proinsulin cleavage in the pancreatic β cell via two distinct site-specific endopeptidases. Nature 333, 93–96.

De Vos, A., Heimberg, H., Quartier, E. Huypens, P., Bouwens, L., Pipeleers, D. and Schuit, F. (1995). Human and rat beta cells differ in glucose transporter but not in glucokinase gene expression. J. Clin. Invest. 96, 2489–2495.

Efrat, S., Linde, S., Kofod, H., Spector, D., Delannoy, M., Grant, S., Hanahan, D. and Baekkeskov, S. (1988). Beta-cell lines derived from transgenic mice expressing a hybrid insulin gene-oncogene. Proc. Natl. Acad. Sci. USA 85, 9037–9041.

Efrat, S., Leiser, M., Surana, M., Tal, M., Fusco-DeMane, D. and Fleischer, N. (1993). Murine insulinoma cell line with normal glucose-regulated insulin secretion. Diabetes 42, 901–907.

Efrat, S., Fusco-DeMane, D., Lemberg, H., Al Emran, O. and Wang, X. (1995). Conditional transformation of a pancreatic β-cell line derived from transgenic mice expressing a tetracycline-regulated oncogene. Proc. Natl. Acad. Sci. USA 92, 3576–3580.

Efrat, S. (1996). Genetic engineering of β-cells for cell therapy of diabetes: Cell growth, function and immunogenicity. Diabetes Rev. 4, 224–234.

Ferber, S., BeltrandelRio, H., Johnson, J.H., Noel, R.J., Cassidy, L.E., Clark, S., Becker, T.C., Hughes, S.D. and Newgard, C.B. (1994). GLUT-2 gene transfer into insulinoma cells confers both low and high affinity glucose-stimulated insulin release. Relationship to glucokinase activity. J. Biol. Chem. 269, 11523–11529.

Gazdar, A.F., Chick, W.L., Oie, H.K., Sims, H.L., King, D.L., Weir, G.C. and Lauris, V. (1980). Continuous, clonal, insulin- and somatostatin-secreting cell lines established from a transplantable rat islet cell tumor. Proc. Natl. Acad. Sci. USA 77, 3519–3523.

Groskreutz, D.J., Sliwkowski, M.X. and Gorman, C.M. (1994). Genetically engineered proinsulin constitutively processed and secreted as mature, active insulin. J. Biol. Chem. 269, 6241–6245.

Grupe, A., Hultgren, B., Ryan, A., Ma, Y.H., Bauer, M. and Stewart, T.A. (1995). Transgenic knockouts reveal a critical requirement for pancreatic β cell glucokinase in maintaining glucose homeostasis. Cell 83, 69–78.

Halban, P.A., Praz, G.A. and Wollheim, C.B. (1983). Abnormal glucose metabolism accompanies failure of glucose to stimulate insulin release from a rat pancreatic cell line (RINm5F). Biochem. J. 212, 439–443.

Hayashi, M., Taguchi, M. and Fujita, T. (1997). The IRI release from insulin gene-transduced hepatic cells responds to ambient glucose concentration. Biochem. Biophys. Res. Commun. 233, 470–475.

Hohmeier, H.E., BeltrandelRio, H., Clark, S.A., Henkel-Rieger, R., Normington, K. and Newgard, C.B. (1997). Regulation of insulin secretion from novel engineered insulinoma cell lines. Diabetes 46, 968–977.

Howell, S.L. (1972). Role of ATP in the intracellular translocation of proinsulin and insulin. Nature New Biol. 235, 85–87.

Hughes, S.D., Johnson, J.H., Quaade, C. and Newgard, C.B. (1992). Engineering of glucose-stimulated insulin secretion and biosynthesis in non-islet cells. Proc. Natl. Acad. Sci. USA 89, 688–692.

Ishihara, H., Asano, T., Tsukuda, K., Katagiri, H., Inukai, K., Anai, M., Kikuchi, M., Yazaki, Y., Miyazaki, J.-I. and Oka, Y. (1993). Pancreatic beta cell line MIN6 exhibits characetristics of glucose metabolism and glucose-stimulated insulin secretion similar to those of normal islets. Diabetologia 36, 1139–1145.

Ishihara, H., Asano, T., Tsukuda, K., Katagiri, H., Inukai, K., Anai, M., Kikuchi, M., Yazaki, Y., Miyazaki, J.-I. and Oka, Y. (1994). Overexpression of hexokinase I but not GLUT1 glucose transporter alters concentration dependence of glucose-stimulated insulin secretion in pancreatic β-cell line MIN6. J. Biol. Chem. 269, 3081–3087.

Kaufmann, J.E., Irminger, J.-C., Mungall, J. and Halban, P.A. (1997). Proinsulin conversion in GH3 cells after coexpression of human proinsulin with the endoproteases PC2 and/or PC3. Diabetes 46, 978–982.

Kawakami, Y., Yamaoka, T., Hirochika, R., Yamashita, K., Itakura, M. and Nakauchi, H. (1992). Somatic gene therapy for diabetes with an immunological safety system for complete removal of transplanted cells. Diabetes 41, 956–961.

Knaack, D., Fiore, D.M., Surana, M., Leiser, M., Laurance, M., Fusco-DeMane, D., Hegre, O.D., Fleischer, N. and Efrat, S. (1994). Clonal insulinoma cell line that stably maintains correct glucose responsiveness. Diabetes 43, 1413–1417.

Liang, Y., Bai, G., Doliba, N., Buettiger, C., Wang, L., Berner, D.K. and Matschinsky, F.M. (1996). Glucose metabolism and insulin release in mouse βHC9 cells, as model for wild-type pancreatic β-cells. Am. J. Physiol. 270, E846–E857.

Lu, D., Hoshino, H. and Takeuchi, T. (1996). Regulatable production of mature insulin from a hepatocyte cell line: insulin production is up-regulated by cAMP and glucocorticoids, and down-regulated by insulin. FEBS Lett. 399, 37–42.

Mashima, H., Shibata, H., Mine, T. and Kojima, I. (1996). Formation of insulin-producing cells from pancreatic acinar AR42J cells by hepatocyte growth factor. Endocrinology 137, 3969–3976.

Matschinsky, F., Liang Y., Kesavan, P., Wang, L., Froguel, P., Velho, G., Cohen, D., Permutt, M.A., Tanizawa, Y., Jetton, T.L., Niswender, K. and Magnusson, M.A. (1993). Glucokinase as pancreatic β-cell glucose sensor and diabetes gene. J. Clin. Invest. 92, 2092–2098.

McClenaghan, N.H., Barnett, C.R., Ah-Sing, E., Abdel-Wahab, Y.H.A., O'Harte, F.P.M., Yoon, T.-W., Swanston-Flatt, S.K. and Flatt, P.R. (1996a). Characterization of a novel glucose-responsive insulin-secreting cell line, BRIN-BD11, produced by electrofusion. Diabetes 45, 1132–1140.

McClenaghan, N.H., Barnett, C.R., O'Harte, F.P.M., Swanston-Flatt, S.K. Ah-Sing, E. and Flatt, P.R. (1996b). Characteristics of BRIN-BG5 and BRIN-BG7, two novel glucose-responsive insulin-secreting cell lines produced by electrofusion. J. Endocrinol. 148, 409–417.

McClenaghan, N.H., Gray, A.M., Barnett, C.R. and Flatt, P.R. (1996c). Hexose recognition by insulin-secreting BRIN-BD11 cells. Biochem. Biophys. Res. Commun. 223, 724–728.

Miyazaki, J.-I., Araki, K., Yamato, E., Ikegami, H., Asano, T., Shibasaki, Y., Oka, Y. and Yamamura, K.-I. (1990). Establishment of a pancreatic β cell line that retains glucose-inducible insulin secretion: special reference to expression of glucose transporter isoforms. Endocrinology 127, 126–132.

Moore, H.-P.H., Walker, M.D., Lee, F. and Kelly, R.B. (1983). Expressing a human proinsulin cDNA in a mouse ACTH-secreting cell. Intracellular storage, proteolytic processing, and secretion on stimulation. Cell 35, 531–538.

Newgard, C.B., Clark, S., BeltrandelRio, H., Hohmeier, H.E., Quaade, C. and Normington, K. (1997). Engineered cell lines for insulin replacement in diabetes: current status and future prospects. Diabetologia 40, S42–S47.

Niesor, E.J., Wollheim, C.B., Mintz, D.H., Blondel, B., Renold, A.E. and Weil, R. (1979). Establishment of rat pancreatic endocrine cell lines by infection with simian virus 40. Biochem. J. 178, 559–568.

Noda, M., Komatsu, M. and Sharp, G.W.G. (1996). The βHC-9 pancreatic β-cell line preserves the characteristics of progenitor mouse islets. Diabetes 45, 1766–1773.

Ogawa, Y., Noma, Y., Davalli, A.M., Wu, Y.-J., Thorens, B., Bonner-Weir, S. and Weir, G.C. (1995). Loss of glucose-induced insulin secretion and GLUT2 expression in transplanted β-cells. Diabetes 44, 75–79.

Pipeleers, D.G., Marichal, M. and Malaisse, W.J. (1973). The stimulus-secretion coupling of glucose-induced insulin release. XIV. Glucose regulation of insular biosynthetic activity. Endocrinology 93, 1001–1011.

Praz, G.A., Halban, P.A., Wollheim, C.B., Blondel, B., Strauss, A.J. and Renold, A.E. (1983). Regulation of immunoreactive-insulin release from a rat cell line (RINm5F). Biochem. J. 210, 345–352.

Radvanyi, F., Christgau, S., Baekkeskov, S., Jolicoeur, C. and Hanahan, D. (1993). Pancreatic β cells cultured from individual preneoplastic foci in a multistage tumorigenesis pathway: A potentially general technique for isolating physiologically representative cell lines. Mol. Cell. Biol. 13, 4223–4232.

Rhodes, C.J. and Halban, P.A. (1987). Newly synthesized proinsulin/insulin and stored insulin are released from pancreatic B cells predominantly via a regulated, rather than a constitutive, pathway. J. Cell Biol. 105, 145–153.

Santerre, R.F., Cook, R.A., Crisel, R.M.D., Sharp, J.D., Schmidt, R.J., Williams, D.C. and Wilson, C.P. (1981). Insulin synthesis in a clonal cell line of simian virus 40-transformed hamster pancreatic beta cells. Proc. Natl. Acad. Sci. USA 78, 4339–4343.

Selbie, L.A., Scmitz-Peiffer, C., Sheng Y and Biden, T.J. (1993). Molecular cloning and characterization of PKCι, an atypical isoform of protein kinase C derived from insulin-secreting cells. J. Biol. Chem. 268, 24296–24302.

Selden, R.F., Skoskiewicz, M.J., Russell, P.S. and Goodman, H.M. (1987). Regulation of insulin-gene expression: implication for gene therapy. N. Eng. J. Med. 317, 1067–1076.

Sodoyez, J.C., Luyckx, A.S. and Lefebvre, P.J. (1967). Biological properties of a transplantable islet-cell tumor of the golden hamster. II. Insulin content of the tumor and some metabolic characteristics of the tumor-bearing animals. Diabetes 16, 415–417.

Soldevila, G., Buscema, M., Marini, V., Sutton, R., James, R.F.L., Bloom, S.R., Robertson, R.P., Mirakian, R., Pujol-Borrell, R. and Bottazzo, G.F. (1991). Transfection with SV40 gene of human pancreatic endocrine cells. J. Autoimmun. 4, 381–396.

Tal, M., Matschinsky, F., Liang, Y., Najafi, H. and Lodish, H. (1992). Expression and function of GLUT-1 and GLUT-2 glucose transporter isoforms in cells of cultured rat pancreatic islets. J. Biol. Chem. 267, 17241–17247.

Thivolet, C.H., Demidem, A., Haftek, M., Durand, A. and Bertrand, J. (1988). Structure, function, and immunogenicity of human insulinoma cells. Diabetes 37, 1279–1286.

Thorens, B., Sarkar, H.K., Kaback, H.R. and Lodish, H.F. (1988). Cloning and functional expression in bacteria of a novel glucose transporter present in liver, intestine, kidney and beta-pancreatic islet cells. Cell 55, 281–290.

Trautmann, M.E. and Wollheim, C.B. (1987). Characterization of glucose transport in an insulin-secreting cell line. Biochem. J. 242, 625–630.

Wang, S., Beattie, G.M., Mally, M.I., Lopez, A.D., Hayek, A. and Levine, F. (1997). Analysis of a human fetal pancreatic islet cell line. Transplantation Proc. 29, 2219.

Wang, H. and Iynedjian, P.B. (1997). Modulation of glucose-responsiveness of insulinoma β-cells by graded overexpression of glucokinase. Proc. Natl. Acad. Sci. USA 94, 4372–4377.

Welsh, N., Bendtzen, K. and Welsh, M. (1995). Expression of an insulin/interleukin-1 receptor antagonist hybrid gene in insulin-producing cell lines (HIT-T15 and NIT-1) confers resistance against interleukin-1-induced nitric oxide production. J. Clin. Invest. 95, 1717–1722.

Whitesell, R.R., Powers, A.C., Regen, D.M. and Abumrad, N.A. (1991). Transport and metabolism of glucose in an insulin-secreting cell line, βTC-1. Biochemistry 30, 11560–11566.

β-CELL DYSFUNCTION AND DEATH

Décio L. Eizirik and Anne Hoorens

I. INTRODUCTION

The different forms of diabetes mellitus are characterized by deficient insulin release in relation to the insulin demand. This can be caused either by immune-, toxin-, or viral-mediated β-cell damage or by a site-specific defect in β-cell substrate metabolism. Moreover, there is evidence that chronic demand for high insulin production may lead to β-cell "exhaustion" and prolonged dysfunction.

A large amount of research in noninsulin-dependent diabetes mellitus (type 2 diabetes) has focused on β-cell biochemical defects, either innate or induced by

Advances in Molecular and Cell Biology
Volume 29, pages 47-73.

ISBN: 0-7623-0547-9

metabolical stress (reviewed in Malaisse, 1994; Polonsky et al., 1996); however, most research in insulin-dependent diabetes mellitus (type 1 diabetes) is directed to immune mediators of β-cell destruction. Indeed, the pancreatic β-cell in type 1 diabetes is often seen as "a corpse in a funeral—the object of much attention, but playing no active role in the proceedings" (adapted from Hyans, 1996). β-Cell dysfunction and damage in early type 1 diabetes may result from a direct contact with islet-infiltrating mononuclear cells (reviewed in Bach, 1995) and/or exposure to inflammatory mediators produced by these cells, such as cytokines and toxic free radicals (reviewed in Rabinovitch, 1993; Mandrup-Poulsen, 1996; Eizirik and Pavlovic, 1997). Moreover, as the β-cell mass decreases, the surviving insulin-producing cells will be submitted to an increased functional demand to maintain normoglycemia, which may further contribute to β-cell dysfunction. Recent evidence indicates that, far from being "a corpse in a funeral," the β-cell is an active participant in all the steps leading to its death or survival in early type 1 diabetes. Thus, following immune- or toxin-mediated β-cell damage, there is induction of diverse intracellular signals, leading to transcription of different genes which may either contribute to β-cell repair and/or defense or lead to cell death (reviewed in Eizirik et al., 1993a; Eizirik et al., 1996b).

In the present chapter, we discuss some of the inflammatory and metabolic agents potentially responsible for β-cell dysfunction and death in type 1 diabetes, the repair/defense mechanisms triggered by these cells following injury, and the different possible β-cell outcomes following damage; namely functional recovery, permanent dysfunction, necrosis, or apoptosis. Special emphasis is devoted to the β-cell intracellular mechanisms participating in the previously described phenomena.

II. INFLAMMATORY AND METABOLICAL AGENTS INDUCING β-CELL DYSFUNCTION

A. Cytokines

Cytokines are polypeptide cell regulators, produced by a wide variety of cells and involved in cellular communication, mostly related to immune system function. Data obtained from animal models of type 1 diabetes suggest that cytokines such as interleukin-1 (IL-1), interferon-γ (IFN-γ), and tumor necrosis factor-α (TNF-α) are produced in and around the islets during insulitis (Jiang and Woda, 1991; Rabinovitch et al., 1996a; Hirasawa et al., 1997). Under in vitro conditions, these cytokines induce β-cell dysfunction and death (reviewed in Mandrup-Poulsen, 1996). The most potent cytokine in this respect is interleukin-1β (IL-1β), which, alone or in combination with IFN-γ and/or TNF-α, inhibits rat islet total protein and (pro)insulin biosynthesis, decreases glucose-induced insulin release and insulin mRNA content, and reduces total islet DNA and insulin content (reviewed in Sandler et al., 1991; Mandrup-Poulsen 1996). This is at least in part secondary to cytokine-induced inhibition of glucose oxidation at the mitochondrial level and consequent decrease in adenosine triphosphate (ATP) production and Ca^{2+} uptake (Sandler et al., 1991). IL-1β–exposed β-cells preserve a better secretory response to arginine and cyclic adenosine monophosphate (cAMP) generators (Eizirik et al., 1989), agents that do not depend on cellular metabolism for the induction of insulin release. Most of the inhibitory effects of cytokines on rat islets are mediated by expression of the inducible form of nitric oxide synthase (iNOS) and consequent synthesis of the radical nitric oxide (NO; reviewed by Eizirik et al., 1996b; Eizirik and Pavlovic, 1997) (see the discussion that follows). IL-1–induced iNOS mRNA expression in rodent insulin-producing cells depends on gene transcription, protein synthesis, and activation of the nuclear transcription factors NF-κB, Stat1α, and interferon regulatory factor-1 (IRF-1) (Eizirik et al., 1996b; Darville and Eizirik, 1998). Cytokine-induced NO production inhibits iNOS mRNA expression in these cells (Niemann et al., 1994), an effect probably secondary to blockage of NF-κB activation (Peng et al., 1995). Considering the potential cellular toxicity of elevated concentrations of NO (see later discussion), this NO-induced negative feedback may represent a protective mechanism against excessive radical production.

IL-1β also inhibits mouse islets insulin release, (pro)insulin biosynthesis, and insulin mRNA expression (Eizirik et al., 1990; Eizirik, 1991), but in this species IL-1β effects are independent of inhibition of glucose oxidation and ATP generation (Eizirik et al., 1990; Eizirik, 1991). However, combinations of IL-1β plus TNF-γ plus IFN-γ impair mouse islet glucose oxidation, a situation associated with increased NO production and decreased islet cell viability (Cetkovic-Cvrlje and Eizirik, 1994). In the case of human islets, IL-1β alone neither induces iNOS expression nor decreases islet function or viability (Eizirik et al., 1993b). On the other hand, combinations of cytokines containing IL-1β plus IFN-γ plus TNF-α or IL-1β plus IFN-γ lead to decreased insulin release (Corbett et al., 1993; Eizirik et al., 1994b), increased iNOS mRNA expression and NO production (Eizirik et al., 1994b), and, after 6 to 9 days of exposure, DNA strand breaks and death by apoptosis (Delaney et al., 1997). In these cells activation of NF-κB is a necessary but not sufficient signal for cytokine-induced NO formation (Eizirik et al., 1996b), indicating a potential role for other transcription factors, such as interferon regulatory factor-1 (Flodström and Eizirik, 1997). As described for mouse islets, a decrease in mitochondrial function is a late event in human islets exposed to cytokines (Eizirik et al., 1994b). Inhibition of iNOS expression fails to protect human islets against the deleterious effects of combinations of cytokines on insulin release, DNA integrity, and cell viability (Eizirik et al., 1994b; Delaney et al., 1997), suggesting that other intracellular mediators contribute to these actions of cytokines. The nature of these mediators remains to be determined, but they may be related to generation of free oxygen radicals and aldehyde production (Rabinovitch et al., 1996b) and/or increased expression of proapoptotic genes and proteins, such as the interleukin-1 converting enzyme (ICE or caspase 1) (Karlsen et al., 1998).

Cytokines induce the expression of several other genes in whole pancreatic islets and insulin-producing cell lines. These include argininosuccinate synthetase (AS), cyclooxygenase (COX), manganese superoxide dismutase (MnSOD), heme oxygenase, and heat shock protein 70 (hsp70) (reviewed in Eizirik et al., 1996b). While proteins such as iNOS, AS (which supports NO formation by regenerating arginine; see later discussion), and COX may exacerbate the inflammatory reaction and thus contribute to β-cell destruction, MnSOD and hsp70 are probably involved in β-cell defense and/or repair. Interestingly, both blockers of protein synthesis and inhibitors of NF-κB activation prevent IL-1β-induced iNOS expression without preventing the increase in cellular MnSOD mRNA (Eizirik et al., 1996b). This suggests the possibility of preserving the "protective" effects of cytokines, such as MnSOD induction, while at the same time blocking the transduction of "deleterious" signals, such as those leading to iNOS expression. Further characterization of the molecular regulation of iNOS, MnSOD, and other cytokine-related genes in β-cells is required to achieve this goal.

B. Reactive Nitrogen and Oxygen Species

The naturally occurring free radicals contain oxygen- or nitrogen-based unpaired electrons, and comprise molecules such as NO, the superoxide anion, and the hydroxyl radical (Lander, 1997). At high concentrations, these radicals are part of the immunological host defense, acting as bactericidal and antiviral agents (MacMicking et al., 1997). The formation of large amounts of free radicals can also be deleterious for the cells producing them, and there are indications that both nitrogen (NO and derived molecules) and oxygen (superoxide and hydroxyl) radicals contribute to tissue damage in diverse autoimmune diseases, including type 1 diabetes (reviewed in Rabinovitch 1993; Eizirik and Pavlovic, 1997; Kröncke et al., 1998). At lower concentrations, these molecules play a role as physiological intracellular messengers (Lander, 1997). For instances, NO participates in such diverse processes as regulation of blood flow and pressure, platelet aggregation and adhesion, and neurogenic vasodilation (Moncada and Higgs, 1993; Moncada et al., 1997).

NO is formed by a family of enzymes named NO synthases, which use the amino acid L-arginine as the main substrate. Three NOS isoforms have been characterized, with two constitutive (eNOS and nNOS) and one inducible (iNOS) form (for reviews see Xie and Nathan, 1994; Griffith and Stuher, 1995; Eizirik et al., 1996b). In the context of β-cell damage, iNOS is clearly the most relevant. Exposure of rat and mouse islets, and rodent insulinoma cell lines, to IL-1β alone, or rodent and human islets to IL-1β plus IFN-γ or IL-1β plus IFN-γ plus TNF-α leads to iNOS expression and production of large amounts of iNOS (reviewed in Eizirik et al., 1996b). In the reaction catalysed by iNOS, arginine is converted into NO and citrulline. Cellular NO formation depends on the availability of arginine, and the amino acid can be provided both extracellularly, gaining access to the cell via the cationic amino acid transport system y + CAT, or intracellularly, either by synthesis from citrulline (the citrulline–NO cycle) or by protein degradation (reviewed by Morris and Billiar, 1994). The rate-limiting enzyme for arginine regeneration from citrulline is AS, and iNOS and AS are co-induced by cytokines in human and rat pancreatic islets, enabling these cells to synthesize NO in the absence of arginine but presence of citrulline (Flodström et al., 1995). Arginine transport into rat β-cells is also increased by IL-1β (Flodström et al., 1999), while arginase, an enzyme that catalyzes arginine convertion into ornithine and urea, is inhibited (Cunningham et al., 1997). Thus, cytokine-induced iNOS expression in pancreatic islets is paralleled by activation of other biochemical pathways which contribute to effective arginine uptake, arginine regeneration from citrulline, and its use for NO production. This is probably of relevance, since low arginine concentration often prevails in sites of inflammation (Morris and Billiar, 1994). Thus, it seems that β-cells developed an efficient way to support continuous NO generation under cytokine stimulation. It remains to be clarified which is the physiological relevance of this complex system for the β-cell, but one possibility is that NO participates in an innate cellular defense system against viral infections (MacMicking et al., 1997).

Excessive NO production may lead to both rat and human β-cell death (reviewed in Eizirik and Pavlovic, 1997). Some of the intracellular targets for NO are the β-cell mitochondria and nucleus. NO reacts with iron/sulphur centers in enzymes, and the Krebs cycle enzyme aconitase contains such a center in its catalitical site. IL-1β–induced NO formation severely decreases aconitase activity in rat pancreatic islets (Welsh et al., 1991), contributing to the above-described inhibitory effect of IL-1β on glucose oxidation. NO also damages mitochondrial DNA (Wilson et al., 1997), which may further aggravate the organelle dysfunction.

NO, or products of its reaction with oxygen free radicals (see later discussion), induce strand breaks in both rat (Delaney et al., 1993) and human pancreatic islets (Eizirik et al., 1996a; Delaney et al., 1996). NO-induced nuclear DNA strand breaks may activate poly(ADP-ribose) polymerase (PARP), an enzyme that participates in DNA repair but consumes nicotinamide adenine nucleotide (NAD) in this process, contributing to the NO-induced impairment in mitochondrial function. Inhibitors of PARP activation, such as nicotinamide, reduce cell loss in rat (Radons et al., 1994) and human (Eizirik et al., 1996a) islets exposed to NO donors. Moreover, islets isolated from PARP-deficient mice are less sensitive to both NO-and oxygen-free radical–induced cell death (Heller et al., 1995).

NO reacts with a high rate constant with superoxide, forming the toxic radical peroxynitrite (for a review see Beckman and Koppenol, 1996). Peroxynitrite may contribute to some of the deleterious effects attributed to NO, such as aconitase blockage and DNA damage. Peroxynitrite leads to nitration of tyrosine residues in proteins, a lesion detectable by immunohistochemical techniques using antibodies to nitrotyrosine. Using this approach, the production of peroxynitrite was demonstrated in different inflammatory conditions, including islets of Langerhans isolated from diabetes-prone nonobese diabetic (NOD) mice already infiltrated by mononuclear cells (Suarez-Pinzon et al., 1997). In these experiments, around 70% of the β-cells isolated from acutely diabetic NOD mice stained positively for nitrotyrosine, suggesting that the majority of β-cells were exposed to both nitrogen and oxygen reactive species during insulitis. Due to obvious ethical limitations, similar information is not available for islets isolated from human patients during the early stages of type 1 diabetes. However, there is evidence that human islets exposed in vitro to peroxynitrite present acute DNA strand breaks and decreased glucose metabolism, evolving to cell death after 1 to 24 hours (Delaney et al., 1996).

C. Chronic Hyperglycemia

Chronic hyperglycemia may contribute to the defective β-cell function observed both in type 2 diabetes and in early type 1 diabetes. Experimental support for this concept relies mostly on animal models, such as the neonatal injection of streptozotocin, partial pancreatectomy, and in vivo infusion of glucose in rats (reviewed in Leahy et al., 1992; Robertson et al., 1994; Leahy 1996). Genetic differences between experimental models are a relevant factor for β-cell susceptibility to sustained hyperglycemia (Korsgren et al., 1990). Thus, when islets obtained from C57/BL6 (B6) and C57BL/Ks (BKs) mice were syngeneically transplanted under the kidney capsule of alloxan diabetic mice, in insufficient amounts to normalize glycemia, BKs islets presented a 50% reduction of graft volume and a severe defect in glucose-induced insulin release after 6 weeks in vivo. This impaired function was not restored after a subsequent 4 to 12 weeks of normoglycemia. On the other hand, B6 islets did not present a decrease in graft volume, and, although these islets failed to release insulin in response to glucose after 6 weeks of hyperglycemia, their function was restored after a subsequent 4 to 12 weeks of normoglycemia (Korsgren et al., 1990). Interestingly, when glucose is infused in rats in parallel with diazoxide, an agent that inhibits insulin secretion by opening ATP-dependent K^+ channels at the cell membrane, the hyperglycemia-induced defect in insulin release is prevented (Sako and Grill, 1990). This finding suggests that excessive β-cell stimulation, rather than a direct "toxic" effect of glucose, underlies hyperglycemia-induced β-cell dysfunction.

There are several hypothesis to explain the in vivo deleterious effects of the diabetic state on rodent β-cells. These hypotheses include depletion of insulin reserves, glycogen deposition, decreased insulin mRNA expression and consequent impaired (pro)insulin biosynthesis, and defective glucose transport and oxidation (Malaisse et al., 1993; Robertson et al., 1994; Leahy 1996). Furthermore, there are data indicating that elevated levels of free fatty acids and ketone bodies, often present in decompensated diabetes, may impair glucose metabolism and further decrease insulin release (Zhou and Grill, 1994). Unfortunately, none of these hypothesis appears to provide a general explanation for the phenomenon and, as discussed in the next paragraph, they do not seem of major relevance for human islets.

Studies in patients at the clinical onset of type 1 diabetes and type 2 diabetes indicate that intensive insulin therapy and strict glycemic control improves β-cell function (Shah et al., 1989; Ilkova et al., 1997), suggesting that hyperglycemia may also be deleterious for human β-cells. In line with this suggestion, in vitro studies demonstrated that isolated human islets exposed for 7 days to 11 or 28 mM glucose fail to respond to a short-term glucose stimulation (Eizirik et al., 1992). In subsequent experiments, human islets were transplanted under the kidney capsule of normoglycemic or diabetic nude mice (Jansson et al., 1995; Eizirik et al., 1997). The amount of transplanted islets was insufficient to cure the diabetic state, and the β-cells were consequently exposed to severe hyperglycemia (>20 mM glucose) for 4 to 6 weeks. In some groups, a second intrasplenic implant of mouse islets restored normoglycemia for 2 extra weeks. Exposure of human islets to hyperglycemia for 4 to 6 weeks induced a severe impairment of the insulin release in response to both glucose and arginine, a defect that was not restored by a subsequent 2 weeks of normoglycemia (Jansson et al., 1995; Eizirik et al., 1997). The decrease in insulin release was independent of glycogen accumulation and depletion of insulin reserves, and it was accompanied neither by a reduction in glucose oxidation nor by decreased expression of insulin mRNA and (pro)insulin biosynthesis (Janson et al., 1995; Eizirik et al., 1997). This finding, and the observation that the diabetic environment inhibits both glucose- and arginine-induced insulin release (Eizirik et al., 1997), suggest that chronic β-cell stimulation leads to a dysfunction distal to ATP production and closure of ATP-sensitive K^+-channels. The nature of this defect remains to be clarified.

III. β-CELL DEFENSE AND REPAIR MECHANISMS

Rats injected with the β-cytotoxic agent streptozotocin (SZ) present hyperglycemia after 24 to 48 hours. The hyperglycemia is permanent, and results from widespread β-cell destruction. However, when SZ-treated rats were fed a high-protein diet for 2 weeks, there was a 20-fold increase in pancreatic insulin content, restoration of islet morphology, and improvement of the diabetic state (Eizirik and Migliorini, 1984). These results, and the subsequent in vitro findings that both purified rat β-cells (Pipeleers and Van de Winkel, 1986) and whole pancreatic islets (reviewed in Eizirik et al., 1993a) recover their function after some types of injury, suggested the existence of putative repair mechanisms in β-cells. Furthermore, islets isolated from NOD mice in the prediabetic state had impaired glucose oxidation and insulin release, but recovered normal function after 1 week in tissue culture, a treatment that removes the infiltrating mononuclear cells (Strandell et al., 1990). Similar observations were made in vivo, where treatment with monoclonal antibodies against infiltrating T cells improved the glucose metabolism in islets isolated from prediabetic NOD mice (Strandell et al., 1992). Taken together, these in vitro and in vivo observations suggest the presence of a population of suppressed and/or partially damaged β-cells in early type 1 diabetes, which may repair themselves and restore function once the immune assault is arrested.

β-Cell repair and/or defense mechanisms also seem to be present in human islets. Thus, cultured human islets are more resistant than rodent islets to damage induced by cytokines, hydrogen peroxide, SZ and alloxan, NO donors, and cycasin (reviewed in Eizirik, 1996). These in vitro observations are in line with in vivo data, indicating that SZ is not diabetogenic in humans (Rerup, 1970). Moreover, human islets grafted into nude mice are not affected by an alloxan injection, whereas the endogenous murine islets are destroyed by the toxin (Eizirik et al., 1994a).

Available evidence suggests the presence of at least two groups of defense proteins in rodent and human β-cells: (1) the stress proteins hsp70 and heme oxygenase; and (2) the antioxidant enzymes superoxide dismutase (SOD), catalase, and reduced glutathione (GSH).

Hsp70 and heme oxygenase are well-defined stress response proteins in different mammalian cells (Welch, 1992). The inducible form of hsp70 is present in pancreatic islets, and its expression increases several fold following rodent islet exposure to thermal stress, SZ, and the cytokine IL-1β (Eizirik et al., 1993a). Hsp70 may play a relevant role for β-cell protection, as suggested by the observations that increased hsp70 content in rat β-cells or insulin-producing cell lines protects them against subsequent damage by cytokines, SZ, NO, and oxygen radicals (Margulis et al., 1991; Bellman et al., 1995; Bellman et al., 1996).

Oxygen-free radicals are probably generated in the vicinity of the β-cells during insulitis (Rabinovitch, 1993; Mandrup-Poulsen, 1996). Catalase, SOD, and GSH are oxygen radical scavengers, and low expression of these enzymes in rat islets may explain their susceptibility to free radical–generating agents, such as alloxan (Grankvist et al., 1981; Malaisse et al., 1982). In agreement with this possibility, liposomal delivery of catalase, SOD, and GSH partially protects these cells against oxidative damage (Welsh et al., 1994); targeted expression of copper/zinc superoxide dismutase (Cu/ZnSOD) in mouse β-cells enhances their resistance against alloxan-induced diabetes (Kubisch et al., 1997); and stable overexpression of MnSOD prevents IL-1β–induced cytotoxicity in insulinoma cell lines (Hohmeier et al., 1998).

Human pancreatic islets are more resistant than rodent islets to damage induced by different β-cell toxins (see preceding discussion). This finding is paralleled by higher expression of hsp70, catalase, and SOD, but not heme oxygenase, as compared with rat and mouse islets (Welsh et al., 1995). Human β-cells express very low levels of the glucose transporter type 2 (GLUT-2) (De Vos et al., 1995), which may contribute to decreased uptake of SZ and alloxan, agents that gain access to the β-cells via the glucose transporter system. However, a slower glucose transport cannot explain the observed differences between human and rodent islets regarding sensitivity to cytokines or hydrogen peroxide. Thus, it is conceivable that increased cellular contents of hsp70, SOD, and catalase contribute to the increased resistance of human islets against toxins. However, while the resistance to SZ, alloxan, and cytokines is human > mouse > rat (Eizirik et al., 1994a), the observed activity of catalase and SOD is human > rat > mouse and the hsp70 expression is human > rat = mouse (Welsh et al., 1995). This discrepancy between the rat and mouse data, and the recent observation that human islets are as susceptible to peroxynitrite as rat islets (Delaney et al., 1996), suggests that other genes and proteins, still to be identified, are of relevance for the observed species differences in the β-cell defense mechanisms.

IV. β-CELL OUTCOME FOLLOWING DAMAGE

A. Functional Recovery or Permanent Dysfunction

A model for different β-cell outcomes following injury is provided in Figure 1. Immediately after damage these cells present decreased function, with impaired mitochondrial substrate oxidation and defective glucose-induced insulin release. In parallel, different defense/repair mechanisms are activated and, depending on the intensity and type of injury, and of the cell capacity to repair the inflicted damage, the β-cell will either survive or die. As will be discussed in the next section, β-cell death may occur via necrosis or apoptosis. In most cases, once the assault is arrested the surviving β-cells are able to recover and regain normal function (reviewed in Eizirik et al., 1993a). However, alkylating agents such as SZ induce long-term β-cell dysfunction, with selective impairment in glucose-induced insulin secretion, defective mitochondrial function, and decreased expression of mitochondrial DNA encoded genes (reviewed in Eizirik et al., 1996c). One of the mechanisms responsible for SZ-induced cell damage involves the formation of highly reactive carbonium ions, causing both nuclear and mitochondrial DNA alkylation (LeDoux et al., 1986; Pettepher et al., 1991). It has been suggested that mutations in mitochondrial DNA contribute to aging and degenerative diseases, such as type 2 diabetes (Gerbitz et al., 1996), and it is conceivable that the chronic decrease in mitochondrial gene expression induced by SZ contributes to the observed long-term β-cell dysfunction (Eizirik et al., 1996c). There is a positive correlation between formation and persistence of specific alkylation products, such as O^6-methyldeoxyguanoside in DNA, and mutagenesis, carcinogenesis, and defective gene transcription (Day et al., 1987). Levels of O^{6}-methylguanine methyltransferase, the enzyme responsible for repairing this lesion, are low in β-cells (Nelson et al., 1993), which could contribute to the inability of these cells to repair the DNA damage induced by SZ. However, this is not the whole explanation for the chronic effects of SZ. Thus, methylnitrosourea (MNU), the aglucone moiety

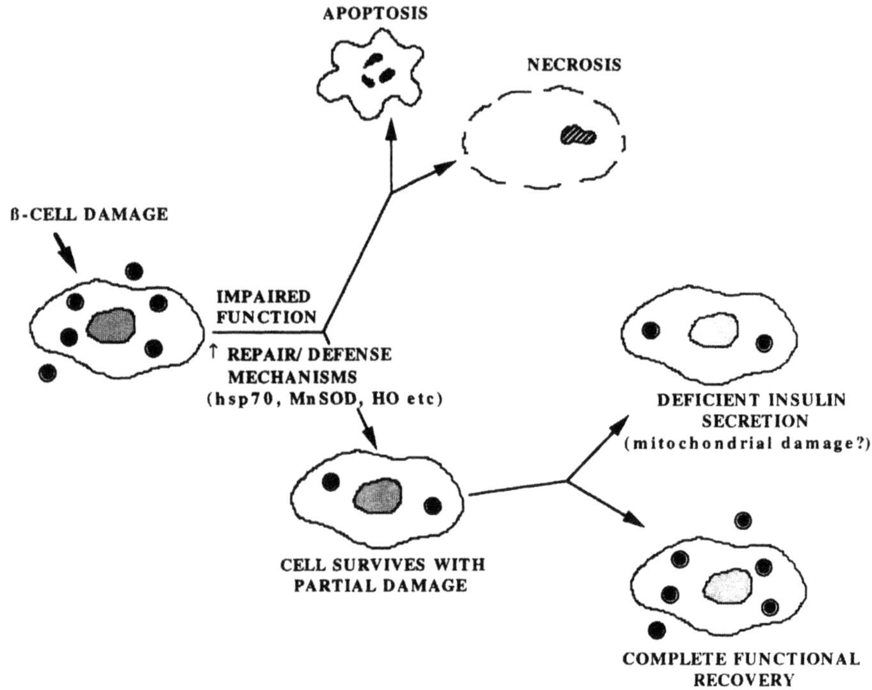

Figure 1. Possible β-cell outcomes following damage (modified from Eizirik et al., 1993a).

of SZ, also induces prolonged damage to β-cell mitochondrial DNA, but this is not paralleled by decreased glucose-induced insulin release (Eizirik et al., 1991). It can be speculated that the glucose moiety of SZ targets the methylation to other intracellular components besides mitochondrial DNA, such as nuclear DNA and/or cytosolic proteins of relevance for the signal-transduction, leading to hexose-induced insulin release. The nature of these extramitochondrial sites, and their role for insulin release, are issues that remain to be clarified.

There is indirect evidence that environmental alkylating agents contribute to some forms of diabetes mellitus (reviewed by Eizirik et al., 1996c). Taking into account the issues described previously, it is conceivable that repeated damage to the nucleus and mitochondria of β-cells, a cell with a putative long life and limited capacity for proliferation, will eventually lead to irreversible mitochondrial dysfunction. Since mitochondrial metabolism and ATP production are crucial for glucose-induced insulin release, this would lead to defective insulin secretion and a consequent rise in serum glucose. As discussed previously, chronic hyperglycemia may be deleterious for the β-cell, further worsening β-cell function and eventually culminating in overt type 2 diabetes. Alternatively, long-term alkylating damage could induce the expression of modified proteins and lead to recognition of the β-cell as "non-self" by the immune system. Depending on the genetic background, this could lead to immune-mediated β-cell destruction and type 1 diabetes. In line with this concept, repeated subdiabetogenic injections of SZ induce mononuclear cell infiltration and β-cell destruction in some, but not all, strains of mice (reviewed in Kolb and Kröncke, 1993). Further epidemiological and experimental data are required to test this hypothesis.

B. β-Cell Death—The Role for Apoptosis

Cell death is an essential event in the development, maintenance, and function of multicellular organisms. Two different forms of cell death have been described, necrosis and apoptosis. They may be induced by distinct stimuli, and differ in their underlying signaling and effector mechanisms, morphological features, and the degree by which they may be affected by therapeutical intervention.

Necrosis is usually present in cases of overwhelming cell injury, such as sudden severe ischemia, leading to loss of selective permeability of the cytoplasmic membrane (reviewed in Buja et al., 1993; Majno and Joris, 1995). This result can occur either due to direct damage to the cytoplasmic membrane, or indirectly, as the result of severe energy depletion. As a consequence, water and calcium ions enter the cell, leading to cellular swelling and subsequent rupture of plasma and organelle membranes and dissolution of organized structures (Table 1, Figure 2). Necrosis is accompanied by a release of cytoplasmic material, triggering an acute inflammatory response (Table 2).

Apoptosis is a genetically controlled and metabolically active type of cell death (reviewed in Steller, 1995; Cummings et al., 1997). It plays a significant role in many physiological and developmental processes, occurring during embryogenesis,

Table 1. Ultrastructural Features of Necrosis and Apoptosis

Necrosis	*Apoptosis*
• Severe swelling of cell organelles (mitochondria) and of the whole cell with formation of blister-like, fluid filled structures at the cell membrane ("blebbing") • Appearance of flocculent and granular densities in the matrix of the mitochondria • Rupture of organelle and plasma membranes • Dissolution of internal structures • Condensation of chromatin (never as homogeneous and as sharply delineated as in apoptosis), with disintegration of the nucleus *Observation*: most of the characteristics described above are based on studies in cells undergoing ischemic cell death	• Earliest visible stage: condensation and margination of chromatin towards the nuclear membrane (crescent-shaped) *Late stage:* • Shrinkage of total cell volume; condensation of the cytoplasmic organelles • Dilatation of the endoplasmic reticulum • Appearance of invaginations and protuberances on the cell surface ("budding") • Fragmentation of nucleus and cytoplasm into multiple, small, membrane-bound apoptotic bodies

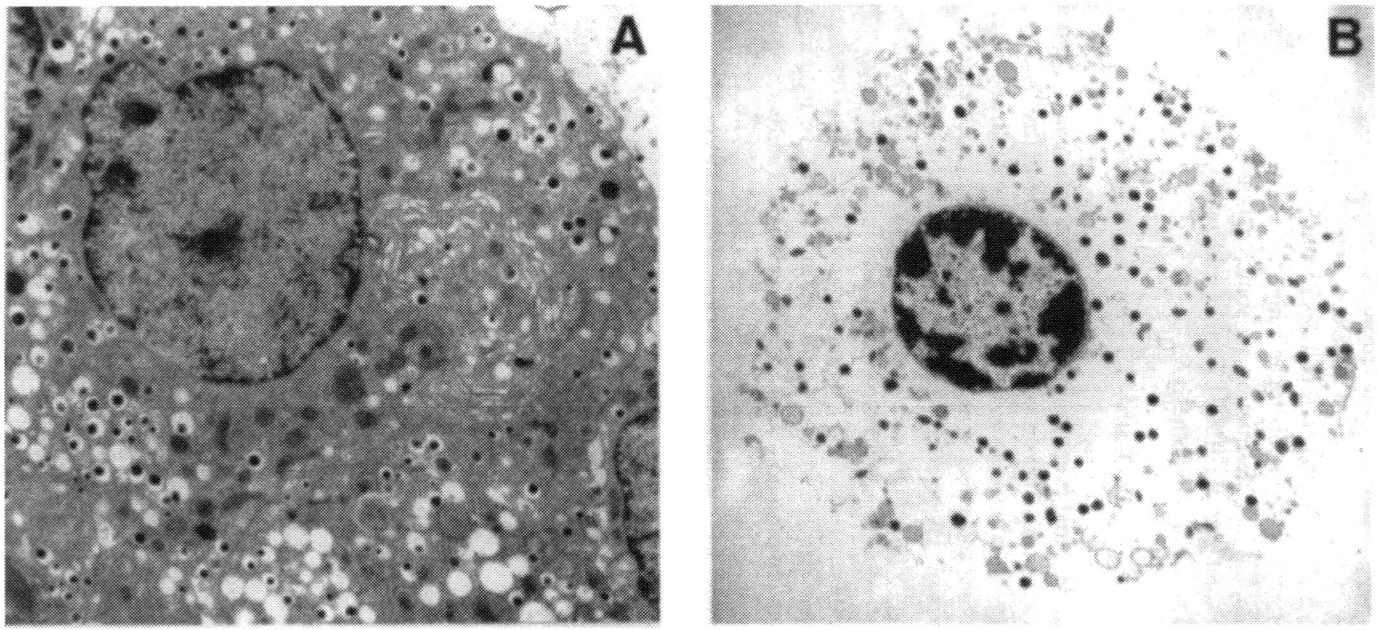

Figure 2. Electron micrograph of a normal rat β-cell cultured under control conditions **(A)** and of a necrotic rat β-cell following treatment with the respiratory poison antimycin A **(B)**. The necrotic β-cell shows condensed chromatin, disruption of the plasma membrane, and dissolution of internal structures.

endocrine-dependent tissue atrophy, and in normal tissue turnover. Apoptosis is also a mechanism to remove cells whose continued existence presents danger for the whole organism, such as cells that are immunologically reactive against self, viral infected, tumoral, and genetically damaged. It also accounts for the cell death occurring under a variety of pathological circumstances (reviewed in Thompson, 1995). For instances, apoptosis is the way target cells are killed by cytotoxic T cells or natural killer (NK) cells. This destruction is effected via Fas/FasL interaction, perforin/granzyme-based mechanisms, and/or release of cytokines (reviewed in Nagata and Golstein, 1995; Nagata, 1997). Although many genes and proteins involved in apoptosis have been identified (reviewed in Nagata, 1997; Thornberry and Lazebnik, 1998; Adams and Cory, 1998), several of the steps in this form of cell death remain to be clarified. Apoptosis can be distinguished from necrosis by its characteristic morphological features (see Tables 1 and 2; reviewed in Buja et al., 1993; Cummings et al., 1997). The earliest event observed in apoptosis is condensation and margination of chromatin toward the nuclear membrane, with formation of sharply delineated uniformly dense crescentic masses abutting the nuclear membrane (Figure 3A). The cell subsequently shrinks, with condensation of the cytoplasm, and detaches from its neighboring cells. Convolutions of the nuclear outlines appear, evolving to fragmentation of the nucleus (Figure 3B). At a later stage, invaginations and protrusions are usually formed at the plasma membrane ("cell budding"), leading to cell fragmentation and formation of apoptotic bodies. These bodies are vesicles lined by an intact cytoplasmic membrane containing condensed but intact organelles, with or without nuclear fragments. Apoptotic cells can also display dilatation of the endoplasmic reticulum. In vivo apoptotic bodies are almost immediately phagocytosed by neighboring epithelial cells and macrophages, preventing leakage of the noxious cellular content. This process explains the absence of acute inflammation in association with apoptotic cells.

At the light microscopic level (Table 2) both apoptotic and necrotic cells show a condensed eosinophilic cytoplasm and condensation of chromatin, proceeding to karyorrhexis and eventual karyolysis (Majno and Joris, 1995). Apoptosis can run its course very fast—in minutes to hours. This, together with the asynchronous nature of the process and the absence of acute inflammation, explains why apoptosis, in contrast to necrosis, is remarkably unobtrusive in tissue sections.

One of the methods to facilitate recognition of apoptotic cells in tissue sections is in situ end labelling of DNA breaks (ISEL) (Gavrieli et al., 1992; Wijsman et al., 1993). This method is based on one of the hallmarks of apoptosis, namely the breakdown of DNA into oligonucleosomal fragments leading to a characteristic ladder pattern in DNA gel electrophoresis (Wyllie, 1980). In late stages of necrosis, DNA breaks also occur, but at random, giving rise to a smear. In ISEL, DNA polymerase I or terminal deoxynucleotidyl transferase (TdT) is used to incorporate biotinylated nucleotides at sites of DNA breaks in the nucleus of apoptotic cells. However, DNA breaks can also occur in late necrotic cells, indicating that ISEL should be combined with some other evidence for apoptosis (Grasl-Kraupp

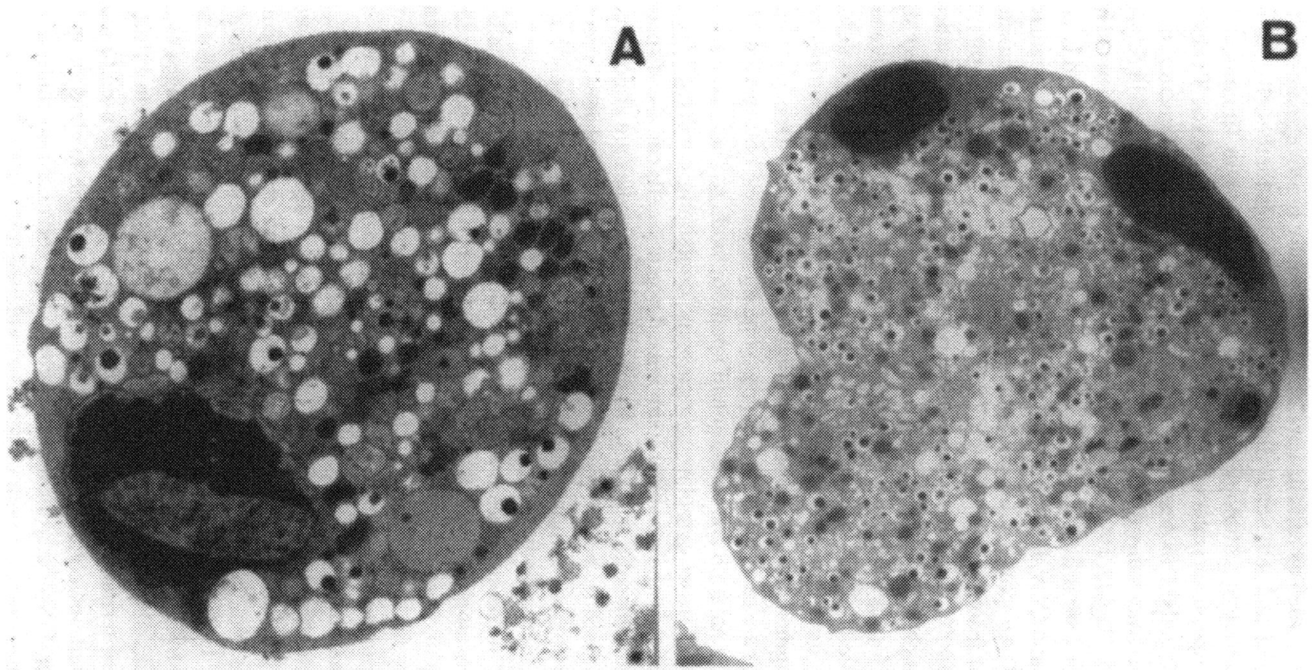

Figure 3. **A.** Immunogold labelling for insulin identifying apoptotic rat β-cells after culture in the presence of cycloheximide. Typical features are condensation and margination of chromatin against the nuclear membrane, condensation of the cytoplasm, and dilatation of the endoplasmic reticulum. **B.** Apoptotic rat β-cell exhibiting fragmentation of chromatin into spheres.

Table 2. Light Microscopic Features of Necrosis and Apoptosis

Necrosis	*Apoptosis*
• Condensed eosinophilic cytoplasm • Karyorrhexis • Karyolysis	
• Massive cell death occurring as large sheets of contiguous cells • Associated with an acute inflammatory infiltrate	• Single dispersed cells • No acute inflammatory infiltrate

et al., 1995). Additional criteria to distinguish different forms of cell death in tissue sections are the occurrence of apoptotic cells as single dispersed cells—in contrast to necrotic cells, which are usually in large contiguous sheets of dead cells—and the absence of an acute inflammatory infiltrate in apoptosis.

The presence of apoptosis in cultured cells can be identified either by their ultrastructural characteristics or by techniques that detect fragmentation of cellular DNA, such as ISEL at the single cell level (Gold et al., 1994) or DNA gel electrophoresis for groups of cells (Wyllie, 1980). Apoptosis in cells cultures can also be detected by flow cytometry, based on the reduced cellular DNA content after fixation or permeabilization (detection of a subdiploid peak) (Nicoletti et al., 1991).

Initial evidence for apoptosis in insulin-producing cells was obtained under in vitro conditions. Thus, it was observed that apoptosis is induced in RINm5F cells by the protein kinase C inhibitor staurosporine (Sánchez-Margalet et al., 1993), and that β-amyloid fibrils (Lorenzo et al., 1994) induce apoptosis in rat β-cells. Overexpression of the anti-apoptotic gene Bcl-2 in the insulin-producing cell line βTC1 partially protected these cells against cytokine-induced cell death (Iwahashi et al., 1996). In RINm5F cells, it was shown that different prooxidant levels may stimulate growth, trigger apoptosis, or lead to cell necrosis (Dypbukt et al., 1994), and it was reported that IgM antibodies present in serum from IDDM patients can induce apoptosis (Juntti-Berggren et al., 1993).

Many of the apoptosis-inducing conditions described previously have only been studied in insulinoma cell lines. Tumor cells on the one hand may develop ways to escape the normal deletion mechanisms and on the other hand may enter apoptosis in the absence of supporting growth factors. Thus, caution is warranted when extrapolating findings obtained in these tumoral cells to primary β-cells. It should be noted that formation of the nucleosome ladder is a common but not universal feature of apoptosis (Cohen et al., 1992; Tomei et al., 1993; Zakeri et al., 1993), and that "ladderlike" DNA degradation patterns have been described in some cells undergoing necrosis (Shen et al., 1991; Bicknell et al., 1995). Moreover, small amounts of cells undergoing apoptosis may lead to a ladder pattern in homogenates from a whole cell population. Thus, contaminating nonendocrine cells undergoing apoptosis can lead to erroneous extrapolations to β-cells when work-

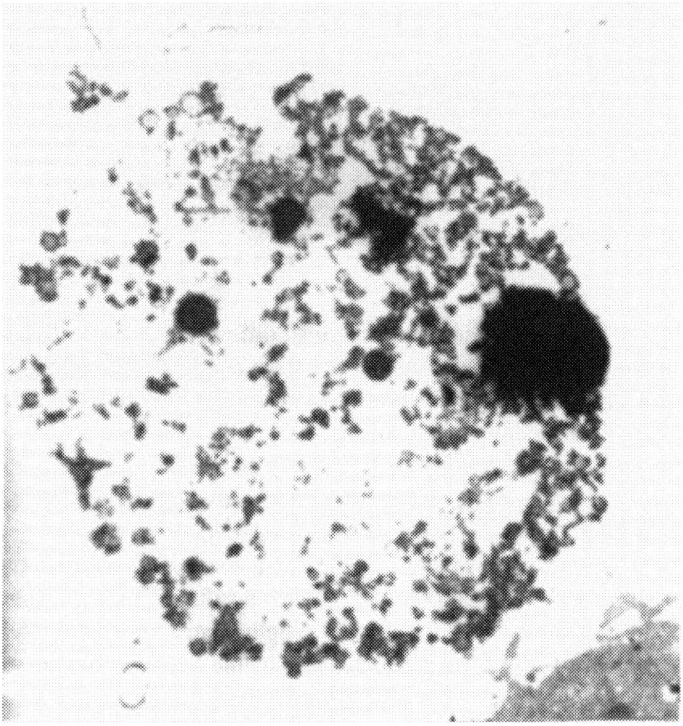

Figure 4. Late apoptotic β-cell ("secondary necrotic cell") following culture with cycloheximide. There is both rupture of cytoplasmic membranes and nuclear fragmentation, a feature reminiscent of apoptosis.

ing with whole islet preparations. Detection and quantification of the number of β-cells undergoing apoptosis in cell culture presents further difficulties. For instances, when apoptosis is not synchronized, longer observation periods may be required to assess the incidence of apoptotic cells. This delay can lead both to loss of apoptotic characteristics and/or considerable overlap between necrosis and apoptosis features (Mundle and Raza, 1995). Indeed, apoptotic cells in vitro, in the absence of phagocytosis, rapidly undergo further degenerative changes that are similar to those occurring in necrosis (Cummings et al., 1997). After about 1 hour they lose their membrane integrity, resulting in swelling, loss of density, membrane rupture, and organelle dispersal, a process referred to as secondary necrosis. Some of the secondary necrotic cells can still be identified ultrastructurally as apoptotic cells because they display features reminiscent of apoptosis, such as the presence of two or more condensed chromatin spheres (Hoorens et al., 1996; Figure 4). However, when karyolysis occurs, it is no longer possible to distinguish between primary and secondary necrosis.

To overcome these obstacles, a new approach based on the membrane and nuclear characteristics of living, apoptotic, and necrotic β-cells was developed, (Hoorens et al., 1996). With this method it is possible to assess, by simultaneous staining with the DNA binding dyes Hoechst 33342 and propidium iodide, the percentage of viable, apoptotic, and necrotic cells in single-cell preparations of primary β-cells cultured in serum free medium. HO 342 freely crosses the plasma membrane, entering both cells with damaged and intact membranes and stains the DNA blue. PI is a highly polar dye, which can only penetrate cells with damaged membranes, staining their nuclei red. After a 15-minute incubation, cells are examined in an inverted fluorescence microscope with UV excitation at 340 to 380 nm. Viable cells are identified by their intact nuclei with blue fluorescence (HO 342); necrotic cells by their intact nuclei with yellow fluorescence (HO 342 + PI); and apoptotic cells by their fragmented nuclei, exhibiting either a blue (HO 342; early apoptosis) or yellow (HO 342 + PI; late apoptosis) fluorescence. The results of the assay were validated by parallel observations of electron micrographs (Hoorens et al., 1996).

Using this method, we studied the role for both protein synthesis and cytokines on β-cell apoptosis. It was observed that rat islet β-cells constitutively express an apoptotic program, which is activated when RNA or protein synthesis are blocked with, respectively, actinomycin D or cycloheximide (Hoorens et al., 1996). Survival of β-cells seems dependent on synthesis of proteins that suppress an endogenous apoptotic program. This observation may explain the beneficial effects of glucose, a well-known stimulator of protein synthesis, on the survival of cultured β-cells (Ling et al., 1994). In line with this possibility, the dose-dependent cytoprotective effect of glucose results from a dose-dependent reduction in the percent of β-cells dying in apoptosis (Hoorens et al., 1996). Moreover, intercellular differences in β-cell sensitivity to glucose (Kiekens et al., 1992) may explain the heterogeneity in β-cell sensitivity to apoptotic conditions (Hoorens et al., 1996). Prolonged treatment of single human pancreatic β-cells in vitro with a combination of three cytokines (IL-1β + TNF-α + IFN-γ) also caused cell death by apoptosis (Delaney et al., 1997).

Most of the data discussed to this point indicate that apoptosis is a frequent form of β-cell death in vitro. However, when these cells are exposed to severe damage by oxygen-free radicals (Dypbukt et al., 1994; Di Matteo et al., 1997), SZ (Saini et al., 1996), or peroxynitrite (Delaney et al., 1996), necrosis is the most prevalent form of cell death. This suggests that the pattern of β-cell death may depend on the type, intensity, and time-course of the cell assault, and also of the β-cell antioxidant defenses and ATP content. Indeed, evidence obtained in human T cells suggests that similar insults may lead either to necrosis or apoptosis, depending on the cellular content of ATP (Leist et al., 1997).

Under in vivo conditions, apoptosis apparently contributes to maintain a β-cell mass well adapted to the insulin demand. For instances, the rat pancreatic β-cell mass nearly doubles during pregnancy, and this is followed by involution after delivery, a consequence of β-cell apoptosis (Scaglia et al., 1995). Apoptosis also participates in the remodelling of the endocrine pancreas in the neonatal rat (Scaglia et al., 1997). In addition, the prolonged hypoglycemia caused by transplanted rat insulinomas induces apoptosis of normal rat pancreatic β-cells (Blume et al., 1995). On the other hand, sucrose feeding (Koyama et al., 1998) or prolonged exposure to high levels of long-chain free fatty acids (FFA) (Shimabukuro et al., 1998) causes β-cell apoptosis in rodents genetically predisposed to develop type 2 diabetes. In the later experiments, apoptosis was prevented by leptin, which apparently acts by preventing FFA-induced decrease in the antiapoptotic gene Bcl-2, which was clearly expressed in the rat islets under basal conditions (Shimabukuro et al., 1998). The fact that Bcl-2 expression is not detectable in human pancreatic islets (Welsh et al., 1995) makes it unclear whether these observations in rodent models can be extrapolated to human type 2 diabetes.

Apoptosis seems also to be the form of β-cell death in two mouse models of type 1 diabetes, namely mice injected with multiple low doses of SZ (O'Brien et al., 1996) and the NOD mice (O'Brien et al., 1997; Augstein et al., 1998). In NOD mice, interaction between the Fas receptor induced on pancreatic β-cells and FasL, present on activated T lymphocytes, may be an important mechanism for β-cell apoptosis (Chervonsky et al., 1997; Itoh et al., 1997). Fas (APO-1,CD95) is a receptor belonging to the TNF/NGF receptor family, whereas its ligand, FasL, belongs to the TNF family (reviewed in Nagata and Golstein, 1995; Nagata, 1997). When Fas-L binds to Fas, it induces apoptosis of the Fas-bearing target cells, and this is one of the mechanisms by which cytotoxic T lymphocytes lyse target cells. It is noteworthy that pancreatic β-cells also die by apoptosis in an accelerated model of diabetes in NOD mice devoid of CD8 + T cells, but containing CD4 + T cells bearing transgenic T-cell receptors (Kurrer et al., 1997). This suggests that cytotoxic mechanisms other than signalling trough Fas may contribute to β-cell apoptosis, a possibility supported by the recent observation that islets isolated from Fas-deficient mice are only marginally protected from the immune attack when transplanted into diabetic NOD mice (Allison and Strasser, 1998). In line with this observation, it has recently been shown that cytokines kill purified human β-cells, cultured in serum-free medium, by apoptosis (Delaney et al., 1997).

It has been suggested that Fas also plays a role for human β-cell death in type 1 diabetes. Stassi et al. (1997) showed that β-cells from pancreata obtained from two young type 1 diabetes patients who died soon after diagnosis expressed Fas and presented apoptosis (as measured by the TUNEL assay) among the β-cells located near Fas ligand-expressing T lymphocytes infiltrating the islets. Based on subsequent in vitro data, these authors suggested that locally produced IL-1β upregulates Fas expression via NO formation (Stassi et al., 1997). Moreover, Loweth et al. (1998) observed that human islets constitutively express FasL and upregulated Fas following IL-1β exposure, leading to apoptosis in around 4% of these cells. These data are intriguing, but they should be interpreted with caution. Thus, IL-1β alone induces neither iNOS mRNA expression nor NO formation by human pancreatic islets (Eizirik et al., 1993; Corbett et al., 1993; Eizirik et al., 1994b; Karlsen and Eizirik, unpublished data), making it unlikely that IL-1β upregulates Fas via NO formation, as suggested by Stassi et al. (1997). Moreover, although combinations of cytokines induce apoptosis in FACS-purified (around 80% purity) human β-cells (Delaney et al., 1997), IL-1β alone neither induces dysfunction nor kills human β-cells (Eizirik et al., 1993b; Hoorens et al., manuscript in preparation). In the study by Loweth et al. (1998), it was not determined whether the small proportion of islet cells killed by IL-1β were indeed β-cells, and no information was provided on the cellular composition of the islet preparations. Collagenase-isolated human islets usually contain, among other cells, 50% to 70% β-cells and 20% to 40% ductal cells (Keymelen et al., 1998; Pavlovic et al., 1999). It is thus conceivable that the islet cells undergoing cell death following exposure to IL-1β (Loweth et al., 1998) are not β-cells, but the nonendocrine cells present in these human islet preparations.

Expression of FasL in target tissues may contribute to protection of immune privileged sites, such as the eye or testis, by deleting infiltrating Fas-expressing T cells (Bellgrau et al., 1995; Griffith et al., 1995). It was thus hypothesized that ectopically expressed FasL could interfere with immune destruction of target cells. Indeed, both transplantation of FasL-expressing myoblasts together with pancreatic islets in mice (Lau et al., 1996), or co-transplantation of allogeneic islets with allogeneic testicular cell aggregates in rats (Korbutt et al., 1997) enabled long-term islet survival. This raised the possibility that FasL expression by the β-cells themselves could protect them against an immune-mediated assault by killing activated autoreactive Fas-expressing T cells. In order to test this hypothesis, different groups developed transgenic mice expressing FasL at the β-cell level (Allison et al., 1997; Chervonsky et al., 1997). Unexpectedly, some transgenic NOD animals presented higher rates of spontaneous diabetes than their nontransgenic litter mates (Chervonsky et al., 1997). This finding may be either due to cytokine-induced Fas expression by the β-cells in the course of insulitis (Yamada et al., 1996; Chervonsky et al., 1997; Stassi et al., 1997), leading them to commit "suicide" by the concomitant expression of Fas and FasL; or to other potential proinflammatory effects of FasL, such as increasing local IL-8 secretion (Allison et al., 1997). In conclusion, the role for Fas and its ligand for human and rodent β-cell death in type 1 diabetes, and following islet transplantation, remains an issue for debate, and additional studies are required to clarify this question.

β-Cell apoptosis was discussed earlier as a form of β-Cell death induced by cytokines and/or cytotoxic T cells during insulitis. Another possibility is that β-cell apoptosis may precede massive mononuclear cell invasion, and thus contribute to triggering the autoimmune process. Such a mechanism was suggested for apoptotic keratinocytes in systemic lupus erythematosus (Casciola-Rosen et al., 1994), and data obtained in mice injected with multiple low doses of SZ (O'Brien et al., 1996) and in NOD mice indicate that increased β-cell apoptosis precedes detectable insulitis (O'Brien et al., 1997). This intriguing possibility deserves further studies.

REFERENCES

Adams, J.M. and Cory, S. (1998). The Bcl-2 protein family: Arbiters of cell survival. Science 281, 1322-1326.

Allison, J., Georgiou, H.M., Strasser, A. and Vaux, D.L. (1997). Transgenic expression of CD95 ligand on islet β cells induces a granulocytic infiltration but does not confer immune privilege upon islet allografts. Proc. Natl. Acad. Sci. USA 94, 3943-3947.

Allison, J. and Strasser, A. (1998). Mechanisms of β-cell death in diabetes: A minor role for CD95. Proc. Natl. Acad. Sci. USA 95, 13818-13822.

Augstein, P., Stephens, L.A., Allison, J., Elefanty, A.G., Ekberg, M., Kay, T.W.H. and Harrison, L.C. (1998). Beta-cell apoptosis in an accelerated model of autoimmune diabetes. Mol. Med. 4, 495-501.

Bach, J.F. (1995). Insulin-dependent diabetes mellitus as a beta-cell targeted disease of immunoregulation. J. Autoimmun. 8, 439-463.

Beckman, J.S. and Koppenol, W.H. (1996). Nitric oxide, superoxide, and peroxynitrite: the good, the bad, and the ugly. Am. J. Physiol. 271, C1424-1437.

Bellgrau, D., Gold, D., Selawry, H., Moore, J., Franzusoff, A. and Duke, R.C. (1995) A role of CD95 ligand in preventing graft rejection. Nature 377, 630-632.

Bellmann, K., Jaattela, M., Wissing, D., Burkart, V. and Kolb, H. (1996). Heat shock protein hsp70 overexpression confers resistance against nitric oxide. FEBS Lett. 391, 185-188.

Bellmann, K., Wenz, A., Radons, J., Burkart, V., Kleemann, R. and Kolb, H. (1995). Heat shock induces resistance in rat pancreatic islet cells against nitric oxide, oxygen radicals and streptozotocin toxicity in vitro. J. Clin. Invest. 95, 2840-2845.

Bicknell, G.R. and Cohen G.M. (1995). Cleavage of DNA to large kilobase pair fragments occurs in some forms of necrosis as well as apoptosis. Biochem. Biophys. Res. Comm. 207, 40-47.

Blume, N., Skouv, J., Larsson, L.-I., Holst, J.J. and Madsen, O.D. (1995). Potent inhibitory effects of transplantable rat glucagonomas and insulinomas on the respective endogenous islet cells are associated with pancreatic apoptosis. J. Clin. Invest. 96, 2227-2235.

Buja, L.M., Eigenbrodt, M.L. and Eigenbrodt, E.H. (1993). Apoptosis and necrosis. Basic types and mechanisms of cell death. Arch. Pathol. Lab. Med. 117, 1208-1214.

Casciola-Rosen, L.A., Anhalt, G. and Rosen A. (1994). Autoantigens targeted in systemic lupus erythematosus are clustered in two populations of surface structures on apoptotic keratinocytes. J. Exp. Med. 179, 1317-1330.

Cetkovic-Cvrlje M. and Eizirik, D.L. (1994). TNF-α and IFN-γ potentiate the deleterious effects of IL-1β on mouse pancreatic islets mainly via generation of nitric oxide. Cytokine 6, 399-406.

Chervonsky, A.V., Wang, Y, Wong, F.S., Visintin, I, Flavell, R.A., Janeway, C.A. Jr., and Matis, L.A. (1997). The role of Fas in autoimmune diabetes. Cell 89, 17-24.

Cohen, G.M., Sun, X.-M., Snowden, R.T., Dinsdale, D. and Skilleter, D.N. (1992). Key morphological features of apoptosis may occur in the absence of internucleosomal DNA fragmentation. Biochem. J. 286, 331-334.

Corbett, J.A., Sweetland, M.A., Wang, J.L., Lancaster, J.R. Jr, and McDaniel, M.L. (1993). Nitric oxide mediates cytokine-induced inhibition of insulin secretion by human islets of Langerhans. Proc. Natl. Acad. Sci. USA 90, 1731-1735.

Cummings, M.C., Winterford, C.M. and Walker N.I. (1997). Apoptosis. Am. J. Surg. Pathol. 21, 88-101.

Cunningham, J.M., Mabley, J.G. and Green, I.C. (1997). Interleukin-1β-mediated inhibition of arginase in RINm5F cells. Cytokine 9, 570-576.

Darville, M.I. and Eizirik, D.L. (1998). Regulation by cytokines of the inducible nitric oxide synthase promoter in insulin-producing cells. Diabetologia 41, 1101-1108.

Day, R.S., Babich, M.A., Yarosh, D.B. and Scudiero, D.A. (1987). The role of O^6-methylguanine in human cell killing, sister chromatid exchange induction and mutagenesis: A review. J. Cell. Sci. 6 (Suppl.), 333-353.

Delaney, C.A., Green, M.H., Lowe, J.E. and Green, I.C. (1993). Endogenous nitric oxide induced by interleukin-1β in rat islets of Langerhans and HIT-T15 cells causes significant DNA damage as measured by the 'comet' assay. FEBS Lett. 333, 291-295.

Delaney, C.A., Pavlovic, D., Hoorens, A., Pipeleers, D.G. and Eizirik, D.L. (1997). Cytokines induce deoxyribonucleic acid strand breaks and apoptosis in human pancreatic islet cells. Endocrinology 138, 2610-2614.

Delaney, C.A., Tyrberg, B., Bouwens, L., Vaghef, H., Hellman, B. and Eizirik, D.L. (1996). Sensitivity of human pancreatic islets to peroxynitrite-induced cell dysfunction and death. FEBS Lett. 394, 300-306.

De Vos, A., Heimberg, H., Quartier, E., Huypens, P., Bouwens, L., Pipeleers D. and Schuit, F. (1995). Human and rat beta cells differ in glucose transporter but not in glucokinase gene expression. J. Clin. Invest. 96, 2489-2495.

Di Matteo, M.A., Loweth, A.C., Thomas, S., Mabley, J.G., Morgan, N.G., Thorpe, J.R. and Green, I.C. (1997). Superoxide, nitric oxide, peroxynitrite and cytokine combinations all cause functional impairment and morphological changes in rat islets of Langerhans and insulin secreting cell lines, but dictate cell death by different mechanisms. Apoptosis 2, 164-177.

Dypbukt, J.M., Ankarcrona, M., Burkitt, M., Sjöholm, Å., Ström, K., Orrenius, S. and Nicotera, P. (1994). Different prooxidant levels stimulate growth, trigger apoptosis, or produce necrosis of insulin-secreting RINm5F cells. J. Biol. Chem. 269, 30553-30560.

Eizirik, D.L. (1991). Interleukin-1β induces an early decrease in insulin release, (pro)insulin biosynthesis and insulin mRNA in mouse pancreatic islets by a mechanism dependent on gene transcription and protein synthesis. Autoimmunity 10, 107-114.

Eizirik, D.L. (1996). Beta-cell defence and repair mechanisms in human pancreatic islets. Horm. Metab. Res. 28, 302-305.

Eizirik, D.L., Delaney, C.A., Green, M.H., Cunningham, J.M., Thorpe, J.R., Pipeleers, D.G., Hellerström, C. and Green, I.C. (1996a). Nitric oxide donors decrease the function and survival of human pancreatic islets. Mol. Cell. Endocrinol. 118, 71-83.

Eizirik, D.L., Flodström, M., Karlsen, A.E. and Welsh, N. (1996b). The harmony of the spheres: inducible nitric oxide synthase and related genes in pancreatic beta-cells. Diabetologia 39, 875-890.

Eizirik, D.L., Jansson, L., Flodström, M., Hellerström, C. and Andersson, A. (1997). Mechanisms of defective glucose-induced insulin release in human pancreatic islets transplanted to diabetic nude mice. J. Clin. Endocrinol. Metab. 82, 2660-2663.

Eizirik, D.L., Korbutt, G.S. and Hellerström, C. (1992). Prolonged exposure of human pancreatic islets to high glucose concentrations in vitro impairs the β-cell function. J. Clin. Invest. 90, 1263-1268.

Eizirik, D.L. and Migliorini, R.H. (1984). Reduced diabetogenic effect of streptozotocin in rats previously adapted to a high-protein carbohydrate-free diet. Diabetes 33, 383-388.

Eizirik, D.L. and Pavlovic, D. (1997). Is there a role for nitric oxide in β-cell dysfunction and damage in IDDM? Diabetes Metab. Rev. 13, 293-307.

Eizirik, D.L., Pipeleers, D.G., Ling, Z., Welsh, N., Hellerström, C. and Andersson, A. (1994a). Major species differences between humans and rodents in the susceptibility to pancreatic β-cell injury. Proc. Natl. Acad. Sci. USA 91, 9253-9256.

Eizirik, D.L., Sandler, S., Ahnström, G. and Welsh, M. (1991). Exposure of pancreatic islets to different alkylating agents decreases mitochondrial DNA content but only streptozotocin induces long-lasting functional impairment of β-cells. Biochem. Pharmacol. 42, 2275-2282.

Eizirik, D.L., Sandler, S., Hallberg, A., Bendtzen, K., Sener, A. and Malaisse, W.J. (1989). Differential sensitivity to β-cell secretagogues in rat pancreatic islets exposed to human interleukin-1β. Endocrinology 125, 752-759.

Eizirik, D.L., Sandler, S. and Palmer J.P. (1993a). Repair of pancreatic b-cells. A relevant phenomenon in early IDDM? Diabetes 42, 1383-1391.

Eizirik, D.L., Sandler, S., Welsh, N., Cetkovic-Cvrlje, M., Nieman, A., Geller, D.A., Pipeleers, D.G., Bendtzen, K. and Hellerström C. (1994b). Cytokines suppress human islet function irrespective of their effects on nitric oxide generation. J. Clin. Invest. 93, 1968-1974.

Eizirik, D.L., Spencer, P. and Kisby, G.E. (1996c). Potential role of environmental genotoxic agents in diabetes mellitus and neurodegenerative diseases. Biochem. Pharmacol. 51, 1585-1591.

Eizirik, D.L., Welsh, N. and Hellerström, C. (1993b). Predominance of stimulatory effects of interleukin-1β on isolated human pancreatic islets. J. Clin. Endocrinol. Metab. 76, 399-403.

Eizirik, D.L., Welsh, M., Strandell, E., Welsh, N. and Sandler, S. (1990). Interleukin-1b depletes insulin messenger ribonucleic acid and increases the heat shock protein hsp70 in mouse pancreatic islets without impairing the glucose metabolism. Endocrinology 127, 2290-2297.

Flodström, M., Chen, M.-C., Smismans, A., Pipeleers, DG., Schuit, F. and Eizirik, D.L. (1999). Interleukin-1β increases arginine accumulation and activates the citrulline-nitric oxide cycle in rat pancreatic β-cells. Cytokine (in press).

Flodström, M. and Eizirik, D.L. (1997). Interferon-γ-induced interferon regulatory factor-1 (IRF-1) expression in rodent and human islet cells precedes nitric oxide formation. Endocrinology 138, 2747-2753.

Flodström, M., Niemann, A., Bedoya, F.J., Morris, S.M. Jr. and Eizirik, D.L. (1995). Expression of the citrulline-nitric oxide cycle in rodent and human pancreatic β-cells: Induction of argininosuccinate synthetase by cytokines. Endocrinology 136, 3200-3206.

Gavrieli, Y., Sherman, Y. and Ben-Sasson, S.A. (1992). Identification of programmed cell death in situ via specific labelling of nuclear DNA fragmentation. J. Cell. Biol. 119, 493-501.

Gerbitz, K.-D., Gempel, K. and Brdiczka, D. (1996). Mitochondria and diabetes. Genetic, biochemical, and clinical implications of the cellular energy circuit. Diabetes 45, 113-126.

Gold, R., Schmied, M., Giegerich, G., Breitschopf, H., Hartung, H.P., Toyka K.V. and Lassmann, H. (1994). Differentiation between cellular apoptosis and necrosis by the combined use of in situ tailing and nick translation techniques. Lab. Invest. 71, 219-225.

Grankvist, K., Marklund, S.L. and Taljedal, I.B. (1981). CuZn-superoxide dismutase, Mn-superoxide dismutase, catalase and glutathione peroxidase in pancreatic islets and other tissues in the mouse. Biochem. J. 199, 393-398.

Grasl-Kraupp, B., Ruttkay-Nedecky, B., Koudelka, H., Bukowska, K., Bursch, W. and Schulte-Hermann, R. (1995) In situ detection of fragmented DNA (TUNEL assay) fails to discriminate among apoptosis, necrosis, and autolytic cell death: a cautionary note. Hepatology 21, 1465-1468.

Griffith, O.W. and Stuehr, D.J. (1995). Nitric oxide synthases: properties and catalytic mechanism. Annu. Rev. Physiol. 57: 707-736.

Griffith. T.S., Brunner, T., Fletcher, S.M., Green D.R. and Ferguson, T.A. (1995). Fas ligand-induced apoptosis as a mechanism of immune privilege. Science 270, 1189-1192.

Heller, B., Wang, Z.Q., Wagner, E.F., Radons, J., Burkle, A., Fehsel, K., Burkart, V. and Kolb, H. (1995). Inactivation of the poly(ADP-ribose) polymerase gene affects oxygen radical and nitric oxide toxicity in islet cells. J. Biol. Chem. 270, 11176-11180.

Hirasawa, K., Jun, H.S., Maeda, K., Kawaguchi, Y., Itagaki, S., Mikami, T., Baek, H.S., Doi K. and Yoon, J.W. (1997). Possible role of macrophage-derived soluble mediators in the pathogenesis of encephalomyocarditis virus-induced diabetes in mice. J. Virology 71, 4024-4031.

Hohmeier, H.-E., Thigpen, A., Tran, V.V., Davis, R. and Newgard, C.B. (1998). Stable expression of manganese superoxide dismutase (MnSOD) in insulinoma cells prevents IL-1β-induced cytotoxicity and reduces nitric oxide production. J. Clin. Invest. 101, 1811-1820.

Hoorens, A., Van de Casteele, M., Klöppel, G. and Pipeleers, D. (1996). Glucose promotes survival of rat pancreatic β cells by activating synthesis of proteins which suppress a constitutive apoptotic program. J. Clin. Invest. 98,1568-1574.

Hyans, J. (1996). Look Ma, no chromosomes! Nature 382, 397-398.

Ilkova, H., Glaser, B., Tunckale, A., Bagriaçik, N. and Cerasi, E. (1997). Induction of long-term glycemic control in newly diagnosed type 2 diabetic patients by transient intensive insulin treatment. Diabetes Care 20, 1353-1356.

Itoh, N., Imagawa, A., Hanafusa, T., Waguri, M., Yamamoto, K., Iwahashi, H., Moriwaki, M., Nakajima, H., Miyagawa, J., Namba, M., Makino, S., Nagata, S., Kono, N. and Matsuzawa, Y. (1997). Requirement of Fas for the development of autoimmune diabetes in nonobese diabetic mice. J. Exp. Med. 186, 613-618.

Iwahashi, H., Hanafusa, T., Eguchi, Y., Nakajima, H., Miyagawa, J., Itoh, N., Tomita, K., Namba, M., Kuwajima, M., Noguchi, T., Tsujimoto, Y. and Matsuzawa, Y. (1996). Cytokine-induced apoptotic cell death in a mouse pancreatic beta-cell line: Inhibition by Bcl-2. Diabetologia 39, 530-536.

Jansson, L., Eizirik, D.L., Pipeleers, D.G., Borg, L.A.H., Hellerström, C. and Andersson, A. (1995). Impairment of glucose-induced insulin secretion in human pancreatic islets transplanted to diabetic nude mice. J. Clin. Invest. 96, 721-726.

Jiang, Z. and Woda, B.A. (1991). Cytokine gene expression in the islets of the diabetic Biobreeding/ Worcester rat. J. Immunol. 146, 2990-2994.

Juntti-Berggren, L., Larsson, O., Rorsman, P., Änmälä, C., Bokvist, K., Wåhlander, K., Nicotera, P., Dypbukt, J., Orrenius, S., Hallberg, A. and Berggren, P.-O. (1993). Increased activity of L-type Ca^{2+} channels exposed to serum from patients with type I diabetes. Science 261, 86-90.

Karlsen, A., Pavlovic, D., Nielsen, K., Jensen, J., Andersen, H.U., Pociot, F., Mandrup-Poulsen, T., Eizirik, D.L. and Nerup, J. (1998). Cytokine induced nitric oxide and interleukin-1 converting enzyme may represent different pathways of cytokine induced β-cell destruction. Diabetologia 41 (Suppl. 1), A56.

Keymelen, B., Ling, Z., Gorus, F.K., Delvaux, G., Bouwens, L., Grupping, A., Hendrieckx, C., Pipeleers-Marichal, M., Van Schravendijk, C., Salmela, K. and Pipeleers, D.G. (1998). Implantation of standardized beta-cell grafts in a liver segment of IDDM patients: Graft and recipient characteristics in two cases of insulin-independence under maintenance immunosuppression for prior kidney graft. Diabetologia 41, 452-459.

Kiekens, R., In't Veld, P., Mahler, T., Schuit, F., Van De Winkel, M. and Pipeleers, D. (1992). Differences in glucose recognition by individual rat pancreatic B cells are associated with intercellular differences in glucose-induced biosynthetic activity. J. Clin. Invest. 89, 117-125.

Kolb, H. and Kröncke, K.-D. (1993). IDDM—Lessons from the low-dose streptozotocin model in mice. Diabetes Rev. 1, 116-126.

Korbutt, G.S., Elliott, J.F. and Rajotte, R.V. (1997). Cotransplantation of allogeneic islets with allogeneic testicular cell aggregates allows long-term graft survival without systemic immunosuppression. Diabetes 46, 317-322.

Korsgren, O., Jansson, L., Sandler, S. and Andersson, A. (1990). Hyperglycemia-induced B-cell toxicity. The fate of pancreatic islets transplanted into diabetic mice is dependent on their genetic background. J. Clin. Invest. 86, 2161-2168.

Koyama, M., Wada, R., Sakuraba, H., Mizukami, H. and Yagihashi, S. (1998). Accelerated loss of islet beta cells in sucrose-fed Goto-Kakizaki rats, a genetic model of non-insulin dependent diabetes mellitus. Am. J. Pathol. 153, 537-545.

Kröncke, K.-D., Fehsel, K. and Kolb-Bachofen, V. (1998). Inducible nitric oxide synthase in human diseases. Clin Exp Immunol. 113, 147-156.

Kubish, M.-H., Wang, J., Bray, T.M. and Philipps, J.P. (1997). Targeted overexpression of Cu/Zn superoxide dismutase protects pancreatic β-cells against oxidative stress. Diabetes 46, 1563-1566.

Kurrer, M.O., Pakala, S.V., Hanson, H.L. and Katz, J.D. (1997). β cell apoptosis in T cell-mediated autoimmune diabetes. Proc. Natl. Acad. Sci. USA 94, 213-218.

Lander, H.M. (1997). An essential role for free radicals and derived species in signal transduction. FASEB J. 11, 118-124.

Lau, H.T., Yu, M., Fontana, A. and Stoeckert C.J. Jr. (1996). Prevention of islet allograft rejection with engineered myoblasts expressing FasL in mice. Science 273, 109-112.

Leahy, J.L. (1996). Impaired β-cell function with chronic hyperglycemia: "Overworked β-cell" hypothesis. Diabetes Rev. 4, 298-319.

Leahy, J.L., Bonner-Weir, S. and Weir, G.C. (1992). β-cell dysfunction induced by chronic hyperglycemia. Diabetes Care 15, 442-455.

LeDoux, S.P., Woodley, S.E., Patton, N.J. and Wilson, G.L. (1986). Mechanisms of nitrosourea-induced B-cell damage. Alterations of DNA. Diabetes 35, 866-872.

Leist, M., Single, B., Castoldi, A.F., Kühlne, S. and Nicotera, P. (1997). Intracellular adenosine triphosphate (ATP) concentration: A switch in the decision between apoptosis and necrosis. J. Exp. Med. 185, 1481-1486.

Ling, Z., Hannaert, J.C. and Pipeleers, D. (1994). Effect of nutrients, hormones and serum on survival of rat islet beta cells in culture. Diabetologia 37, 15-21.

Lorenzo, A., Razzaboni, B., Weir, G.C. and Yankner, B.A. (1994). Pancreatic islet cell toxicity of amylin associated with type-2 diabetes mellitus. Nature 368, 756-760.

Loweth, A.C., Williams, G.T., James, R.F.L., Scarpello, J.H.B. and Morgan, N.G. (1998). Human islets of Langerhans express Fas ligand and undergo apoptosis in response to IL-1β and Fas ligation. Diabetes 47, 727-732.

MacMicking, J., Xie, Q.W. and Nathan, C. (1997). Nitric oxide and macrophage function. Annu. Rev. Immunol. 15, 323-350.

Majno, G. and Joris, I. (1995). Apoptosis, oncosis, and necrosis. An overview of cell death. Am. J. Pathol. 146, 3-15.

Malaisse, W.J. (1994). The beta cell in non-insulin-dependent diabetes: giving light to the blind. Diabetologia 37 (Suppl. 2), S36–S42.

Malaisse, W.J., Maggetto, C., Leclerk-Meyer, V. and Sener, A. (1993). Interference of glycogenolysis with glycolysis in pancreatic islets from glucose-infused rats. J. Clin. Invest. 91, 432-436.

Malaisse, W.J., Malaisse-Lagae, F., Sener, A. and Pipeleers, D.G. (1982). Determinants of the selective toxicity of alloxan to the pancreatic B cell. Proc. Natl. Acad. Sci. U S A 79, 927-930.

Mandrup-Poulsen, T. (1996). The role of interleukin-1 in the pathogenesis of IDDM. Diabetologia 39, 1005-1029.

Margulis, B.A., Sandler, S., Eizirik, D.L., Welsh, N., and Welsh, M. (1991). Liposomal delivery of purified heat shock protein hsp70 into rat pancreatic islets as protection against interleukin 1β-induced impaired β-cell function. Diabetes 40, 1418-1422.

Moncada, S. and Higgs, A. (1993). The L-arginine-nitric oxide pathway. N. Engl. J. Med. 329, 2002-2012.

Moncada, S., Higgs, A. and Furchgott, R. (1997). International Union of Pharmacology nomenclature in nitric oxide research. Pharmacol. Rev. 49, 137-142.

Morris, S.M. Jr and Billiar, T.R. (1994). New insights into the regulation of inducible nitric oxide synthesis. Am. J. Physiol. 266, E829-E839.

Mundle, S.D. and Raza A. (1995). The two in situ techniques do not differentiate between apoptosis and necrosis but rather reveal distinct patterns of DNA fragmentation in apoptosis. Lab. Invest. 72, 611-613.

Nagata, S. (1997). Apoptosis by death factor. Cell 88, 355-365.

Nagata, S., and Golstein, P. (1995). The Fas death factor. Science 267, 1449-1456.

Nelson, J.W., LeDoux, S.P. and Wilson, G.L. (1993). Repair of O^{6-}methylguanine in rat pancreatic β-cells after exposure to N-methyl-N-nitrosourea. Diabetes 42, 1187-1194.

Nicoletti, I., Migliorati, G., Pagliacci, M.C., Grignani, F. and Riccardi, C. (1991). A rapid and simple method for measuring thymocyte apoptosis by propidium iodide staining and flow cytometry. J. Immunol. Methods 139, 271-279.

Niemann, A., Björklund, A. and Eizirik, D.L. (1994). Studies on the molecular regulation of the inducible form of nitric oxide synthase (iNOS) in insulin-producing cells. Mol. Cell. Endocrinol. 106, 151-155.

O'Brien, B.A., Harmon, B.V., Cameron, D.P. and Allan, D.J. (1996). Beta cell apoptosis is responsible for the development of IDDM in the multiple low-dose streptozotocin model. J. Pathol. 178, 176-181.

O'Brien, B.A., Harmon, B.V., Cameron, D.P. and Allan, D.J. (1997). Apoptosis is the mode of β-cell death responsible for the development of IDDM in the nonobese diabetic (NOD) mouse. Diabetes 46, 750-757.

Pavlovic, D., Chen, M.-C., Bouwens, L., Eizirik, D.L. and Pipeleers, D. (1999). Contribution of ductal cells to cytokine responses by human pancreatic islets. Diabetes 48, 29–33.

Peng, H.B., Libby, P. and Liao, J.K. (1995). Induction and stabilization of IκB alpha by nitric oxide mediates inhibition of NF-κB. J. Biol. Chem. 270, 14214-14219.

Pettepher, C.C., LeDoux, S.P., Bohr, V.A. and Wilson, G.L. (1991). Repair of alkali-labile sites within the mitochondrial DNA of RINr38 cells after exposure to the nitrosourea streptozotocin. J. Biol. Chem. 266, 3113-3117.

Pipeleers, D. and Van de Winkel, M. (1986). Pancreatic B cell possess defense mechanisms against cell-specific toxicity. Proc. Natl. Acad. Sci. USA 83, 5267-5271.

Polonsky, K.S., Sturis, J. and Bell, G.I. (1996). Non-insulin dependent diabetes mellitus—a genetically programmed failure of the beta cell to compensate for insulin resistance. N. Engl. J. Med. 334, 777-783.

Rabinovitch, A. (1993). Roles of cytokines in IDDM pathogenesis and islet β-cell destruction. Diabetes Rev. 1, 215-240.

Rabinovitch, A., Suarez-Pinzon, W.L., Sorensen, O. and Bleackley, R.C. (1996a). Inducible nitric oxide synthase (iNOS) in pancreatic islets of nonobese diabetic mice: identification of iNOS-expressing cells and relationships to cytokines expressed in the islets. Endocrinology 137, 2093-2099.

Rabinovitch, A., Suarez-Pinzon, W.L., Strynadka, K., Lakey, J.R.T., Rajotte, R.V. (1996b). Human pancreatic islet β-cell destruction by cytokines involves oxygen free radicals and aldehyde production. J. Clin. Endocrinol. Metab. 81, 3197-3202.

Radons, J., Heller, B., Burkle, A., Hartmann, B., Rodriguez, M.L., Kroncke, K.D., Burkart, V. and Kolb, H. (1994). Nitric oxide toxicity in islet cells involves poly(ADP-ribose) polymerase activation and concomitant NAD+ depletion. Biochem. Biophys. Res. Commun. 199, 1270-1277.

Rerup, C.C. (1970). Drugs producing diabetes through damage of the insulin secreting cells. Pharmacol. Rev. 22, 485-518.

Robertson, R.P., Olson, L.K. and Zhang, H.-J. (1994). Differentiating glucose toxicity from glucose desensitization: a new message from the insulin gene. Diabetes 43, 1085-1089.

Saini, K.S., Thompson, C., Winterford, C.M., Walker, N.I. and Cameron, D.P. (1996). Streptozotocin at low doses induces apoptosis and at high doses causes necrosis in a murine pancreatic beta cell line, INS-1. Biochem. Mol. Biol. Internat. 39, 1229-1236.

Sako, Y. and Grill, V.E. (1990). Coupling of β-cell desensitization by hyperglycemia to excessive stimulation and circulating insulin in glucose-infused rats. Diabetes 39, 1580-1583.

Sánchez-Margalet, V., Lucas, M., Solano, F. and Goberna, R. (1993). Sensitivity of insulin-secreting RIN m5F cells to undergoing apoptosis by the protein kinase C inhibitor staurosporine. Exp. Cell Res. 209,160-163.

Sandler, S., Eizirik, D.L., Svensson, C., Strandell, E., Welsh, M. and Welsh, N. (1991). Biochemical and molecular actions of interleukin-1 on pancreatic β-cells. Autoimmunity 10, 241-253.

Scaglia, L., Cahill, C.J., Finegood, D.T. and Bonner-Weir, S. (1997). Apoptosis participates in the remodelling of the endocrine pancreas in the neonatal rat. Endocrinology 138, 1736-1741.

Scaglia, L., Smith, F.E. and Bonner-Weir, S. (1995). Apoptosis contributes to the involution of the β cell mass in the post partum rat pancreas. Endocrinology 136, 5461-5468.

Shah, S., Malone, J.I. and Simpson, N.E. (1989). A randomized trial of intensive insulin therapy in newly diagnosed insulin-dependent diabetes mellitus. N. Engl. J. Med. 320, 550-554.

Shen, W., Kamendulis, L.M., Ray, S.D. and Corcoran, G.B. (1991). Acetaminophen-induced cytotoxicity in cultured mouse hepatocytes: Correlation of nuclear Ca^{2+} accumulation and early DNA fragmentation with cell death. Toxicol. Appl. Pharmacol. 111, 242-254.

Shimabukuro, M., Wang, M.-Y., Zhou, Y.-T., Newgard, C.B. and Unger, R.H. (1998). Protection against lipoapoptosis of β cells trough leptin dependent maintenance of Bcl-2 expression. Proc. Natl. Acad. Sci. USA 95, 9558-9561.

Stassi, G., De Mari, R., Trucco, G., Rudert, W., Testi, R., Galluzzo, A., Giordano, C. and Trucco, M. (1997). Nitric oxide primes pancreatic β-cells for Fas-mediated destruction in insulin-dependent diabetes mellitus. J. Exp. Med. 186, 1193-1200.

Steller, H. (1995). Mechanisms and genes of cellular suicide. Science 267,1445-1449.

Strandell, E., Eizirik, D.L. and Sandler, S. (1990). Reversal of β-cell suppression in vitro in pancreatic islets isolated from nonobese diabetic mice during the phase preceding insulin-dependent diabetes mellitus. J. Clin. Invest. 85, 1944-1950.

Strandell, E., Sandler, S., Boitard, C. and Eizirik, D.L. (1992). Role of infiltrating T cells for impaired glucose metabolism in pancreatic islets isolated from non-obese diabetic mice. Diabetologia 35, 924-931.

Suarez-Pinzon, W.L., Szabo, C. and Rabinovitch, A. (1997). Development of autoimmune diabetes in NOD mice is associated with the formation of peroxynitrite in pancreatic islet β-cells. Diabetes 46, 907-911.

Thompson, C.B. (1995). Apoptosis in the pathogenesis and treatment of disease. Science 267,1456-1462.

Thornberry, N.A. and Lazebnik, Y. (1998). Caspases: enemies within. Science 281, 312-316.

Tomei, L.D., Shapiro, J.P. and Cope, F.O. (1993). Apoptosis in C3H/10T1/2 mouse embryonic cells: Evidence for internucleosomal DNA modification in the absence of double-strand cleavage. Proc. Natl. Acad. Sci. USA 90, 853-857.

Welch, M.J. (1992). Mammalian stress response: cell physiology, structure/function of stress proteins, and implications for medicine and disease. Physiol. Rev. 72, 1063-1081.

Welsh, N., Eizirik, D.L., Bendtzen, K. and Sandler, S. (1991). Interleukin-1β-induced nitric oxide production in isolated rat pancreatic islets requires gene transcription and may lead to inhibition of the Krebs cycle enzyme aconitase. Endocrinology 129, 3167-3173.

Welsh, N., Margulis, B., Bendtzen, K. and Sandler, S. (1994). Liposomal delivery of antioxidant enzymes protects against hydrogen peroxide—but not interleukin-1β-induced inhibition of glucose metabolism in rat pancreatic islets. J. Endocrinol. 143, 151-156.

Welsh, N., Margulis, B., Borg, L.A., Wiklund, H.J., Saldeen, J., Flodström, M., Mello, M.A., Andersson, A., Pipeleers, D.G., Hellerström, C. and Eizirik, D.L. (1995). Differences in the expression of heat-shock proteins and antioxidant enzymes between human and rodent pancreatic islets: Implications for the pathogenesis of insulin-dependent diabetes mellitus. Mol. Med. 1, 806-820.

Wijsman, J.H., Jonker, R.R., Keijzer, R., Van De Velde, C.J.H., Cornelisse, C.J. and Van Dierendonck, J.H. (1993). A new method to detect apoptosis in paraffin sections: In situ end-labelling of fragmented DNA. J. Histochem. Cytochem. 41, 7-12.

Wilson, G.L., Patton, N. and LeDoux, S.P. (1997). Mitochondrial DNA in β-cells is a sensitive target for damage by nitric oxide. Diabetes 46, 1291-1295.

Wyllie A.H. (1980). Glucocorticoid-induced thymocyte apoptosis is associated with endogenous endonuclease activation. Nature 284, 555-556.

Xie, Q. and Nathan, C (1994). The high-output nitric oxide pathway: role and regulation. J. Leukoc. Biol. 56, 576-582.

Yamada, K., Takane-Gyotoku, N., Yuan, X., Ichikawa, F., Inada, C. and Nonaka, K. (1996). Mouse islet cell lysis mediated by interleukin-1-induced Fas. Diabetologia 39, 1306-1312.

Zakeri, Z.F., Quaglino, D., Latham, T. and Lockshin, R.A. (1993). Delayed internucleosomal DNA fragmentation in programmed cell death. FASEB J. 7, 470-478.

Zhou, Y.-P. and Grill V.E. (1994). Long-term exposure of rat pancreatic islets to fatty acids inhibits glucose-induced insulin secretion and biosynthesis through a glucose fatty acid cycle. J. Clin. Invest. 93, 870-876.

ISLET PROTEINS IMPLICATED IN PATHOGENESIS OF TYPE 1 DIABETES MELLITUS

Michael R. Christie

Advances in Molecular and Cell Biology
Volume 29, pages 75-100.

ISBN: 0-7623-0547-9

I. INTRODUCTION

Type 1 diabetes mellitus is caused by the specific destruction of the pancreatic β-cells within the pancreatic islet by a mechanism in which autoimmune responses to islet components is implicated, but the precise mechanism whereby the β-cells are destroyed in the disease are not well understood. Much of recent research aimed at understanding the mechanisms of disease in Type 1 diabetes has focused on defects in immune regulation that may lead to autoimmune responses to β-cell antigens (Bach, 1994). It is also recognized that the pancreatic β-cell itself very likely plays a part in its own destruction. This may take the form of:

- Upregulation of factors that promote cell destruction
- Provision of tissue-specific or tissue-restricted target molecules that focus an immune response on the pancreatic islet
- Failure to provide or activate appropriate defence mechanisms

Alterations in protein expression within the islets occur in the disease state, and characterization of these changes may provide clues as to mechanisms of β-cell destruction in the disease. Genetic factors that influence protein expression may contribute to diabetes susceptibility and identification of the genes involved may provide important markers for the identification of individuals at risk. Furthermore, an understanding of islet proteins that are involved in the disease would be valuable for the design of procedures to block interactions between the immune system and the pancreatic islet to effectively prevent the course of the disease in those at risk. This chapter reviews the changes that take place within the islet in Type 1 diabetes and discusses the proteins involved in the disease process.

II. THE PANCREAS IN TYPE 1 DIABETES MELLITUS

A. The Pancreas in Human Type 1 Diabetes

Pathological investigations of pancreases obtained from patients with Type 1 diabetes demonstrated the absence, or severe reduction, in numbers of insulin-containing β-cells that is responsible for the insulin deficiency in the disease (Gepts, 1965). Other endocrine cell types within the islet, secreting glucagon, somatostatin, and pancreatic polypeptide, appear to survive. The reason for this highly specific loss of β-cells in the disease has been the subject of investigation for many years. Investigators at the beginning of the 20th century described inflammatory infiltrates into the islets of a few cases of juvenile diabetes (Warren and Root, 1925) and the term "insulitis" was coined for this phenomenon. A more extensive analysis of pancreases from Type 1 diabetic patients diagnosed before the age of 40 and with varying disease duration was performed by Gepts (1965). This work clearly established that insulitis was not a rare phenomenon, but a common feature of the diabetic pancreas around the time of disease onset. Gepts noted that insulitis was absent in pancreases of patients with a long duration of disease and in recent-onset Type 1 diabetes inflammation was only seen within islets containing residual β-cells. Later studies demonstrated a strong correlation between the degree of insulitis and the proportion of islet cells expressing insulin (Somoza et al., 1994). These observations imply a link between the inflammatory lesion and β-cell destruction. LeCompte (1958) suggested a number of reasons for the observed insulitis. First, he proposed that the infiltration may be a consequence of an inflammatory response to viral or bacterial infection within the islet. The common association of diabetes with an infectious illness was quoted as evidence for this hypothesis. Alternatively, the infiltrate might be a secondary consequence of overactivity of the β-cell, or an immune reaction to islet components perhaps released as a consequence of islet cell injury by unknown agents. The presence in the blood of diabetic patients before insulin therapy of circulating antibodies that recognize insulin (Pav et al., 1963) or islet cells on frozen sections (Bottazzo et al., 1974) gave support to the view that the immune response in Type 1 diabetes is directed, at least in part, to islet self components. However, 40 years after the LeCompte paper there is still considerable debate as to the relative contribution of infectious agents and islet self proteins as targets of the islet inflammation.

Improvements in the diagnosis and treatment of Type 1 diabetes mean that pancreases from diabetic patients close to the time of disease onset are now rarely available for detailed investigations into the precise nature of the infiltrating cells and the targets of the inflammatory response. However, immunohistochemical analysis of postmortem and biopsy samples of pancreases that have become available have established that the majority of the cells within the infiltrate are T cells, and that in most patients T cells expressing CD8, a cell surface marker for cytotoxic/suppressor cells, predominate (Bottazzo et al., 1985; Hanninen et al., 1992; Itoh et al., 1993; Somoza et al., 1994). Smaller numbers of CD4 positive (helper) T cells, macrophages, and B lymphocytes are present. Upregulation of the expression by endocrine cells of major histocompatibility complex (MHC) class I molecules, which bind peptides of antigens for the recognition by the antigen receptor of cytotoxic T cells, is also a characteristic feature of the diabetic pancreas (Foulis et al., 1987a). The immunohistochemical observations in the human pancreas are therefore consistent with an involvement of cytotoxic T cell recognition of foreign or self antigen on β-cells, which may be promoted by the upregulation of MHC class I molecules on the cell surface.

B. Animal Models of Type 1 Diabetes

Our understanding of potential mechanisms of β-cell destruction in Type 1 diabetes has been greatly advanced by the study of various animal models of the disease, of which two rodent strains, the nonobese diabetic (NOD) mouse and the Bio-Breeding (BB) rat, have been particularly valuable. The NOD mouse and the BB rat both spontaneously develop diabetes with autoimmune characteristics similar to the human disease; in particular, there is progressive insulitis from around 3 weeks of age in the mice, which is generally more severe than that usually observed in the human disease. Disease generally starts to develop at around 12 weeks of age. The earliest cell type observed infiltrating the islet is the macrophage, followed initially by CD8 T cells and, subsequently, by those expressing CD4 (Hanenberg et al., 1989; Jarpe et al., 1990; O'Reilly et al., 1991). "Peri-insulitis," in which intact islets are surrounded by an accumulation of mononuclear cells that do not enter the islet, is a common feature of the prediabetic period. This observation implies that there are multiple steps in the course of inflammation, which include the homing of lymphocytes into the pancreas to surround the islet and the subsequent entry of lymphocytes into the islets, which is associated with β-cell destruction. By the time of disease onset, T cells, predominantly expressing CD4, form the majority of the infiltrate. A combination of CD4 and CD8 T cell subsets from diabetic NOD mice are capable of transferring diabetes to irradiated (immunodeficient) nondiabetic animals, thus demonstrating the essential role of T cells in the disease process (Miller et al., 1988). Islet reactive CD4+ and CD8+ T cells have been cloned from the islets of NOD mice and shown to be capable of inducing disease in severe combined immunodeficient NOD (NOD–SCID) mice, which themselves lack T cells (Wong et al., 1996). Thus, both CD4+ and CD8+ T cells have been implicated directly in the destruction of β-cells, and there is considerable debate concerning the relative importance of each of these subsets in the human disease.

III. FACTORS INVOLVED IN PROMOTING β-CELL DESTRUCTION

A. Perforin and Fas

Perforin and Fas are the two main mediators of target cell lysis by T lymphocytes. Perforin is a pore-forming molecule that is secreted predominantly by CD8+ T cells and incorporated into the membrane of target cells. Increased expression of perforin mRNA has been detected by reverse transcriptase-polymerase chain reaction (RT-PCR) in the pancreases of patients with IDDM (Somoza et al., 1994), and by immunocytochemistry in NOD mice (Young et al., 1989), consistent with a role for perforin-mediated lysis in β-cell destruction. The role of perforin in autoimmune diabetes has been explored further by the generation of perforin-deficient NOD mice (Kagi et al., 1997). These mice show no defect in the activation and proliferation of T cells and display insulitis similar to control animals. However, the perforin-deficient mice show a dramatically reduced incidence and delayed onset of diabetes compared to control animals.

These observations support a crucial role for perforin-mediated cytotoxity in the NOD mouse model, but indicate that other, perhaps less efficient, mechanisms can also operate.

Fas is a cell surface receptor that, following interaction with Fas ligand, induces cell death by apoptosis. Fas-mediated apoptosis plays an important role in a number of processes, including deletion of self-reactive T cells during development in the thymus and tissue homeostasis, and is a major mechanism for cytolysis of target cells by CD4+ T cells. Fas ligand expression in immune privileged sites, such as the testis and some tumors, may assist in protecting them from immune attack by causing the deletion of Fas-expressing activated T cells. Induction of Fas ligand on β-cells could potentially prevent diabetes by deleting autoreactive T cells involved in β-cell destruction, and form the basis of an effective gene therapy for the disease. However, Fas ligand expression on β-cells of NOD mice, induced by transfection or germ line transfer, resulted in an acceleration of the disease process rather than its prevention. (Chervonsky et al., 1997) These results imply that immune-mediated induction of Fas on β-cells already expressing Fas ligand results in death by apoptosis and that the Fas pathway may play a role in spontaneous diabetes in the NOD mouse. In support of this observation, Fas-deficient NOD mice do not develop diabetes and are resistant to disease induction after transfer of diabetogenic T-cell lines (Itoh et al., 1997). Fas deficiency therefore operates at the level of the islet to prevent islet cell death, rather than affecting lymphocyte development and generation of effector immune cells responsible for disease.

Normal β-cells do not express detectable levels of Fas. However, Fas expression can be induced in vitro on both mouse and human islets by the inflammatory cytokine, interleukin-1 (Stassi et al., 1997). This cytokine is known to be cytotoxic to β-cells in vitro by mechanisms that include the induction of nitric oxide synthase (Nerup et al., 1994). Fas expression induced by interleukin-1 is apparently mediated by nitric oxide production (Stassi et al., 1997). Recent investigations on pancreases of two patients who died shortly after development of Type 1 diabetes indicated that Fas can be strongly upregulated specifically on β-cells in the diabetic pancreas. β-Cells expressing Fas and having characteristics of apoptosing cells were localized in close proximity to T cells expressing Fas-ligand. Apoptosis following induction of Fas expression by β-cells may therefore be an important element of the disease process in both human and mouse Type 1 diabetes.

B. Adhesion Molecules

Immune surveillance, the process of patrolling the body in search of foreign antigens, is achieved by the circulation of lymphocytes through the blood system and the lymphatics. Localized inflammation at the site of infection involves the migration of lymphocytes from the blood across the vascular epithelium in a multistep process initiated by the attachment of lymphocytes to endothelial cells. Attachment is mediated by specific receptors on lymphocytes which bind ligands

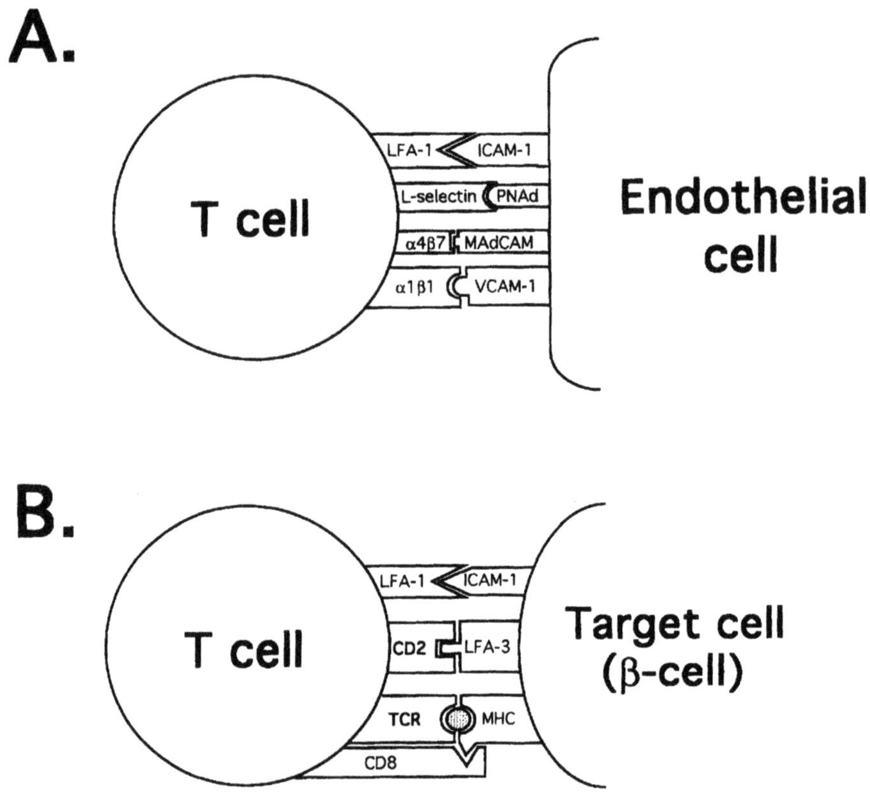

Figure 1. Adhesion receptors on T cells involved in **(A)** binding to vascular endothelium, and **(B)** antigen-specific recognition of target cells.

on endothelial cells that are upregulated in response to cytokine stimulation. Multiple adhesion receptors have been described on lymphocytes (Springer, 1990), including L-selectin, α4β7- and α4β1-integrins and lymphocyte function associated antigen-1 (LFA-1). These receptors interact with a variety of ligands on vascular endothelium, including peripheral node addressin (PNAd), mucosal addressin cell adhesion molecule (MAdCAM-1), vascular adhesion molecules (VCAM-1), and intercellular cell adhesion molecules (ICAM-1, ICAM-2) (Figure 1A). Adhesion molecules also have important roles in facilitating antigen-specific recognition of target cells by lymphocytes. Thus, adhesion receptors LFA-1 and CD2 on lymphocytes interact with ICAM-1 and LFA-3 on the target cell, strengthening the interaction of the T cell antigen receptor with antigenic peptide presented by MHC class I molecules (Figure 1B). These interactions induce T cell activation and target cell killing. In Type 1 diabetes, increased expression of ligands for adhesion receptors on vascular epithelium and pancreatic β-cells may be critical for the homing of lymphocytes to the pancreatic islet and β-cell destruction. Blocking of the adhesion molecule interactions may therefore represent a potential site for therapy of the disease.

Expression of adhesion molecules has been examined in the pancreas in both human and mouse diabetes. In the NOD mouse, a progressive increase in expression of MAdCAM-1 on pancreatic vascular endothelium in the region of islets has been observed from 4 weeks of age (Yang et al., 1996), followed by upregulation of PNAd at 8 weeks. Increased expression of ICAM-1 has been detected on many vessels within NOD mouse islets, both with and without insulitis (Hanninen et al., 1993b). Lymphocytes infiltrating the pancreas express high levels of cell adhesion molecules (Faveeuw et al., 1994). In human diabetes, fewer changes in adhesion molecules have been reported, but ICAM-1 is hyperexpressed on vascular epithelium of inflamed islets and there is also evidence of weak expression of ICAM-1 on pancreatic β-cells (Hanninen et al., 1993a; Somoza et al., 1994).

Increased expression of MHC class I molecules on islet cells is established as a very common feature of the pancreas in Type 1 diabetes (Foulis et al., 1987a) and, in combination with increased ICAM-1 expression, could promote activation of islet specific T lymphocytes and target cell killing. Increased MHC class I expression is detected in both inflamed and noninflamed islets and therefore is likely to precede infiltration of inflammatory cells and may represent an early event in the course of the disease. In vitro studies have shown that inflammatory cytokines interferon-γ and tumor necrosis factor-α can cause upregulation of both MHC class I molecules and ICAM-1 on pancreatic islet cells (Campbell et al., 1985; Campbell et al., 1989). Hyperexpression of MHC class I molecules specifically in the pancreatic β-cells has been shown to cause diabetes in transgenic mice (Harrison et al., 1989). Diabetes in these animals is not, however, immune mediated; rather the upregulation of MHC class I appears to interfere with the normal insulin-secretory function of the β-cells. Whether this phenomenon can contribute to insulin deficiency in human diabetes, or is merely an artefact of transgenic hyperexpression, is unclear.

The observed effects of interferon-γ and tumor necrosis factor-α on MHC class I and adhesion protein expression in pancreatic islets in vitro, as well as the specific cytotoxicity of certain cytokines (Nerup et al., 1994), particularly interleukin-1, on pancreatic β-cells has prompted investigations of cytokine expression within the pancreas of diabetic patients. These studies have not revealed increases in the expression of interleukin-1, interferon-γ, or tumor necrosis factor-α around the time of diabetes onset, although increased expression of interferon-α and interleukin-6, frequently seen in viral infection of cells, was observed (Foulis et al., 1987b; Somoza, et al., 1994; Huang et al., 1995). The pattern of cytokine expression is consistent with the observed predominance of CD8-positive T cells, which are poor producers of cytokines. These observations, however, are necessarily restricted to the overt diabetic pancreas, and reflects cytokine expression at a very late stage of the disease process. The situation at the critical period of initiation of the inflammatory process may be very different but will clearly be difficult, if not impossible, to characterize.

IV. TARGETS OF THE IMMUNE RESPONSE IN TYPE 1 DIABETES MELLITUS

A. Viruses

The low concordance rate of diabetes in identical twins has emphasised the importance of environmental factors in the pathogenesis of Type 1 diabetes. There has been particular interest in viral infection as a contributor to disease development. The seasonal variation in disease, with peak incidence during seasons with a high prevalence of infectious disease, together with the reports of respiratory infections shortly before diabetes development, support a role for viral infection in precipitation of disease. Viruses may act to cause direct lysis of infected β-cells during infection or trigger autoimmune responses to islet components, either by release of autoantigens during cell damage or by immunological cross-reactivity between viral antigens to which immune responses have been elicited and sequences on β-cell self proteins (molecular mimicry). Associations between specific viral infections and diabetes development have been reported, including congenital rubella, mumps and coxsackievirus (Yoon, 1991). Furthermore, a direct role for coxsackievirus infection in β-cell destruction was demonstrated by the isolation of coxsackievirus from the pancreas of a child who died of acute onset diabetes (Yoon et al., 1979), that was capable of inducing diabetes after infection of mice (Toniolo et al., 1982). However, there is conflicting data concerning the significance of these observations to the majority of cases of Type 1 diabetes. Thus, examination of pancreases from diabetic patients provided no evidence of coxsackie, mumps, cytomegalovirus, or Epstein-Barr virus infection (Foulis et al., 1997). Similarly, there is no clear evidence of increased viral antibody titres in Type 1 diabetes. Analysis of islet autoimmunity by the detection of islet cell antibodies suggests that the disease is a chronic process that can occur over many years. If viral infection acts as a trigger for an autoimmune process, evidence of infection may have disappeared by the time disease develops, and would therefore be difficult to detect except in very young patients. Recent studies have indeed shown increased enteroviral RNA in the serum of very young patients at the time of IDDM onset (Clements et al., 1995).

The characteristics of the antigen receptors of T cells within pancreases of two children who died with acute-onset IDDM do suggest that T cells within the islet inflammation may be driven, at least in part, by a persistent viral infection. The structure of T cell receptors is defined by rearrangements of genes encoding variable domains (Vβ and Vα) of the receptor, which interact with both the antigenic peptide and regions of MHC molecules to form a trimolecular complex (see Figure 1B). Each T-cell clone has a unique combination of Vβ, Vα, and other hypervariable elements, which determines the antigen specificity of the particular clone. Antigen stimulation can result in an expansion of T cells with a restricted usage of Vβ and Vα genes, so the characterization of T cell receptor genes within an inflammation can be valuable in assessing the nature of the target antigen. In addition, some components of viruses and bacteria, termed superantigens, can interact directly with Vβ regions of T cell receptors, leading to a polyclonal expansion of T cells. The stimulated T cell population shares common Vβ sequences, but differs in usage of other hypervariable elements. Polymerase chain reaction amplification and sequence analysis of T cell receptor variable domains from isolated islets from two diabetic patients indicated an over-representation of Vβ7 sequences but diversity of other elements, suggestive of a response to superantigen (Conrad et al., 1994). T cells expanded from the pancreases of the patients responded only to membrane preparations of islets from the two diabetic patients, and not to islets from nondiabetic donors, indicating that the target antigen is restricted to the diabetic pancreas.

Further investigations to search for potential viral superantigens led to the isolation of a retrovirus from supernatants of cultures of islets isolated from these patients (Conrad et al., 1997). The retrovirus was localized in lymphocytes, rather than the islets themselves. The retrovirus was novel, but sequence analysis indicated a relationship with mouse mammary tumor virus. The viral envelope gene was found to encode for a protein with superantigenic properties, which stimulated T cells specifically expressing Vβ7 T cell receptors. Furthermore, retroviral mRNA was detected in plasma from 10 patients with Type 1 diabetes, but was absent in 10 control individuals, indicating that activation of the retrovirus may be common in human Type 1 diabetes. Interestingly, endogenous retrovirus expression has also been reported in the β-cells of NOD mice (Gaskins et al., 1992; Leiter et al., 1986). Activation of a retrovirus may therefore be a general feature of diabetes in both mouse and human.

A number of explanations for the involvement of endogenous retroviruses in mouse and human diabetes have been proposed (Benoist and Mathis, 1997). Thus, expression of a retroviral superantigen in the pancreas or elsewhere may stimulate the activation and proliferation of T cells expressing specific Vβ receptors (e.g., Vβ7), some of which may be reactive with β-cell antigens. Activation of retrovirus may therefore represent the trigger for an autoimmune reaction. Alternatively, retroviral expression may be a secondary consequence of inflammatory responses or metabolic changes associated with diabetes development. The retrovirus may therefore not play a direct role in diabetes pathogenesis, but rather be a marker of the disease process. Clearly a more extensive analysis of expression of T cell receptor gene expression in the diseased pancreas, and of retroviral transcripts expressed in diabetes, will be important to assess the significance of retroviral activation in the disease process.

B. Glutamic Acid Decarboxylase

The vast majority of patients with Type 1 diabetes have, at the time of disease diagnosis, circulating serum antibodies (islet cell antibodies, or ICA) that bind to islets on sections of normal human pancreas, indicating that the generation of antibodies to antigens present in the undiseased pancreas is an important component of the immune response. Considerable effort has been applied to identifying the target molecules for these ICA. It is now clear that there is no single specific target of ICA (Table 1). Nevertheless, the immune response does appear to be directed predominantly to a few specific islet cell antigens. The identification of these antigens has increased our understanding of the molecular basis of disease, which will be valuable in both disease prediction and the design of procedures to block the autoimmune process and prevent disease.

Early experiments to determine the nature of islet components recognized by the immune system in Type 1 diabetes involved the immunoprecipitation of proteins in detergent solubilized extracts of islets with antibodies in sera of Type 1 diabetic patients. These studies showed that 64,000 Mr proteins were the major proteins recognized by antibodies in the sera of patients with recent onset Type 1 diabetes (Baekkeskov et al., 1982). Antibodies to these 64,000 Mr antigens were subsequently shown to be present in approximately 70% of patients but absent in control individuals (Christie et al., 1988) and to precede diabetes onset by many years (Baekkeskov et al., 1987). The 64,000 Mr antigen recognized by patients in these early studies is now known to represent multiple proteins co-migrating on SDS-PAGE (Christie et al., 1993). Following the observation that patients with stiff-man syndrome (SMS), a rare neurological disorder that is commonly accompanied by IDDM development, have antibodies to a 64,000 Mr protein in γ-aminobutyric acid (GABA)-ergic neurons and pancreatic islets (Solimena et al.,

Table 1. Dominant Targets for Autoimmune Responses in IDDM

Antigen	*Reference*
GAD65	Baekkeskov et al. (1990)
GAD67	Hagopian et al. (1993)
IA-2 (ICA512)	Payton et al. (1995)
IA-2β (Phogrin)	Lu et al. (1996)
Insulin	Pav et al. (1963)
38kDa antigen (Glima 38)	Aanstoot et al. (1996)
38kDa antigen (Imogen)	Roep et al. (1991)
Ganglioside GM2-1	Dotta et al. (1996)
Sulphatide	Buschard et al. (1993)
ICA69	Pietropaolo et al. (1993)

Table 2. Tissue Expression of GAD Isoforms

Tissue	*GAD65*	*GAD67*
Brain	++	++
Human β-cells	++	–
Rat β-cells	++	++
Mouse β-cells	+/–	+
Testis	+	+
Spleen	–	+
Liver	–	+

1990), the target protein for autoantibodies in both of these disorders was identified as glutamic acid decarboxylase (GAD) (Baekkeskov et al., 1990). GAD is the enzyme responsible for the biosynthesis of GABA, and GAD activity is found in the GABA-ergic neurons of the central nervous system and in a number of peripheral tissues, particularly the pancreatic islets, but also at lower levels in testis, ovary, kidney, and liver. GAD is known to be expressed as two isoforms (Erlander et al., 1991), GAD65 and GAD67, which are the products of different genes localized on human chromosome 10 and chromosome 2, respectively. The two enzymes show different tissue distribution and subcellular localization (Table 2), and within islets there are species-specific differences in the expression of the two isoforms. Thus, human islets express predominantly GAD65 (Karlsen et al., 1992), rat islets express both isoforms (Kim et al., 1993), and in mouse islets the expression of GAD65 is very low (Velloso et al., 1994b).

The synthesis of GABA by GAD is proposed to have a dual function in pancreatic islets and other tissues. Thus, GAD has a metabolic role in several tissues by allowing energy formation from glutamic acid via the synthesis of succinate through the "GABA shunt" (Figure 2). However, in islets and the nervous system, GABA has also developed as a signaling molecule and GAD has a role in the biosynthesis of GABA for storage in synapticlike vesicles in these tissues. The different subcellular properties of the two GAD isoforms may be related to their relative importance in these two cellular functions. Thus, within islet cells GAD67 is predominantly cytoplasmic, whereas GAD65 is membrane associated and localized on the cytoplasmic face of the synapticlike vesicles (Reetz et al., 1991). The nature of membrane anchoring is unclear, as the amino acid sequence of GAD65 does not reveal any hydrophobic regions typical of transmembrane domains. GAD65 is palmitoylated on cysteine residues located close to the amino terminus of the molecule (amino acids 30 and 45) and palmitoylation appears to promote membrane association, but is not essential for it (Shi et al., 1994). The integrity of amino acids 24–31 of the molecule appears to be necessary for membrane association, and this may occur via the interaction with other as yet unidentified membrane bound proteins. A proportion of GAD67 is also associated with membrane structures, but this appears to occur as a consequence of heterodimer formation with GAD65

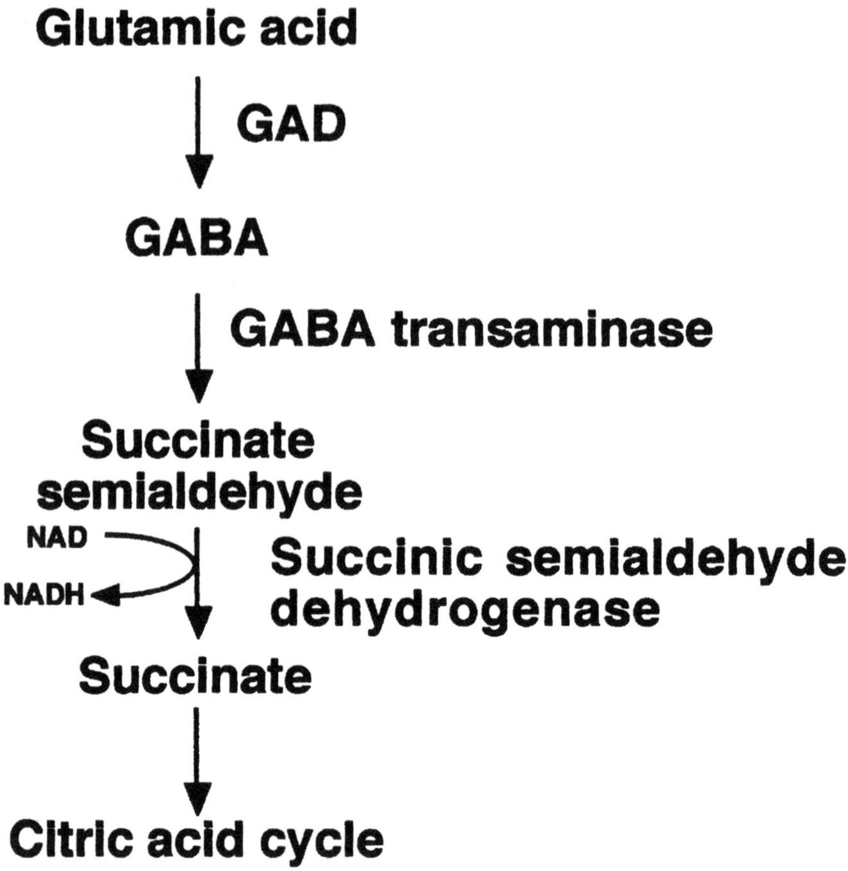

Figure 2. The γ-aminobutyric acid (GABA) shunt pathway. GAD, glutamic acid decarboxylase.

(Lernmark, 1996). Indeed, the major proportion of GAD in cells is multimeric, predominantly dimeric, and a dimerization domain has been localized to the first 83 amino acids of the GAD65 molecule. The significance of homo- and heterodimer formation for enzyme activity is not understood.

Within pancreatic islets, GAD expression is upregulated by glucose, implicating changes in GABA synthesis in islet cell function (Velloso et al., 1994a). The synapticlike vesicles within islets very likely act as stores of GABA for subsequent exocytosis. GABA has, indeed, been demonstrated to be secreted from islets and in some studies glucose stimulates GABA release (Gaskins et al., 1996). However, GABA secretion has clearly been shown to be dissociated from insulin release, indicating that there are different mechanisms of stimulus-secretion coupling for products within the two types of secretory vesicles, insulin secretory granules and synapticlike vesicles (Smismans et al., 1997). The release of GABA appears to be directly related to cellular GABA content, indicating that the major factor regulating GABA release is the rate of biosynthesis. GABA receptors have been shown to be expressed within the islet on α- and δ-cells, suggesting that GABA may have a paracrine role to modulate hormone release within the islet (Rorsman et al., 1989). Exogenous GABA does, indeed, inhibit glucagon secretion from isolated pancreatic islets in vitro consistent with such a role for GABA in the islet. Situations that result in an inhibition of GABA synthesis by GAD may, therefore, lead to an inbalance in hormone secretion by the pancreatic islet.

Of the two GAD isoforms, GAD65 is the predominant target of the antibody response in Type 1 diabetes. Thus, approximately 70% of recent-onset Type 1 diabetic patients possess antibody reactivity to GAD65, whereas antibodies to GAD-67 are detectable in around 15% (Hagopian et al., 1993). Although the amino acid sequences of GAD65 and GAD67 are most divergent within the first 100 amino acids, the majority of antibody reactivity to GAD65 is directed to the more homologous C-terminal region. GAD antibodies in the disease appear to be quite diverse in their epitope recognition, and no immunodominant region has yet been identified (Ujihara et al., 1994). Investigators have remarked that there is a region on GAD65 that has a strong sequence similarity to part of the PC-2 protein of coxsackievirus, and have proposed that autoimmunity to GAD might initially arise following an immune response to coxsackievirus infection, already implicated in diabetes development (Kaufman et al., 1992). In support of this hypothesis, T cells from NOD mice that respond to peptides representing the PC-2 sequence have been shown cross-react with the equivalent sequence of GAD, and T cell reactivity to this region appears to be unique to mice who express diabetes susceptible NOD MHC genes (Tian et al., 1994).

T cell responses to GAD have been demonstrated using spleen cells in NOD mice and peripheral blood lymphocytes in human Type 1 diabetes (Atkinson et al., 1992; Kaufman et al., 1993; Tisch et al., 1993; Atkinson et al., 1994; Endl et al., 1997;), but the regions recognized are quite diverse, and may not always include the coxsackievirus homology domain. Studies of NOD mice of differing ages suggest that the T cell response in the early phase of the disease, at the time insulitis is first detected, may be restricted to a region close to the C-terminus of the molecule, and may subsequently spread to other regions, including the PC-2 homology sequence (Kaufman et al., 1993). The fact that the region similar to the coxsackievirus sequence is not the earliest target of the T cell response, argues against a role for coxsackievirus infection in initiation of GAD autoimmunity, at least in the NOD mouse model. Nevertheless, autoimmunity to GAD does appear to be an early and important event in disease progression in the NOD mouse model. Thus, spontaneous T cell responses to GAD are first detectable at 3 weeks of age and precede immune responses to other potential islet cell antigens in the disease (Kaufman et al., 1993; Tisch et al., 1993). Furthermore, induction of immunological tolerance to GAD blocks islet inflammation and diabetes development, as well as immune responses to other islet autoantigens. GAD-specific T cell lines generated from NOD mice immunized with GAD have been shown to be capable of inducing diabetes in NOD SCID mice (Zekzer et al., 1998). Together these data suggest that autoimmunity to GAD in the NOD mouse is an essential component of a progressively diverse immune response to islets that culminates in diabetes development. Since autoantibodies and T cells to GAD are detected at high frequency in human Type 1 diabetes, the findings in the mouse may also have significance to the human disease.

C. Protein Tyrosine Phosphataselike Proteins IA-2 and IA-2β

While GAD was clearly shown to be a major 64,000 Mr islet cell autoantigen in Type 1 diabetes, other evidence suggested the presence of at least two 64,000 Mr proteins distinct from GAD recognized by antibodies in the disease (Christie et al., 1993). Analysis of antibodies in a number of populations at risk for diabetes suggested that autoimmunity to these proteins may be even more closely linked to diabetes development than autoimmunity to GAD (Christie et al., 1992; Bingley et al., 1994; Christie et al., 1994). The two proteins have recently been identified as closely related protein tyrosine phosphatase (PTP)-like proteins designated IA-2 (or ICA512) and IA-2β (or phogrin) respectively (Bonifacio et al., 1995; Payton et al., 1995; Hawkes et al., 1996; Lu et al., 1996). IA-2 appears to be the dominant PTP-like target for autoantibodies in Type 1 diabetes, and most of antibody reactivity to IA-2β appears to be the result of cross-reactivity to the structurally similar IA-2 (Hatfield et al., 1997).

IA-2 and IA-2β are the products of two distinct genes localized on human chromosomes 2 and 7, repectively, but the two proteins have a number of features in common (Figure 3). The predicted amino acid sequences of the two gene products encode for proteins of 105kDa and 110kDa, respectively, with putative signal peptide sequences, transmembrane domains, and regions at the C-terminus that have high degrees of sequence homology with known protein tyrosine phosphatases (Lan et al., 1994; Wasmeier and Hutton, 1996). Both IA-2 and IA-2β are expressed in neuroendocrine tissues, including regions of the brain, pituitary and the endocrine pancreas; both are integral membrane components of secretory granules in these tissues, and are subject to post-translational proteolytic processing to form mature proteins of approximately 64,000 Mr (Solimena et al., 1996). IA-2 and IA-2β have more than 80% amino acid sequence similarity in their cytoplasmic part, which encompasses the PTP-like domain, but sequence similarity is less than 50% in the region located within the secretory granule lumen (Wasmeier and Hutton, 1996). Although possessing PTP-like domains, neither IA-2 nor IA-2β have been shown to have enzymatic activity using peptides or proteins that are common substrates for these enzymes. Furthermore, both IA-2 and IA-2β have a number of amino acid substitutions at residues that are highly conserved in other

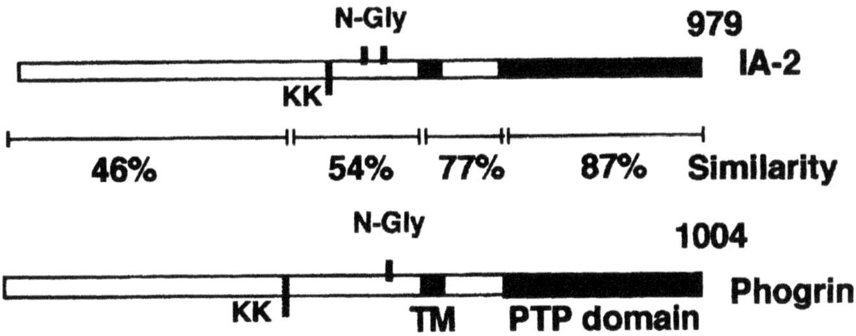

Figure 3. Schematic representation of IA-2 and phogrin (IA-2β) showing regions of amino acid homology and sites of post-translational proteolysis at double lysine motif (KK) and glycosylation (N-Gly). Transmembrane (TM) domain is shaded and protein tyrosine phosphatase (PTP) domain is in black.

tyrosine phosphatases, and which in other PTPs inactivate phosphatase activity (Streuli et al., 1990; Magistrelli et al., 1996). It has therefore been suggested that these proteins either have a very restricted substrate specificity, or that they may indeed be inactive as phosphatase enzymes. Instead, the PTP regions may bind other tyrosine phosphorylated proteins either to mediate protein–protein interactions or to act as a competitive inhibitors.

Complementary DNAs for IA-2 and IA-2β have been cloned independently by a number of research groups studying various aspects of neuroendocrine cell development, cell proliferation, and tumorigenesis. Thus, an equivalent of IA-2β was identified as a PTP-like protein showing expression early in neuronal development of the mouse (Chiang and Flanagan, 1996). Further investigation indicated that IA-2β was also expressed very early in the development of the pancreas, before the formation of the pancreatic rudiment and the expression of known markers of the endocrine and exocrine lineages. A mouse equivalent of IA-2 was cloned as a major PTP-like protein showing increased expression in proliferating 3T3 fibroblasts, a model system for studying regulation of cell proliferation, whereas expression of the protein was strongly downregulated in resting cells (Magistrelli et al., 1995). Furthermore, IA-2 and IA-2β have been shown to be expressed in a number of human cancers, including colon carcinoma (Cui et al., 1996), lung cancer (Xie et al., 1996), and breast tumors. Many of the tumors expressing IA-2 and IA-2β had a neuroendocrine phenotype. These latter findings indicate that, as well as having a potential role in neuroendocrine secretory granule biogenesis and function, IA-2 and IA-2β are also implicated in neuroendocrine cell differentiation and proliferation and may provide important markers for neuroendocrine type tumors.

Antibodies to IA-2 are detected in approximately 70% of new-onset Type 1 diabetic patients and appear to be most common in younger patients (Bingley et al.,

1997; Christie et al., 1997), and in patients expressing the diabetes susceptibity HLA-DR4 alleles (Genovese et al., 1996). A high proportion of both Type 1 diabetic patients and first-degree relatives of Type 1 diabetic patients with high risk for disease (islet antibody positive) were found to have peripheral blood T cell reactivity to IA-2 preparations, whereas T cells from most control subjects did not respond (Durinovic Bello et al., 1996). Unfortunately, further investigations into the potential role of IA-2 in disease pathogenesis have been hampered by the lack of a suitable animal model to study; in contrast to GAD, neither the NOD mouse nor the BB rat have evidence of an autoimmune responses to this autoantigen.

D. Insulin

The generation of antibodies to insulin is a common response to therapeutic insulin injections in Type 1 diabetes, which in very rare cases can lead to immunological insulin resistance. Insulin antibodies were also detected more than 30 years ago in a proportion of patients who had not received insulin treatment (Pav et al., 1963). Subsequent studies have shown that insulin antibodies are commonly detected in young diabetic patients at the time of disease diagnosis, but are less common in older patients (Vardi et al., 1988). Insulin is thus a third major target of autoimmunity in Type 1 diabetes, and is indeed the only autoantigen that is specifically expressed by the pancreatic β-cell.

T cell reactivity to insulin has been described in both human Type 1 diabetic patients (Scheinin et al., 1990) and in NOD mice (Kaufman et al., 1993), although in the latter may be detectable only at a late stage of the disease. Importantly, insulin reactive T cells have been isolated from the pancreases of NOD mice and these cells may represent a considerable proportion of the infiltrating T cell population (Haskins and Wegmann, 1996). T cell lines reactive to insulin have been shown to be capable of inducing disease in transfer experiments (Daniel et al., 1995) and may therefore contribute to β-cell destruction in the animal model.

Further evidence for a crucial role of insulin in the disease process has come from characterization of genes contributing to diabetes susceptibility. A major susceptibility locus has been mapped to variable number of tandem repeats (VNTR) upstream of the insulin gene on chromosome 11 (Undlien et al., 1995). Class I VNTR alleles (short repeats) predispose to diabetes whereas class III alleles (longer repeats) are protective. VNTR expression was found to be correlated with pancreatic insulin expression, but paradoxically, the protective class II alleles were associated with lower insulin expression. A possible explanation for the protective effects of class III VNTR has come from an analysis of autoantigen expression in the thymus. The thymus is an important site of immunological self-tolerance induction. During T cell development, T cells that have high affinity for self proteins expressed in the thymus are deleted. Messenger RNAs for insulin, GAD, and IA-2 have all been detected in the human thymus (Pugliese et al., 1997; Vafiadis et al., 1997) implying that deletion of T lymphocytes to these islet antigens during thymic T cell development is one potential mechanism for avoiding islet autoimmunity. Importantly, thymic expression of insulin was higher in individuals with class III VNTR. Higher levels of insulin in the thymus may result in more efficient deletion of insulin-reactive T cells and protect from diabetes development. Unfortunately, no clear links between insulin gene VNTR and autoimmune responses to the protein have been established. Nevertheless, factors that influence thymic expression of major islet autoantigens could have a profound influence on diabetes susceptibility.

E. Other Antigens

Blocking studies with purified recombinant antigens have indicated that ICA in sera from type 1 diabetic patients detected on frozen sections of pancreas by indirect immunofluorescence cannot be totally accounted for by GAD, IA-2, or insulin. There is evidently at least one other major target for ICA in Type 1 diabetes, and a range of biochemical and molecular techniques have been used to identify potential disease-associated antigens. A number of these have been described (see Table 1), but not all of these have consistently been demonstrated to be disease-specific targets of islet cell antibodies. Early biochemical characterization of the targets of ICA implicated glycolipid molecules as the major components recognized by ICA (Colman et al., 1988). Subsequent studies have implicated a number of glycolipid molecules as autoantigens in Type 1 diabetes, including sulfatides (Buschard et al., 1993) and the sialo-ganglioside GM2-1 (Dotta et al., 1996). Antibodies to GM2-1 have been found to be associated with progression to Type 1 diabetes in relatives with ICA (Dotta et al., 1996), but their relationship to ICA remains unclear.

Other potential islet autoantigens are poorly characterized. Antibodies in approximately 20% of Type 1 diabetic patients immunoprecipitate a 38-kDa membrane glycoprotein (Glima 38) from islet extracts, but the identity or function of the protein is unknown (Aanstoot et al., 1996). Roep and colleagues (1991) have detected T cell reactivity to an apparently distinct 38kDa molecule, which has been identified as an ubituitous mitochondrial protein (Imogen 38), whose function is unclear (Arden et al., 1996). A 69kDa protein target for autoantibodies in Type 1 diabetes (ICA69) raised considerable interest as the protein has a region of similarity with the cow's milk protein bovine albumin (Pietropaolo et al., 1993). However, antibodies to ICA69 are also detected in rheumatoid arthritis (Martin et al., 1995), and in some studies do not discriminate between Type 1 diabetic patients and control subjects (Lampasona et al., 1994).

F. Progressive Responses to Islet Antigens in the Early Phase of Human Disease

Studies on the NOD mouse have demonstrated a progressive immune response to different islet autoantigens, which has implicated induction of autoimmunity to

Age (years)		1	2	3	4	5
Relative 1	IA-2	-	-	+	+	+
	GAD	-	+	+	+	+
	IAA	+	+	+	+	+
Relative 2	IA-2	-	+	+	+	+
	GAD	-	-	-	-	+
	IAA	-	+	+	+	+
Relative 3	IA-2	-	+	+		
	GAD	+	+	+		
	IAA	-	+	+		
Relative 4	IA-2	+				
	GAD	-				
	IAA	-				
Relative 5	IA-2	+				
	GAD	+	Diabetes			
	IAA	+				

Figure 4. Progressive antibody responses to islet antigens in early life. The figure shows progressive development of positive antibody responses (+) to GAD, IA-2, and insulin (IAA) in five offspring of diabetic parents followed prospectively from birth. Relative 5 developed Type 1 diabetes at 1 year of age (see Roll et al., 1996).

GAD as an essential key event in the early phase of disease (Kaufman et al., 1993; Tisch et al., 1993). Prospective studies to evaluate the progression of immune responses to islet antigens in humans have been carried out in Germany (BABY-DIAB study) and the United States (DAISY) by following offspring of diabetic parents and young relatives of Type 1 diabetic patients. These studies indicate that autoimmunity to islet antigens can be detected very early in life, frequently before the age of 5 years, although autoimmunity can also be initiated later (Roll et al., 1996; Yu et al., 1996). Similar to the NOD mouse, antibody responses to different antigens can occur progressively over a period of several years (Figure 4), but no consistent order in the progression of autoimmune responses to the different antigens was observed. Thus, in contrast to the observations in the NOD mouse studies, there is no evidence of a single primary islet cell antigen triggering islet autoimmunity in human Type 1 diabetes.

V. IMPLICATIONS FOR DIABETES PREDICTION AND PREVENTION

Analysis of the immune response and islet pathology in human Type 1 diabetes, and in animal models of the disease, suggests that insulin deficiency associated with the disease is the result of a progressive inflammatory response within the islets in which T cells play a major role. Although cytotoxic T cells form the majority of cells in the infiltrate in most cases of human Type 1 diabetes, cellular destruction may involve both perforin-dependent and apoptotic mechanisms and involve both CD8 and CD4 positive T cells. A few dominant components of the islet cells, including GAD, IA-2, and insulin, may drive an autoimmune response, but involvement of infectious agents also has to be considered. However, circulating antibodies to islet antigens are detectable early in the disease process and are therefore valuable markers for the identification of individuals at risk for disease. No single antibody marker identified to date has sufficient sensitivity and specificity to be used alone for reliable disease prediction. However, existing antibody markers can be used in combination to provide accurate assessments of disease risk (Bingley et al., 1994; Christie et al., 1997). Thus, most Type 1 diabetic patients have antibodies to multiple islet cell antigens at the time of disease diagnosis, and prospective studies of non-diabetic relatives of Type 1 diabetic patients show that possession of two or more antibodies to islet antigens confers a very high risk for subsequent disease development. Screening for multiple antibodies also appears to be a valid strategy to identify individuals with no family history, who represent the major proportion of individuals who develop disease (Bingley et al., 1997).

The ability to identify individuals at risk for Type 1 diabetes with a high degree of accuracy, combined with a better understanding of molecular events important in the disease process, offers the potential to intervene in the disease process. A wide range of therapeutic procedures are effective in animal models of disease, and some of these have been tested in clinical trials in humans. These include general immunosuppressive therapy (e.g., cyclosporin), which has some beneficial effects in human Type 1 diabetes; blockade of adhesion molecule interactions with monoclonal antibodies (reviewed in Yang et al., 1996), to prevent islet inflammation; and procedures to specifically block autoimmune responses to islet cell antigens. Much interest has focussed on the latter, as antigen-specific intervention should avoid immune deficiency, which might occur with more general immunosuppressive treatments. Promising strategies involve induction of immunological tolerance by administering antigen by oral or nasal routes, or "T cell vaccination," in which autoreactive T cells to autoantigens are used as vaccines to prevent an autoimmune reponse. Both approaches have shown promise in other autoimmune disorders. There is, therefore, real hope that characterization of molecules involved in IDDM will have practical benefits in the near future for the treatment and prevention of the disease.

REFERENCES

Aanstoot, H.J., Kang, S.M., Kim, J., Lindsay, L.A., Roll, U., Knip, M., Atkinson, M., Mose Larsen, P., Fey, S., Ludvigsson, J. and Baekkeskov, S. (1996). Identification and characterization of glima

38, a glycosylated islet cell membrane antigen, which together with GAD65 and IA2 marks the early phases of autoimmune response in type 1 diabetes. J. Clin. Invest. 97, 2772–2783.

Arden, S.D., Roep, B.O., Neophytou, P.I., Usac, E.F., Duinkerken, G., de Vries, R.R. and Hutton, J.C. (1996). Imogen 38: A novel 38-kD islet mitochondrial autoantigen recognized by T cells from a newly diagnosed type 1 diabetic patient. J. Clin. Invest 97, 551–561.

Atkinson, M.A., Bowman, M.A., Campbell, L., Darrow, B.L., Kaufman, D.L. and Maclaren, N.K. (1994). Cellular immunity to a determinant common to glutamate decarboxylase and Coxsackie virus in insulin-dependent diabetes. J. Clin. Invest. 94, 2125–2129.

Atkinson, M., Kaufman, D., Campbell, L., Gibbs, K., Shah, S., Bu, D.-F., Erlander, M., Tobin, A. and Maclaren, N. (1992). Response of peripheral-blood mononuclear cells to glutamate decarboxylase in insulin-dependent diabetes. Lancet 339, 458–459.

Bach, J.-F. (1994). Insulin-dependent diabetes as an autoimmune disease. Endocrine Rev. 14, 516–542.

Baekkeskov, S., Aanstoot, H., Christgau, S., Reetz, A., Solimena, M., Cascalho, M., Folli, F., Richter-Olesen, H. and De Camilli, P. (1990). Identification of the 64k autoantigen in insulin-dependent diabetes as the GABA-synthesizing enzyme glutamic acid decarboxylase. Nature 347, 151–156.

Baekkeskov, S., Nielsen, J. H., Marner, B., Bilde, T., Ludvigsson, J. and Lernmark, Å. (1982). Autoantibodies in newly diagnosed diabetic children immunoprecipitate pancreatic islet cell proteins. Nature 298, 167–169.

Baekkeskov S.L.M., Kristensen, J.K., Srikanta, S., Bruining, G.J., Mandrup-Poulsen T., de Beaufort, C., Soeldner, J.S., Eisenbarth, G., Lindgren, F., Sundquist, G., Lernmark, A. (1987). Antibodies to a 64000-Mr human islet cell protein precede the clinical onset of insulin-dependent diabetes. J. Clin. Invest. 79, 926–934.

Benoist, C. and Mathis, D. (1997). Autoimmune diabetes. Retrovirus as trigger, precipitator or marker? Nature 388, 833–834.

Bingley, P.J., Bonifacio, E., Williams, A.J., Genovese, S., Bottazzo, G.F. and Gale, E.A. (1997). Prediction of IDDM in the general population: Strategies based on combinations of autoantibody markers. Diabetes 46, 1701–1710.

Bingley, P.J., Christie, M.R., Bonifacio, E., Bonfanti, R., Shattock, M., Fonte, M.T., Bottazzo, G.F. and Gale, E. (1994). Combined analysis of autoantibodies improves prediction of IDDM in islet cell antibody-positive relatives. Diabetes 43, 1304–1310.

Bonifacio, E., Lampasona, V., Genovese, S., Ferrari, M. and Bosi, E. (1995). Identification of protein tyrosine phosphatase-like IA-2 (islet cell antigen 512) as the insulin-dependent diabetes-related 37/40k autoantigen and a target of islet cell antibodies. J. Immunol. 155, 5419–5426.

Bottazzo, G., Florin Christensen, A. and Doniach, D. (1974). Islet cell antibodies in diabetes mellitus with autoimmune polyendocrine deficiencies. Lancet ii, 1279–1283.

Bottazzo, G.F., Dean, B.M., McNally, J.M., MacKay, E.H., Swift, P.G.F. and Gamble, D.R. (1985). In situ characterization of autoimmune phenomena and expression of HLA molecules in the pancreas in diabetic insulitis. N. Engl. J. Med. 313, 353–360.

Buschard, K., Josefsen, K., Horn, T. and Fredman, P. (1993). Sulphatide and sulphatide antibodies in insulin-dependent diabetes mellitus. Lancet 342, 840.

Campbell, I.L., Cutri, A. , Wilkinson, D., Boyd, A.W. and Harrison L.C. (1989). Intercellular adhesion molecule 1 is induced on isolated endocrine islet cells by cytokines but not by reovirus infection. Proc. Natl. Acad. Sci. 86, 4282–4286.

Campbell, I.L., Wong, G.H.W., Schrader, J.W., Harrison, L.C. (1985). Interferon γ enhances the expression of the major histocompatibility class I antigens on mouse pancreatic beta cells. Diabetes 34, 1205–1209.

Chervonsky, A.V., Wang, Y., Wong, F.S., Visintin, I., Flavell, R.A., Janeway, C.A. Jr., and Matis, L.A. (1997). The role of Fas in autoimmune diabetes. Cell, 89, 17–24.

Chiang, M.K. and Flanagan, J.G. (1996). PTP-NP, a new member of the receptor protein tyrosine phosphatase family, implicated in development of nervous system and pancreatic endocrine cells. Development 122, 2239–2250.

Christie, M., Landin Olsson, M., Sundkvist, G., Dahlquist, G., Lernmark, A. and Baekkeskov, S. (1988). Antibodies to a Mr-64000 islet cell protein in Swedish children with newly diagnosed type 1 (insulin-dependent) diabetes. Diabetologia 31, 597–602.

Christie, M., Roll, U., Payton, M., Hatfield, E. and Ziegler, A. (1997). Validity of screening for individuals at risk for Type I diabetes by combined analysis of autoantibodies to recombinant proteins. Diabetes Care 20, 965–970.

Christie, M.R., Genovese, S., Cassidy, D., Bosi, E., Brown, T.J., Lai, M., Bonifacio, E. and Bottazzo, G.F. (1994). Antibodies to islet 37k-antigen, but not to glutamate decarboxylase, discriminate rapid progression to insulin-dependent diabetes mellitus in endocrine autoimmunity. Diabetes 43, 1254–1259.

Christie, M.R., Hollands, J.A., Brown, T.J., Michelsen, B.K. and Delovitch, T.L. (1993). Detection of pancreatic islet 64,000 Mr autoantigens in insulin-dependent diabetes distinct from glutamate decarboxylase. J. Clin. Invest., 92, 240–248.

Christie, M.R., Tun, R.Y.M., Lo, S.S.S., Cassidy, D., Brown, T.J., Hollands, J., Shattock, M., Bottazzo, G.F. and Leslie, R.D.G. (1992). Antibodies to glutamic acid decarboxylase and tryptic fragments of islet 64kD antigen as distinct markers for the development of insulin-dependent diabetes. Studies with identical twins. Diabetes, 41, 782–787.

Clements, G.B., Galbraith, D.N. and Taylor, K.W. (1995). Coxsackie B virus infection and onset of childhood diabetes. Lancet 346, 221–223.

Colman, P., Nayak, R., Campbell, I. and Eisenbarth, G. (1988). Binding of cytoplasmic islet cell antibodies is blocked by human pancreatic glycolipid extracts. Diabetes 37, 645–652.

Conrad, B., Weidmann, E., Trucco, G., Rudert, W.A., Behboo, R., Ricordi, C., RodriquezRilo, H., Finegold, D. and Trucco, M. (1994). Evidence for superantigen involvement in insulin-dependent diabetes mellitus aetiology. Nature 371, 351–355.

Conrad, B., Weissmahr, R.N., Boni, J., Arcari, R., Schupbach, J. and Mach, B. (1997). A human endogenous retroviral superantigen as candidate autoimmune gene in type I diabetes. Cell 90, 303–313.

Cui, L., Yu, W.P., DeAizpurua, H.J., Schmidli, R.S. and Pallen, C.J. (1996). Cloning and characterization of islet cell antigen-related protein-tyrosine phosphatase (PTP), a novel receptor-like PTP and autoantigen in insulin-dependent diabetes. J. Biol. Chem, 271, 24817–24823.

Daniel, D., Gill, R.G., Schloot, N. and Wegmann, D. (1995). Epitope specificity, cytokine production profile and diabetogenic activity of insulin-specific T cell clones isolated from NOD mice. Eur. J. Immunol. 25, 1056–1062.

Dotta, F., Gianani, R., Previti, M., Lenti, L., Dionisi, S., D'Erme, M., Eisenbarth, G. S. and Di Mario, U. (1996). Autoimmunity to the GM2-1 islet ganglioside before and at the onset of type I diabetes. Diabetes 45, 1193–1196.

Durinovic Bello, I., Hummel, M. and Ziegler, A.G. (1996). Cellular immune response to diverse islet cell antigens in IDDM. Diabetes 45, 795–800.

Endl, J., Otto, H., Jung, G., Dreisbusch, B., Donie, F., Stahl, P., Elbracht, R., Schmitz, G., Meinl, E., Hummel, M., Ziegler, A.G., Wank, R. and Schendel, D.J. (1997). Identification of naturally processed T cell epitopes from glutamic acid decarboxylase presented in the context of HLA-DR alleles by T lymphocytes of recent onset IDDM patients. J. Clin. Invest. 99, 2405–2415.

Erlander, M.G., Tillakaratne, N.J.K., Feldblum, S., Pate, N. and Tobin, A.J. (1991). Two genes encode distinct glutamate decarboxylases. Neuron 7, 91–100.

Faveeuw, C., Gagnerault, M.C. and Lepault, F. (1994). Expression of homing and adhesion molecules in infiltrated islets of Langerhans and salivary glands of nonobese diabetic mice. J. Immunol. 152, 5969–5978.

Foulis, A.K., Farquharson, M.A. and Hardman, R. (1987a). Aberrant expression of class II major histocompatibility complex molecules by B cells and hyperexpression of class I major histocompatibility complex molecules by insulin containing islets in type 1 (insulin-dependent) diabetes mellitus. Diabetologia 30, 333–343.

Foulis, A.K., Farquharson, M.A. and Meager, A. (1987b). Immunoreactive alpha-interferon in insulin-secreting beta cells in type 1 diabetes mellitus. Lancet 2, 1423–1427.

Foulis, A.K., McGill, M., Farquharson, M.A. and Hilton, D.A. (1997). A search for evidence of viral infection in pancreases of newly diagnosed patients with IDDM. Diabetologia 40, 53–61.

Gaskins, R., Baldeon, M., Selassie, L. and Beverly, L. (1996). Glucose modulates gamma-amino butyric acid release from the pancreatic beta TC6 line. J. Biol. Chem. 270, 30286–30289.

Gaskins, H.R., Prochazka, M., Hamaguchi, K., Serreze, D.V. and Leiter, E.H. (1992). Beta cell expression of endogenous xenotropic retrovirus distinguishes diabetes-susceptible NOD/Lt from resistant NON/Lt mice. J. Clin. Invest. 90, 2220–2227.

Genovese, S., Bonfanti, R., Bazzigaluppi, E., Lampasona, V., Benazzi, E., Bosi, E., Chiumello, G. and Bonifacio, E. (1996). Association of IA-2 autoantibodies with HLA DR4 phenotypes in IDDM. Diabetologia 39, 1223–1226.

Gepts, W. (1965). Pathologic anatomy of the pancreas in juvenile diabetes mellitus. Diabetes 14, 619–633.

Hagopian, W.A., Michelsen, B., Karlsen, A.E., Larsen, F., Moody, A., Grubin, C.E., Rowe, R., Petersen, J., McEvoy, R. and Lernmark, Å. (1993). Autoantibodies in IDDM primarily recognize the 65,000-M_r rather than the 67,000-M_r isoform of glutamic acid decarboxylase. Diabetes 42, 631–636.

Hanenberg, H., Kolb Bachofen, V., Kantwerk Funke, G. and Kolb, H. (1989). Macrophage infiltration precedes and is a prerequisite for lymphocytic insulitis in pancreatic islets of pre-diabetic BB rats. Diabetologia 32, 126–134.

Hanninen, A., Jalkanen, S., Salmi, M., Toikkanen, S., Nikolakaros, G. and Simell, O. (1992). Macrophages, T cell receptor usage, and endothelial cell activation in the pancreas at the onset of insulin-dependent diabetes mellitus. J. Clin. Invest. 90, 1901–1910.

Hanninen, A., Salmi, M., Simell, O. and Jalkanen, S. (1993a). Endothelial cell-binding properties of lymphocytes infiltrated into human diabetic pancreas. Implications for pathogenesis of IDDM. Diabetes 42, 1656–1662.

Hanninen, A., Taylor, C., Streeter, P.R., Stark, L.S., Sarte, J.M., Shizuru, J.A., Simell, O. and Michie, S.A. (1993b). Vascular addressins are induced on islet vessels during insulitis in nonobese diabetic mice and are involved in lymphoid cell binding to islet endothelium. J. Clin. Invest. 92, 2509–2515.

Harrison, L.C., Campbell, I.L., Allison, J. and Miller, J.F. (1989). MHC molecules and beta-cell destruction. Immune and nonimmune mechanisms. Diabetes 38, 815–818.

Haskins, K. and Wegmann, D. (1996). Diabetogenic T-cell clones. Diabetes 45, 1299–305.

Hatfield, E., Hawkes, C., Payton, M. and Christie, M. (1997). Cross reactivity between IA-2 and phogrin/IA-2 beta in binding of autoantibodies in IDDM. Diabetologia 40, 1327–1333.

Hawkes, C.J., Wasmeier, C., Christie, M.R. and Hutton, J.C. (1996). Identification of the 37-kDa antigen in IDDM as a tyrosine phosphatase-like protein (phogrin) related to IA-2. Diabetes 45, 1187–1192.

Huang, X., Yuang, J., Goddard, A., Foulis, A., James, R. F., Lernmark, A., Pujol Borrell, R., Rabinovitch, A., Somoza, N. and Stewart, T.A. (1995). Interferon expression in the pancreases of patients with type I diabetes. Diabetes 44, 658–664.

Itoh, N., Hanafusa, T., Miyazaki, A., Miyagawa, J., Yamagata, K., Yamamoto, K., Waguri, M., Imagawa, A., Tamura, S. and Inada, M. (1993). Mononuclear cell infiltration and its relation to the expression of major histocompatibility complex antigens and adhesion molecules in pancreas biopsy specimens from newly diagnosed insulin-dependent diabetes mellitus patients. J. Clin. Invest. 92, 2313–2322.

Itoh, N., Imagawa, A., Hanafusa, T., Waguri, M., Yamamoto, K., Iwahashi, H., Moriwaki, M., Nakajima, H., Miyagawa, J., Namba, M., Makino, S., Nagata, S., Kono, N. and Matsuzawa, Y. (1997). Requirement of Fas for the development of autoimmune diabetes in nonobese diabetic mice. J. Exp. Med. 186, 613–618.

Jarpe, A.J., Hickman, M.R., Anderson, J.T., Winter, W.E. and Peck, A.B. (1990). Flow cytometric enumeration of mononuclear cell populations infiltrating the islets of Langerhans in prediabetic NOD mice: Development of a model of autoimmune insulitis for type I diabetes. Reg. Immunol. 3, 305–317.

Kagi, D., Odermatt, B., Seiler, P., Zinkernagel, R.M., Mak, T.W. and Hengartner, H. (1997). Reduced incidence and delayed onset of diabetes in perforin-deficient nonobese diabetic mice. J. Exp. Med, 186, 989–997.

Karlsen, A.E., Hagopian, W.A., Petersen, J.S., Boel, E., Dyrberg, T., Grubin, C., Michelsen, B., Madsen, O.D. and Lernmark, Å. (1992). Recombinant glutamic acid decarboxylase representing the single isoform expressed in human islets detects IDDM-associated 64k autoantibodies. Diabetes 41, 1335–1339.

Kaufman, D.L., ClareSalzler, M., Tian, J., Forsthuber, T., Ting, G., Robinson, P., Atkinson, M.A., Sercarz, E.E., Tobin, A.J. and Lehmann, P.V. (1993). Spontaneous loss of T-cell tolerance to glutamic acid decarboxylase in murine insulin-dependent diabetes. Nature 366, 69–72.

Kaufman, D.L., Erlander, M.G., ClareSalzler, M., Atkinson, M.A., Maclaren, N.K. and Tobin, A.J. (1992). Autoimmunity to two forms of glutamate decarboxylase in insulin-dependent diabetes mellitus. J. Clin. Invest. 89, 283–292.

Kim, J., Richter, W., Aanstoot, H.J., Shi, Y., Fu, Q., Rajotte, R., Warnock, G. and Baekkeskov, S. (1993). Differential expression of GAD65 and GAD67 in human, rat and mouse pancreatic islets. Diabetes 42, 1799–1808.

Lampasona, V., Ferrari, M., Bosi, E., Pastore, M.R., Bingley, P.J. and Bonifacio, E. (1994). Sera from patients with IDDM and healthy individuals have antibodies to ICA69 on western blots but do not immunoprecipitate liquid phase antigen. J. Autoimmun. 7, 665–674.

Lan, M.S., Lu, J., Goto, Y. and Notkins, A.L. (1994). Molecular cloning and identification of a receptor-type protein tyrosine phosphatase, IA-2, from human insulinoma. DNA Cell. Biol. 13, 505–514.

LeCompte, P. (1958). "Insulitis" in early juvenile diabetes. Arch. Pathol. 66, 450–457.

Leiter, E.H., Fewell, J.W. and Kuff, E.L. (1986). Glucose induces intracisternal type A retroviral gene transcription and translation in pancreatic beta cells. J. Exp. Med. 163, 87–100.

Lernmark, A. (1996). Glutamic acid decarboxylase—gene to antigen to disease. J. Intern. Med. 240, 259–277.

Lu, J., Li, Q., Xie, H., Chen, Z.J., Borovitskaya, A.E., Maclaren, N.K., Notkins, A.L. and Lan, M.S. (1996). Identification of a second transmembrane protein tyrosine phosphatase, IA-2β, as an autoantigen in insulin-dependent diabetes mellitus: Precursor of the 37-kDa tryptic fragment. Proc. Natl. Acad. Sci. 93, 2307–2311.

Magistrelli, G., Covini, N., Mosca, M., Lippoli, G. and Isacchi, A. (1995). Expression of PTP35, the murine homologue of the protein tyrosine phosphatase-related sequence IA-2, is regulated during cell growth and stimulated by mitogens in 3T3 fibroblasts. Biochem. Biophys. Res. Commun. 217, 154–161.

Magistrelli, G., Toma, S. and Isacchi, A. (1996). Substitution of two variant residues in the protein tyrosine phosphatase-like PTP35/IA-2 sequence reconstitutes catalytic activity. Biochem. Biophys. Res. Commun. 227, 581–588.

Martin, S., Kardorf, J., Schulte, B., Lampeter, E.F., Gries, F.A., Melchers, I., Wagner, R., Bertrams, J., Roep, B.O., Pfutzner, A., Pietropaolo, M. and Kolb, H. (1995). Autoantibodies to the islet antigen ICA69 occur in IDDM and in rheumatoid arthritis. Diabetologia 38, 351–355.

Miller, B., Appel, M., O'Neil, J. and Wicker, L. (1988). Both the LYT-2+ and L3T4+ T cell subsets are required for the transfer of diabetes in nonobese diabetic mice. J. Immunol. 140, 52–58.

Nerup, J., Mandrup Poulsen, T., Helqvist, S., Andersen, H. U., Pociot, F., Reimers, J. I., Cuartero, B.G., Karlsen, A.E., Bjerre, U. and Lorenzen, T. (1994). On the pathogenesis of IDDM. Diabetologia 37(Suppl. 2), 82–89.

O'Reilly, L.A., Hutchings, P.R., Crocker, P.R., Simpson, E., Lund, T., Kioussis, D., Takei, F., Baird, J. and Cooke, A. (1991). Characterization of pancreatic islet cell infiltrates in NOD mice: Effect of cell transfer and transgene expression. Eur. J. Immunol. 21, 1171–1180.

Pav, J., Jezkova, Z. and Skrha, F. (1963). Insulin antibodies. Lancet 2, 221–222.

Payton, M.A., Hawkes, C.J. and Christie, M.R. (1995). Relationship of the 37,000- and 40,000-M(r) tryptic fragments of islet antigens in insulin-dependent diabetes to the protein tyrosine phosphatase-like molecule IA-2 (ICA512). J. Clin. Invest. 96, 1506–1511.

Pietropaolo, M., Castano, L., Babu, S., Buelow, R., Kuo, Y.-L.S., Martin, S., Martin, A., Powers, A.C., Prochada, M., Naggert, J., Leiter, E.H. and Eisenbarth, G.S. (1993). Islet cell autoantigen 69kD (ICA 69). Molecular cloning and characterization of a novel diabetes-associated autoantigen. J. Clin. Invest. 92, 359–371.

Pugliese, A., Zeller, M., Fernandez, A., Jr., Zalcberg, L.J., Bartlett, R.J., Ricordi, C., Pietropaolo, M., Eisenbarth, G.S., Bennett, S.T. and Patel, D.D. (1997). The insulin gene is transcribed in the human thymus and transcription levels correlated with allelic variation at the INS VNTR-IDDM2 susceptibility locus for type 1 diabetes. Nat. Genet. 15, 293–297.

Reetz, A., Solimena, M., Matteoli, M., Folli, F., Takei, K. and De Camilli, P. (1991). GABA and pancreatic β-cells: Colocalization of glutamic acid decarboxylase (GAD) and GABA with synaptic-like microvesicles suggests their role in GABA storage and secretion. EMBO J. 10, 1275–1284.

Roep, B.O., Kallan, A.A., Hazenbos, W., Bruining, G.J., Bailyes, E.M., Arden, S.D., Hutton, J.C. and De Vries, R. (1991). T-cell reactivity to 38 kD insulin-secretory-granule protein in patients with recent-onset type 1 diabetes. Lancet 337, 1439–1441.

Roll, U., Christie, M.R., Fuchtenbusch, M., Payton, M.A., Hawkes, C.J. and Ziegler, A.G. (1996). Perinatal autoimmunity in offspring of diabetic parents: The German multicenter BABY-DIAB study: Detection of humoral immune responses to islet antigens in early childhood. Diabetes 45, 967–973.

Rorsman, P., Berggren, P.O., Bokvist, K., Ericson, H., Mohler, H., Ostenson, C.G. and Smith, P.A. (1989). Glucose-inhibition of glucagon secretion involves activation of GABAA-receptor chloride channels. Nature 341, 233–236.

Scheinin, T., Maenpaa, J. and Kontiainen, S. (1990). Immune responses to insulin and lymphocyte subclasses at diagnosis of insulin-dependent diabetes and one year later. Immunobiol. 180, 431–440.

Shi, Y., Veit, B. and Baekkeskov, S. (1994). Amino acid residues 24-31 but not palmitoylation of cysteines 30 and 45 are required for membrane anchoring of glutamic acid decarboxylase, GAD65. J. Cell Biol. 124, 927–934.

Smismans, A., Schuit, F. and Pipeleers, D. (1997). Nutrient regulation of gamma-aminobutyric acid release from islet beta cells. Diabetologia 40, 1411–1415.

Solimena, M., Dirkx, R. Jr., Hermel, J.M., Pleasic Williams, S., Shapiro, J.A., Caron, L. and Rabin, D.U. (1996). ICA 512, an autoantigen of type I diabetes, is an intrinsic membrane protein of neurosecretory granules. EMBO J. 15, 2102–2114.

Solimena, M., Folli, F., Aparisi, R., Pozza, G. and De Camilli, P. (1990). Autoantibodies to GABA-ergic neurons and pancreatic beta cells in stiff-man syndrome. New Engl. J. Med. 322, 1555–1560.

Somoza, N., Vargas, F., Roura Mir, C., Vives Pi, M., Fernandez Figueras, M.T., Ariza, A., Gomis, R., Bragado, R., Marti, M., Jaraquemada, D. and Pujol-Borrell, R. (1994). Pancreas in recent onset insulin-dependent diabetes mellitus. Changes in HLA, adhesion molecules and autoantigens, restricted T cell receptor Vβ usage, and cytokine profile. J. Immunol. 153, 1360–1377.

Springer, T.A. (1990). Adhesion receptors of the immune system. Nature 346, 425–434.

Stassi, G., Maria, R.D., Trucco, G., Rudert, W., Testi, R., Galluzzo, A., Giordano, C. and Trucco, M. (1997). Nitric oxide primes pancreatic beta cells for Fas-mediated destruction in insulin-dependent diabetes mellitus. J. Exp. Med. 186, 1193–1200.

Streuli, M., Krueger, N.X., Thai, T., Tang, M. and Saito, H. (1990). Distinct functional roles of the two intracellular phosphatase like domains of the receptor-linked protein tyrosine phosphatases LCA and LAR. EMBO J. 9, 2399–2407.

Tian, J., Lehmann, P.V. and Kaufman, D.L. (1994). T cell cross-reactivity between coxsackievirus and glutamate decarboxylase is associated with a murine diabetes susceptibility allele. J. Exp. Med. 180, 1979–1984.

Tisch, R., Yang, X.-D., Singer, S.M., Liblau, R.S., Fugger, L. and McDevitt, H.O. (1993). Immune response to glutamic acid decarboxylase correlates with insulitis in non-obese diabetic mice. Nature 366, 72–75.

Toniolo, A., Onodera, T., Jordan, G., Yoon, J.W. and Notkins, A.L. (1982). Virus-induced diabetes mellitus. Glucose abnormalities produced in mice by the six members of the Coxsackie B virus group. Diabetes 31, 496–499.

Ujihara, N., Daw, K., Gianani, R., Boel, E., Yu, L. and Powers, A.C. (1994). Identification of glutamic acid decarboxylase autoantibody heterogeneity and epitope regions in type I diabetes. Diabetes 43, 968–975.

Undlien, D.E., Bennett, S.T., Todd, J.A., Akselsen, H.E., Ikaheimo, I., Reijonen, H., Knip, M., Thorsby, E. and Ronningen, K.S. (1995). Insulin gene region-encoded susceptibility to IDDM maps upstream of the insulin gene. Diabetes 44, 620–625.

Vafiadis, P., Bennett, S.T., Todd, J.A., Nadeau, J., Grabs, R., Goodyer, C.G., Wickramasinghe, S., Colle, E. and Polychronakos, C. (1997). Insulin expression in human thymus is modulated by INS VNTR alleles at the IDDM2 locus. Nat. Genet. 15, 289–292.

Vardi, P., Ziegler, A. G., Mathews, J.H., Dib, S., Keller, R.J., Ricker, A.T., Wolfsdorf, J.I., Herskowitz, R.D., Rabizadeh, A., Eisenbarth, G.S. and Soeldner, J.S. (1988). Concentration of insulin autoantibodies at onset of type I diabetes: Inverse log-linear correlation with age. Diabetes Care 11, 736–739.

Velloso, L.A., Bjork, E., Ballagi, A.E., Funa, K., Andersson, A., Kampe, O., Karlsson, F.A. and Eizirik, D.L. (1994a). Regulation of GAD expression in islets of Langerhans occurs both at the mRNA and protein level. Mol. Cell. Endocrinol. 102, 31–38.

Velloso, L.A., Eizirik, D.L., Karlsson, F.A. and Kampe, O. (1994b). Absence of autoantibodies against glutamate decarboxylase (GAD) in the non-obese diabetic (NOD) mouse and low expression of the enzyme in mouse islets. Clin. Exp. Immunol. 96, 129–137.

Warren, S. and Root, H. (1925). The pathology of diabetes with special reference to pancreatic regeneration. Am. J. Pathol. 1, 415–430.

Wasmeier, C. and Hutton, J.C. (1996). Molecular cloning of phogrin, a protein-tyrosine phosphatase homologue localized to insulin secretory granule membranes. J. Biol. Chem. 271, 18161–18170.

Wong, F.S., Visintin, I., Wen, L., Flavell, R.A. and Janeway, C.A. Jr. (1996). CD8 T cell clones from young nonobese diabetic (NOD) islets can transfer rapid onset of diabetes in NOD mice in the absence of CD4 cells. J. Exp. Med. 183, 67–76.

Xie, H., Notkins, A.L. and Lan, M.S. (1996). IA-2, a transmembrane protein tyrosine phosphatase, is expressed in human lung cancer cell lines with neuroendocrine phenotype. Cancer. Res. 56, 2742–2744.

Yang, X.D., Michie, S.A., Mebius, R.E., Tisch, R., Weissman, I. and McDevitt, H.O. (1996). The role of cell adhesion molecules in the development of IDDM: Implications for pathogenesis and therapy. Diabetes 45, 705–710.

Yoon, J.W. (1991). Role of viruses in the pathogenesis of IDDM. Ann. Med. 23, 437–45.

Yoon, J.W., Austin, M., Onodera, T. and Notkins, A.L. (1979). Isolation of a virus from the pancreas of a child with diabetic ketoacidosis. N. Engl. J. Med. 300, 1173–1179.

Young, L.H., Peterson, L.B., Wicker, L.S., Persechini, P.M. and Young, J.D. (1989). In vivo expression of perforin by CD8+ lymphocytes in autoimmune disease. Studies on spontaneous and adoptively transferred diabetes in nonobese diabetic mice. J. Immunol. 143, 3994–3999.

Yu, L., Rewers, M., Gianani, R., Kawasaki, E., Zhang, Y., Verge, C., Chase, P., Klingensmith, G., Erlich, H., Norris, J. and Eisenbarth, G.S. (1996). Antiislet autoantibodies usually develop sequentially rather than simultaneously. J. Clin. Endocrinol. Metab. 81, 4264–4267.

Zekzer, D., Wong, F., Ayalon, O., Millet, I., Altieri, M., Shintani, S., Solimena, M. and Sherwin, R. (1998). GAD-reactive CD4+ Th1 cells induce diabetes in NOD/SCID mice. J. Clin. Invest. 101, 68–73.

SECTION II

CELLULAR MECHANISMS OF INSULIN SYNTHESIS AND SECRETION

INSULIN GENE EXPRESSION

Kevin Docherty

Advances in Molecular and Cell Biology
Volume 29, pages 103-123.

ISBN: 0-7623-0547-9

I. INTRODUCTION

Insulin is a small globular protein (5.8 kDa) consisting of two peptide chains, designated A and B, linked together by disulphide bridges. Insulin is synthesized in the β-cells of the islets of Langerhans as a larger precursor molecule, preproinsulin, which undergoes post-translational processing during its transit through the secretory pathway. Preproinsulin contains an NH_2-terminal 24-amino acid hydrophobic signal peptide. The signal peptide is removed in the endoplasmic reticulum to generate proinsulin, in which the insulin B chain and A chain are linked by a connecting peptide, C-peptide.

Insulin is a member of a superfamily of structurally related peptides which includes the ovarian hormone relaxin, and insulinlike growth factors I and II (IGF-I and IGF-II) (Steiner et al., 1985). Insulin is present in all vertebrates, while insulinlike molecules have been isolated from invertebrates including the snail *Lymnea* (Smit et al., 1988), the silkworm *Bombyx mori* (Nagasawa et al., 1986), and the locust (Lageux et al., 1990). These members of the insulin-related family probably evolved from a common ancesterl through a process of gene duplication and diversification which has been described elsewhere in some detail (Steiner et al., 1985). This chapter focuses on the molecular mechanisms controlling expression of the insulin gene, and the regulation of proinsulin biosynthesis.

II. STRUCTURE OF THE INSULIN GENE

The insulin gene has been cloned from a wide range of species including primates, rodents, birds, fish, and amphibians. Its structure is highly conserved among species and consists of three exons and two introns (Bell et al., 1980). Exon 1 is located in the 5′ untranslated region of the gene. Exon 2 contains sequences encoding the signal peptide, the insulin B chain, and part of the C-peptide. Exon 3 encodes the remainder of the C-peptide, the insulin A chain, and the 3′ untranslated sequences. The length and sequences of the introns are highly variable among species (Figure 1). However, the relative length (intron 1 is always shorter than intron 2) and position (intron 1 is located in the 5′ untranslated region while intron 2 interrupts the gene between the first and second nucleotides of the codon for amino acid 7 of the C-peptide) are highly conserved.

There is only one insulin gene in all species studied with the exception of rat, mouse, and Xenopus, which have two nonallelic insulin genes. Intron 2 is absent in the rat and mouse insulin I genes. The rodent insulin I gene arose from the ancestral insulin II gene by a process of reverse transcription of a partially spliced germ line transcript (Soares et al., 1985). The RNA transcript involved was initiated from a site approximately 500 base pairs upstream of the transcription start site of the insulin II gene. The rat and mouse insulin I and II genes are expressed equally in normal islets (Clark and Steiner, 1969), though not in the fetal yolk sac

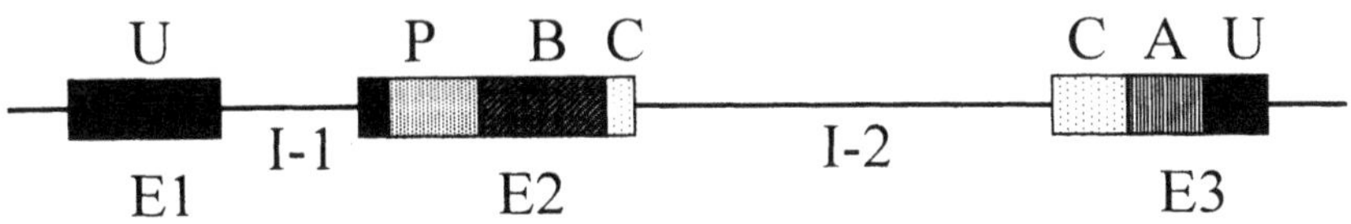

79 bp	786 bp	Human
119 bp		Rat 1
119 bp	499 bp	Rat 2
151 bp	264 bp	Dog
119 bp	580 bp	Guinea Pig
119 bp	~3500 bp	Chicken
393 bp	288 bp	Salmon
>5000 bp	141 bp	Hagfish

Figure 1. Structure of the insulin gene in vertebrates. The exons (E1 to E3) sequences that appear in mature preproinsulin mRNA are shown as bars (U, untranslated region; P, prepeptide coding region; B, B chain coding region; C, C-peptide coding region; and A, A chain coding region). The size of the two introns in various species are shown below. Redrawn from Docherty and Steiner (1997).

(Giddings and Carnaghi, 1989) nor in some insulinoma-derived cells (Fiedorek et al., 1990). Both alleles of the mouse insulin gene are active in embryonic tissue, but only the paternal alleles for both genes are active in the yolk sac, suggesting that maternal alleles undergo selective imprinting in extrapancreatic tissue during embryogenesis (Giddings et al., 1994). Selective imprinting of human insulin and IGF-II alleles, which are located within the same region of chromosome 11, has also been reported (Polychronakos et al.,1995)

III. INSULIN GENE PROMOTER

Like all eukaryotic genes, transcription of the insulin gene is dependent on the assembly of a preinitiation complex (PIC) comprising RNA polymerase II, a number of general transcription factors (designated TFIIA to TFIIJ) and associated proteins (Roeder, 1996). The positioning of the PIC is dependent on DNA sequences, termed the promoter, located adjacent to the transcription start site. In the insulin gene, and most other eukaryotic genes, the sequence TATAA (the TATA) box, is located at about −25 (i.e., 25 base pairs upstream of the start site). The TATA box binds a factor TBP (TATA binding protein), which forms a complex (TFIID) involving a number of transcription associated factors (TAFs). The binding of TFIID to the TATA box causes distortion of the DNA backbone, which brings sequences upstream and downstream of the TATA box into close proximity and permits stepwise formation of the PIC at the start site. Initiation of transcription is dependent on the activation of a kinase activity within the PIC (TFIIH), phosphorylation of the COOH-terminal domain of RNA polymerase II, and activation of a helicase activity that unwinds a small region of DNA and allows the RNA polymerase to commence transcription.

Regulation of the PIC is dependent on sequences that can be located close (about 100 bp) to the start site, upstream promoter elements, or at distances (0.1 to >10 kb) from the promoter, enhancer elements. The defining feature of an enhancer element is that it can operate in a manner independent of orientation and position relative to the promoter. Similar elements can be found within the upstream promoter regions and distant enhancers. Regardless of where they are located, these regulatory elements (usually 8 to 20 bp in length) bind transcription factors that are normally composed of separate domains that are capable of recognizing and binding to specific DNA sequences or involved in protein–protein interactions, to form homo- or heterodimers, to interact with other proteins either on or off the DNA, or to interact with components of the PIC. The situation is complicated by the fact that specific DNA sequences can bind different proteins under different situations, and the resultant effect on transcription can be either positive or negative (Clark and Docherty, 1993).

Transcription of the insulin gene is dependent on regulatory sequences located up to 300 base pairs from the start site (Walker et al., 1983; Edlund et al., 1985). Although individual elements can function as enhancers, for reasons of simplicity, and because the distinction between promoter and enhancer element in this situation is somewhat artificial, this region is usually referred to as the insulin promoter. Until recently the nomenclature for the various *cis*-acting regulatory elements located within the promoter region was to say the least confusing, with most elements having numerous names. However, a rational system has been applied to the major elements located within the promoters of the most extensively studied insulin genes (i.e., the human and the two rat alleles) (German et al., 1995). The transcription factors that bind to these promoter elements play an important role in: (1) restricting expression of the insulin gene to β-cells, (2) silencing the insulin gene in non-β-cells, (3) determining when and where the insulin gene is expressed during islet cell ontogeny, and (4) regulating expression of the insulin gene in response to changes in nutrient status. Transcription factors, some of which bind to the insulin promoter, also play an important role in lineage determination in the developing pancreas (see the first chapter in this volume), while defects in these transcription factors may be related to the development of diabetes mellitus. The following sections describe the promoter elements and their related transcription factors, the manner in which insulin gene transcription and proinsulin biosynthesis are regulated so as to replenish insulin stores, and defects in these regulatory mechanisms in diabetes mellitus.

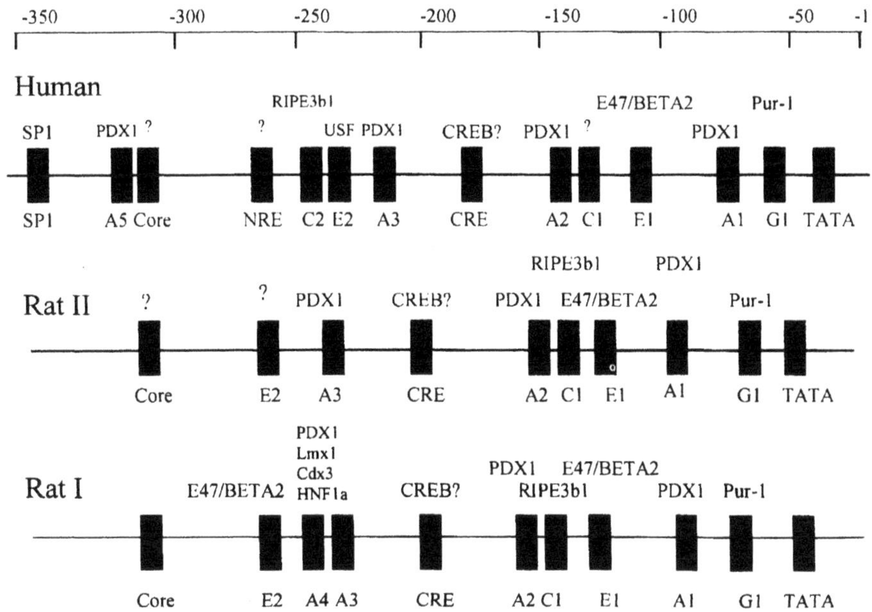

Figure 2. The insulin promoter. Schematic depicting the arrangement of regulatory sequences in human, rat II, and rat I insulin gene promoters. The scale at the top represents base pairs upstream of the transcription start site, where +1 is the first nucleotide of the mRNA. The boxes on the dark line denote sequence elements, where the name of the element (German et al., 1995) is given below each box. The name of the transcription factor(s) that binds at each element is given above each box.

IV. REGULATION OF THE INSULIN PROMOTER

The arrangement of regulatory elements within the insulin promoter and the transcription factors that bind to these sequences are shown schematically in Figure 2. The three genes that have been most extensively studies are the human, the rat I, and the rat II. Although the three promoters exhibit a similar overall arrangement of regulatory elements, there are subtle differences, particularly around the region –200 to –300. Two important sequences that will be described in some detail in this section are the E boxes and the A boxes. About 12 different transcription factors that can bind to sequences within the promoter, and in most cases exhibit transcriptional activity, have been identified. Most of these transcription factors are expressed in a large number of tissues, although some may show restricted expression (e.g., in neuroendocrine cells). Three important transcription factors are PDX-1, E47 ,and BETA2/NeuroD.

A. The E Boxes

E boxes have the consensus sequence CANNTG (where N represents any nucleotide) and are implicated in the tissue-specific regulation of genes in muscle, pancreas, and lymphoid tissue (Weintraub, 1993). Transcription factors that bind E boxes consist of heterodimers between a ubiquitous class A and a tissue-specific class B member of the basic helix loop helix (bHLH) family. These proteins contain a conserved basic DNA-binding domain and two amphipathic helices separated by a loop structure, which are required for dimerization. Class A bHLH members include the E2A gene products E12 and E47, which are also known as Pan1 and Pan2. Class B family members include BETA2/NeuroD, MyoD, and myogenin. Other class B proteins, such as Id, emc, and hairy, which lack a basic DNA-binding domain, can form inactive heterodimers with class A proteins.

The E1 element located at approximately −100 is highly conserved among insulin genes. It is critical for insulin gene expression, since mutagenesis has a drastic effect on insulin promoter activity (Karlsson et al., 1987). The E1 element binds a heterodimer of the ubiquitously expressed *E2A* gene product E47 and BETA2 (which is also known as NeuroD) (Naya et al., 1995; Lee et al., 1995). BETA2/NeuroD is expressed in islets of Langerhans, brain, and intestine (where it is thought to regulate the secretin gene) (Naya et al., 1995). The importance of BETA2/NeuroD is emphasized by inactivation of the gene in mice: mice homozygous for the *BETA2/NeuroD* gene deletion develop severe diabetes and die 3 to 5 days after birth (Naya et al., 1997). On the other hand, mice lacking the *E2A* gene exhibit normal levels of insulin, suggesting functional redundancy among members of the class A bHLH family (Itkin-Ansari, et al., 1996).

The E2 region, located at around −230, is not as well conserved as the E1 region. In the rat insulin I gene, the sequence of the E2 box is identical to that of the E1 box and it binds a similar factor (i.e., IEF1 [E47/BETA2]). The E2 box of the human insulin gene differs from the E1 box. It has an imperfect E box, CACCGG, which binds E47/BETA2 but only weakly. However, it binds the ubiquitously expressed bHLH factor (upstream stimulating factor (USF) with high affinity (Read et al., 1993; Read et al., 1995). On the 5′ side of the E2 box is an element (C2) containing a number of CAGG sequences. This C2 element binds an islet cell–specific factor RIPE3b1 (Read et al., 1997), which was originally shown to bind to the C1 element located between the A1 and E1 boxes of the rat insulin II gene (Shieh and Tsai, 1991). RIPE3b1 has yet to be cloned and characterized.

B. The A Boxes

The A boxes have the consensus sequence C(C/T)TAATG. There are four in the human insulin gene at positions −82 (A1), −140 (A2), −215 (A3), and −319 (A5); three in the rat II insulin gene; and four in the rat I insulin gene. The A box binds a transcription factor PDX-1 which was previously known as IPF-1 (Ohlsson et al., 1993), IDX-1 (Miller et al., 1994), STF-1 (Leonard et al., 1993), and IUF-1 (Boam and Docherty, 1989). Expression of PDX-1 in the adult is restricted to β-cells of the pancreas and to somatostatin-expressing cells of the pancreas and small intestine. PDX-1 belongs to the *Antennapedia* class of homeodomain proteins and is highly homologous to the *XlHbox* 8 from *Xenopus laevis* (Peshavaria et al., 1994). Mice homozygous for a targeted mutation in the PDX-1 gene selectively lack a pancreas (Jonsson et al., 1994). These and other findings (Jonsson et al., 1995; Guz et al., 1995) suggest that PDX-1 functions both in the early specification of the primitive gut to a pancreatic fate and in the maturation of the pancreatic β-cell (see the first chapter in this volume). The human PDX-1 gene, located on chromosome 13q/12 (Inoue et al., 1996), encodes a 283 amino acid protein with a central homeodomain flanked by two proline-rich domains. The transactivation domain is located at the NH_2-terminus of the protein (Peers et al., 1994; Lu et al., 1996; Peshavaria et al., 1997). The mechanisms whereby PDX-1 binding to the A box activates the insulin gene is not well understood. It synergizes with bHLH proteins bound to the E1 element (Peers et al., 1994) and with the homeodomain factor Pbx (Peers et al., 1995). The interaction with Pbx involves a conserved pentapeptide sequence, FPWMK, located amino terminal to the homeodomain. Some preliminary data on the sequences regulating the PDX-1 gene have been published. A regulatory sequence that binds hepatocyte nuclear factor-3β (HNF-3β) has been mapped to a DNase hypersensitive regions at about −2 kb (Wu et al., 1997), while the bHLH protein USF binds to a regulatory site close to the promoter, at −104 (Sharma et al., 1996).

The A4/A3 region of the rat insulin I promoter is complex. There is some form of negative interaction between these boxes, such that mutation of the A3 box leads to a substantial increase in the rate of transcription. The A4/A3 boxes bind PDX-1 and other homeodomain proteins, including cdx-3, lmx-1, HNF-1α, and Isl-1 (German et al., 1992). The role of these and other homeodomain proteins in regulating the promoter are unclear. Isl-1 may be important in development of the dorsal exocrine pancreas, and also in the generation of all endocrine islet cells (Ahlgren et al., 1997).

C. Other Promoter Sequences

Other putative regulatory elements that are conserved among species include a cyclic adenosine monophosphate responsive element (CRE) and an enhancer core sequence that could bind C/EBP. In the rat insulin I gene, a GAGA-rich sequence (G1 box) binds a factor, Pur-1, that is ubiquitously expressed (Kennedy et al., 1992). A STAT5 binding site that may mediate the effects of growth hormone has been mapped to the sequence between −330 and −322 in the rat insulin I gene (Galsgaard et al., 1996). And a complex negative regulatory element has been identified between −258 and −279 of the human insulin gene (Boam et al., 1990). This element binds a number of proteins (Clark et al., 1995) and contains a palindromic repeat sequence that may function as a negative glucocorticoid response element (Goodman et al., 1996).

D. ILPR and Far Upstream Sequences

The human insulin gene contains variable numbers of tandem repeats (VNTRs) of the sequence 5′ TCTGGGGAGAGGGG 3′ located upstream of −365 (Bell et al., 1982; Bell et al., 1984). Termed the insulin-linked polymorphic region (ILPR), this region has been shown to exhibit transcriptional activity (Bennett et al., 1995; Kennedy et al., 1995), possibly through the binding of the transcription factor Pur-1 to G-rich sequences. Interestingly, the ILPR has the ability to adopt an altered DNA structure, which has been characterized as a quadriplex involving interactions between the G residues on the top strand (Hammond-Kosack et al., 1992a; Hammond-Kosack and Docherty 1993; Catasti, et al., 1996), while the C-rich strand forms hairpin structures (Catasti et al., 1997). The quadriplex structure is present in vivo (Hammond-Kosack et al., 1992b) and in DNA assembled in chromosomes in vitro (Hammond-Kosack et al., 1993). The structure is present when the insulin gene is actively transcribed, while mutations that destabilize the structure lower the transcriptional activity of the gene.

Further regulatory sequences, including a steroid responsive element (Clark et al., 1995) and an RIPE3b1 binding site may reside upstream of the ILPR.

V. REGULATION OF INSULIN GENE EXPRESSION AND PROINSULIN BIOSYNTHESIS BY NUTRIENTS

The β-cells of the islets of Langerhans secrete insulin in response to changes in the circulating blood glucose concentrations. Following the release of insulin, it is important that the β-cell replenish its stores of insulin. There are several stages at which this can occur: transcription of the insulin gene, preRNA processing, transport into the cytoplasm and turnover of mRNA, and translation of preproinsulin mRNA (Docherty and Clark, 1994). The major acute (<2 hours) control of insulin production is exerted at the level of translation. This is in keeping with the requirement to replenish insulin stores rapidly following insulin secretion in response to elevated circulating glucose levels. Over longer periods (>6 hours), glucose stimulates transcription of the insulin gene, and, at least in the mouse, it also regulates the level of preRNA splicing (Wang et al., 1997).

In recent years, it has become clear that the cells within islets communicate with each other through connexinlike molecules, such that the overall response to nutrient stimulation reflects the integrated action of β-cells with individually differing behaviors. Thus, an increase in glucose concentrations not only stimulates gene expression (deVargas et al., 1997) and proinsulin biosynthesis (Schuit et al., 1988) within individual β-cells, but also increases the number of cells that respond to glucose. The following account of events should, therefore, be viewed in the context of this synergistic response.

A. Nutrient Effects on Insulin Gene Transcription

Evidence that glucose affects insulin mRNA levels initially came from in vivo experiments in which rats that fasted for 4 days exhibited decreased insulin mRNA levels, which could be restored to normal on refeeding or injection of glucose (Giddings et al., 1981). In vitro studies (Brundstedt and Chan, 1982) confirmed that this occurred through stabilization of insulin mRNA and through effects on the rate of transcription of the insulin gene (Nielsen et al., 1985). Insulin mRNA is relatively stable with a half-life of about 30 hours; however, its half-life was almost threefold longer in islets incubated in 17 mM glucose compared with 3 mM glucose (Welsh et al., 1985). Direct effects of glucose on gene transcription were demonstrated by using transcription run-on assays (Efrat et al., 1991) or transfection of plasmid constructs containing insulin promoter fragments joined to a reporter gene encoding chloramphenicol acetyl transferase (CAT) or firefly luciferase (German et al., 1990; Goodison et al., 1992).

Glucose effects on the insulin gene promoter were originally mapped to the E2/A4/A3 (−196 to −247) region of the rat insulin I gene promoter in transfected neonatal rat islets (German et al., 1990). Part of this response results from the binding at the E2 site of a protein complex that includes the bHLH protein E47or E12. Subsequent studies suggested that multiple regulatory elements interact with each other to illicit a full transcriptional response to glucose (German and Wang, 1994). It has also been shown that the binding of the yet to be characterized factor RIPE3b1 to its recognition sequence (the C1 box) in the rat insulin II promoter may be regulated by glucose (Sharma and Stein, 1994). The fact that RIPE3b1 DNA-binding activity is inhibited by the oxidizing agent diamide and the alkylating agent N-ethylmaleimide suggests that changes in the redox state of the cell may contribute to the regulation of the insulin promoter by nutrients (Read et al., 1997). There is no evidence that the cAMP response element (CRE) located at around −200 plays a role in mediating the effects of glucose and other nutrients on the insulin gene.

There is, however, overwhelming evidence that the homeodomain transcription PDX-1 plays a major role in linking nutrient metabolism in the β-cell to the regulation of the insulin gene (Melloul et al., 1993; Macfarlane et al., 1994; Petersen et al., 1994). Melloul et al. (1993) demonstrated that glucose stimulated the DNA binding activity of a factor (GSF) that was subsequently identified as PDX-1 (Marshak et al., 1996). Glucose activation of PDX-1 occurred rapidly (within 10 to 15 minutes) and involved phosphorylation of the transcription factor (Macfarlane et al., 1994). The effects of glucose could not be mimicked by cyclic AMP–elevating agents or tumor-promoting phorbol esters, suggesting that neither cyclic AMP–dependent protein kinase nor protein kinase C were involved. The mecha-

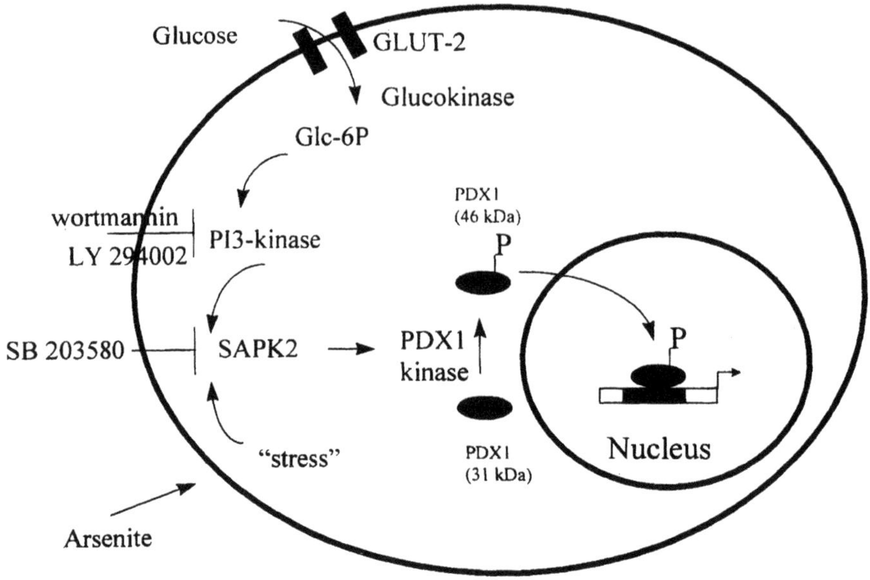

Figure 3. Mechanism whereby glucose regulates PDX1 and insulin promoter activity. The schematic depicts the events that occur following entry of glucose into the β-cell via the high Km GLUT-2 transporter and phosphorylation by glucokinase. The SAPK2 (RK/p38) pathway is activated within 15 minutes. PDX1in the cytoplasm is converted from a 31kDa inactive form to a phosphorylated active 46 kDa form. The 46 kDa form translocates to the nucleus, where it binds to and transactivates the insulin promoter. The effects of glucose on the SAPK2 pathway, PDX1 DNA-binding activity, and insulin promoter activity are inhibited by the SAPK2 inhibitor SB 203580 and by the phosphatidylinositol 3 kinase (PI3-kinase) inhibitors wortmannin and LY 294002. These effects of glucose are mimicked by the stress-inducing agent arsenite; however, the arsenite-stimulated effects are inhibited by SB203508 but not by wortmannin or LY 294002. This suggests that PI3-kinase lies upstream of SAPK2. Insulin (1 to 10 ng/ml) as well as a large variety of metabolites, including mannose, fructose, glyceraldehyde, and α-ketoisocaproate, can also activate PDX1 DNA-binding activity and insulin promoter activity in an SB 203580 inhibitable manner (H. Wu, W.M. Macfarlane, and K. Docherty, unpublished).

nism by which glucose stimulates PDX-1 DNA binding activity and insulin promoter activity has recently been elucidated. It involves a stress-activated protein kinase pathway (Figure 3) in which SAPK2 (also known as RK/p38) plays a critical role (Macfarlane et al., 1997a). SAPK2 is a member of an expanding family of MAP kinase-related kinases that are activated in response to adverse stimuli such as heat, osmotic shock, ultraviolet light, and DNA-damaging reagents, and by proinflammatory cytokines that are produced under conditions of stress

(Cohen, 1997). In β-cells, activation of the SAPK2 pathway by glucose involves phosphatidylinositol 3 kinase (PI3-kinase), while activation by stress (e.g., treatment with sodium arsenite) is independent of PI3-kinase. Interestingly, activation of recombinant PDX-1 by SAPK2 in vitro is associated with an increase in molecular mass from 31 kDa to 46 kDa. This increase in mass is also seen in vivo, where the 31 kDa form of PDX-1 is present in the cytoplasm of low glucose treated islets. Upon transfer to a high glucose concentration, PDX-1 is rapidly (within 10 to 15 minutes) converted to an active phosphorylated 46 kDa form that is present predominantly in the nucleus (Macfarlane et al., 1999).

B. Nutrient Effects on Proinsulin Biosynthesis

Glucose and other nutrients stimulate the biosynthesis of proinsulin (Permutt and Kipnis, 1972; Itoh and Okamoto, 1980) and a number of other proteins, including the secretory granule membrane protein SGM110 (Grimaldi et al., 1987) and chromogranin a (Guest et al., 1989). There may be a general, but relatively weak, stimulatory effect on the synthesis of a large number of islet proteins (Guest et al., 1991), along with a rapid and strong stimulation of proinsulin biosynthesis. The biosynthesis of proinsulin is coupled to secretion in so far as many insulin secretagogues also stimulate proinsulin biosynthesis. However, the transducing systems differ in a number of respects. First, glucose-stimulated insulin release is inhibited in Ca^{2+}-free medium, whereas biosynthesis is still activated. Second, tolbutamide and phosphodiesterase inhibitors, such as isobutylmethylxanthine (IBMX), potentiate glucose-stimulated insulin release by increasing islet cAMP levels, but do not affect proinsulin biosynthesis. Third, the threshold for glucose-stimulated insulin secretion (4.2 to 5.6 mM) is higher than that for proinsulin synthesis (2.5 to 3.9 mM). The point of divergence appears to occur early in the β-cell metabolic signaling pathway (Skelly et al., 1998).

The molecular mechanisms whereby glucose stimulates translation of preproinsulin mRNA are not well understood. Translation initiation factors may be involved. The activity of the guanine exchange factor eIF2B is stimulated rapidly (within 15 minutes) and over the same range of glucose concentrations that stimulate proinsulin synthesis (Gilligian et al., 1996). EIF2B facilitates the exchange of guanosine triphosphate (GTP) between inactive and active forms of eIF-2, the factor involved in mediating the binding of the initiator Met-tRNA to the 40S ribosomal subunit. In other systems, the activity of eIF2B is modulated by phosphorylation of its subunit eIF2α, however, in islets glucose stimulates eIF2B by an alternative mechanism that does not involve phosphorylation of eIF2α. Another factor, eIF4E-BP1, which is also known as PHAS-1, is phosphorylated in response to glucose in β-cells (Xu et al., 1998). EIF4E-BP1 is a protein that binds and inhibits the activity of eIF4E, a component of the mRNA cap site recognition complex. Phosphorylation of eIF4E-BP1 causes it to dissociation from eIF4E. Glucose-stimulated phosphorylation of eIF4E-BP1 is dependent on the presence of amino acids and may be mediated via insulin acting on the β-cell. Thus, insulin release from the β-cells may activate an insulin-receptor signaling pathway to autoregulate its own biosynthesis. In this context, it is noteworthy that insulin also stimulates PDX-1 DNA-binding activity and insulin promoter activity in β-cells (H.Wu, W.M. Macfarlane, and K. Docherty, unpublished). The activated eIF4E forms part of a complex that, in addition to recognizing the mRNA cap site, also unwinds secondary structure within the 5′ untranslated region of the mRNA. The selective effects of glucose on proinsulin may, therefore, be based on the presence of stem loop structures within the 5′ untranslated region of preproinsulin mRNA (Knight and Docherty, 1994).

VI. INSULIN GENE AND DIABETES MELLITUS

There are several aspects of the regulation of insulin gene expression and β-cell transcription factors that may relate to a better understanding of the development of insulin-dependent (type 1) and noninsulin-dependent (type 2) diabetes mellitus.

A. IDDM2 and the ILPR

Resulting from a genomewide search, a number of loci (around 18) that confer susceptibility to type 1 diabetes (IDDM) have been described (Davies et al., 1994). Of these, the IDDM1 locus within the major histocompatibility complex on chromosome 6p21, and the IDDM2 locus within the insulin gene region on chromosome 11p15 have the most powerful effects. The IDDM2 locus has been further mapped and identified as allelic variations in the insulin VNTR. There are two main classes of VNTR allele: class I (26 to 36 repeats) and class III (140 to 200 repeats) (Bell et al., 1982). The class I VNTRs confer an increased risk of type 1 diabetes, while class III VNTRs are dominantly protective (Bennet and Todd, 1996). As described previously, the VNTR (or ILPR) is transcriptionally active in β-cells (Bennet et al., 1995; Kennedy et al., 1995), with the shorter class I allele appearing to exhibit stronger transcriptional activity (Owerbach and Gabbay, 1996) than the class III allele, although not in all studies (Kennedy et al., 1995). The reason why the class III should provide protection against type 1 diabetes has proven puzzling until it was recently discovered that insulin mRNA (along with that of other β-cell autoantigens such as glutamic acid decarboxylase [GAD], islet cell antibody 69 [ICA69], and IA-22) was expressed at low levels in human fetal thymus (Pugliese et al., 1997; Vafiadis et al., 1997). In the thymus, as opposed to the β-cell, the class III allele was associated with higher insulin mRNA levels. Expression of the insulin gene in the thymus may serve to enhance immune tolerance to proinsulin, which is a major autoantigen in type 1 diabetes, by deletion of autoreactive thymocytes or the generation of regulatory T lymphocytes.

Ectopic expression of insulin has been a controversial subject for many years. The major problem has been that insulin receptors are present throughout the body and insulin can be internalized in most cells. However, transgenic mice studies have shown that the insulin upstream region is transcriptionally active in the brain as well as the pancreas (Dandroy-Dron et al., 1991), and insulin mRNA has been detected in isolated neurons of the rat brain (Devaskar et al., 1994). Regardless of the controversy, the reports of insulin gene expression in the thymus are potentially exciting and provide plausible explanations for the protective influence of the class III VNTR allele in type 1 diabetes. It will be interesting to determine which of the known *cis*-acting elements and transcription factors are involved in thymic expression of the gene.

B. PDX1 and Type 2 Diabetes

A number of lines of evidence suggest that PDX1 is a major regulator of β-cell function and a potential factor contributing to the pathogenesis of diabetes mellitus (Stoffers et al., 1997a). The key role of this transcription factor in islet cell ontogeny is emphasized by the demonstration that in mouse gene inactivation models (Jonsson et al., 1994) and in a human subject with a homozygous mutation in the PDX-1 gene, the pancreas fails to develop (Stoffers et al., 1997b). In the adult, PDX-1 is expressed primarily in the β-cell, where it regulates expression of a number of genes that are preferentially expressed in the β-cell, including insulin, glucose transporter type 2 (GLUT-2), glucokinase, and islet amyloid polypeptide. It also mediates the activation of the insulin gene by glucose (see the preceding discussion).

In β-cells cultured for prolonged periods of time in a high glucose concentration, expression of the insulin gene is repressed (Olson et al., 1993), as is expression of PDX1 (Olson et al., 1995) and another insulin gene regulatory protein RIPE3b1 (Sharma et al., 1995). This suggests that the glucotoxic effects of prolonged hyperglycemia, as seen in type 2 diabetes, on the β-cell may be mediated in part by repression of PDX1 (Robertson et al., 1994).

Mutations in PDX1 have been found in patients with maturity onset diabetes of the young (MODY4) (Stoffers et al., 1997c), a monogenic early onset form of Type 2 Diabetes. Three other forms of MODY have been described with mutations in the genes encoding glucokinase (MODY2) (Froguel et al., 1993), and the transcription factors HNF-1α (MODY3) and HNF-4α (MODY 1) (Hattersley et al., 1998). HNF-1α is a homeodomain protein that is expressed in the β-cell and that has been shown to bind to the A4/A3 region of the rat insulin I gene (Emens et al., 1992). However, a number of transcription factors, including cdx-3 and lmx-1 can bind to this element with varying affinity. Whether the binding of HNF-1α or these other factors to this sequence reflects a physiological role for these factors in regulating the insulin gene is unclear. HNF-4α is an orphan member of the steroid receptor family. The ligand for HNF-4α is unknown as is its role in glucose homeostasis. The discovery of mutations in PDX-1, HNF-1α, and HNF-4α in MODY families stresses the importance of transcription factors in the pathogenesis of certain forms of Type 2 Diabetes. This is further emphasized by the finding that heterozygous (+/–) PDX-1 mice injected intraperitoneally with 20 mM glucose remained hyperglycemic for up to 2 hours while normal control mice returned to euglycemia within this time period (Dutta et al., 1998). As well as regulating development of the endocrine pancreas, PDX-1 also appears to play a key role in glucose homeostasis in the adult.

Impaired expression of PDX-1 was also observed in islet cells from a child with persistent hyperinsulinemic hypoglycemia of infancy (PHHI) (Macfarlane et al., 1997b), a rare disorder associated in most cases with mutations in the sulphonyl urea receptor. The reason why a defect in the K_{ATP}-channel should affect PDX-1 expression is unclear. Interestingly, these PHHI-derived cells were able to proliferate in tissue culture where they secreted substantial amounts of insulin, although they lacked a secretory response to glucose. However, a full excitatory and secretory response to glucose in the physiological concentration range was achieved by stably transfecting these cells with cDNAs encoding PDX-1, SUR1 (the sulphonyl urea receptor) and Kir6.2 (the K_{ATP} channel) (Macfarlane et al., 1999b). The resulting cell lines may be of use in the treatment of PHHI or as a vehicle for administering insulin to Type 1 Diabetes and Type 2 Diabetes patients.

Attempts to generate human β-cell lines that proliferate in culture have so far been unsuccessful despite intensive efforts. The ability of the PHHI cells to proliferate may be because these cells have not fully differentiated, the absence of PDX-1 reflecting some intermediate stage in ontogeny, and suggests that PDX-1 may play a role in controlling the cell cycle. What is difficult to explain, however, is the paradoxical high insulin levels in the absence of PDX-1 (Macfarlane et al., 1997b). Similar findings were observed when PDX-1 levels were suppressed by using antisense olgodeoxynucleotides in the mouse β-cell line MIN6 (Kajimoto et al., 1997). These results are at variance with the MODY and heterozygous PDX-1 knock-out mice studies, which support a physiological role for PDX-1 in insulin gene transcription and in maintenance of glucose homeostasis. Clearly this difference must be resolved.

VII. CONCLUSIONS

Relative to many other genes (such as globin or lysozyme), where regulatory sequences may be located at large distances (>10 kb) from the transcription start site, the insulin gene is fairly simple with the major regulatory elements that determine when and where the gene is expressed located in a small region (~350 bp) adjacent to the start site. Within this region, most of the transcriptionally active elements have been mapped and proteins capable of binding at these sites, some of which exhibit transcriptional activity, identified. Despite this progress, we are far from formulating a model describing the molecular mechanisms whereby the insulin gene is regulated. The data to date suggest that the promoter may comprise discrete units involving a combination of adjacent elements (e.g., the E2/A3/A4 unit of the rat insulin I gene or the A2/C1/E1 element of the rat insulin II gene). Within these units, proteins binding at one site may act synergistically with proteins binding at other sites (e.g., E47 and PDX-1 [Peers et al., 1994] or E47 and lmx-1 [German et al., 1992]). As a consequence, it is difficult to examine single elements in isolation. Thus, although PDX-1 binding to the A sites mediates the activation of the promoter by glucose, it is not surprising that mutagenesis of other sites within the promoter also attenuates the effects of glucose (German and Wang, 1994). This complexity is further emphasized by the negative regulatory element (NRE) of the human insulin promoter, which exhibits positive activity when studied in isolation and negative activity within the context of surrounding promoter elements (Clark et al., 1995).

The question also arises as to what is the best system to study the promoter; primary cultures of islet cells (neonatal or adult?) or insulinoma-derived cell lines, since specific elements exhibit different properties in different systems (Odagiri et al., 1996). The answer is that both types of study are invaluable—the insulinoma may be less differentiated but may reflect a combination of transcriptional events that occur during ontogeny and in the adult, whereas isolated islet cells may provide a better model for studying cell-specific aspects of insulin gene transcription.

What has not been covered in this chapter is the progress that has been made in understanding the regulation of the preinitiation complex (PIC) at the transcription start site. The challenge ahead will be to map the interactions of insulin gene regulatory proteins with the components of the PIC and to determine how these interactions are modulated under different conditions.

ACKNOWLEDGMENTS

The author's research has been supported by the British Diabetic Association and The Wellcome Trust.

REFERENCES

Ahlgren, U., Pfaff, S.L., Jessel, T.M., Edlund, T. and Edlund, H. (1997). Independent requirement for ISL1 in formation of the pancreatic mesenchyme and islet cells. Nature 385, 257–260.

Bell, G.I., Horita, S. and Karam, (1984). The highly polymorphic region near the human insulin gene is composed of simple tandemly repeating sequences. Nature 295, 31–35.

Bell, G.I., Pictet, R.L., Rutter, W.J., Cordell, B., Tischer, E. and Goodman, H.M. (1980). Sequence of the human insulin gene. Nature 284, 26–32.

Bell, G.I., Selby, M.J. and Rutter, W.J. (1982). The highly polymorphic region near the human insulin gene is composed of simple tandemly repeating sequences. Nature 295, 31–35.

Bennet, S.T. and Todd, J.A. (1996). Human type 1 diabetes and the insulin gene: Principles of mapping polygenes. 30, 343–370.

Bennett, S.T., Lucassen, A.M., Gough, S.C.L., Powell, E.E., Undlien, D.E., Pritchard, L.E., Merriman, M.E., Kawaguchi, Y., Dronsfield, M.J., Pociot, F., Nerup, J., Bouzekri, N., Cambon-Thomsen, A., Rønningen, Barnett, A.H., Bain, S.C. and Todd, J.A. (1995) Susceptibility to human type 1 diabetes at IDDM2 is determined by tandem repeat variations at the insulin gene minisatellite locus. Nat. Genet. 9, 284–292.

Boam, D.S.W., Clark, A.R. and Docherty, K. (1990). Positive and negative regulation of the human insulin gene by multiple trans-acting factors. J. Biol. Chem. 265, 8285–8296.

Boam, D.S.W. and Docherty, K. (1989). A tissue specific nuclear factor binds to multiple sites in the human insulin gene enhancer. Biochem. J. 264, 233–239.

Brundstedt, J. and Chan, S.J. (1982). Direct effect of glucose on the preproinsulin mRNA level in isolated pancreatic islets. Biochem. Biophys. Res. Commun. 106, 1383–1389.

Catasti, P. Chen, X., Deaven, L.L., Moyzis, R.K., Bradbury, E.M. and Gupta, G. (1997). Cytosine-rich strands of the insulin minisatellite adopt hairpins with intercalated cytosine(+) center dot cytosine pairs. J. Mol. Biol. 272, 369–382.

Catasti,. P., Chen, X., Moyzis, R.K., Bradbury, E.M. and Gupta, G. (1996). Structure-function correlations of the insulin-linked polymorphic region. J. Mol. Biol. 264, 534–54.

Clark A.R., and Docherty K. (1993). Negative regulation of transcription in eukaryotes. Biochem J. 296, 521–541.

Clark, A.R., Wilson, M.E., Leibiger, I., Scott, V. and Docherty, K. (1995). A silencer and an adjacent positive element interact to modulate the activity of the human insulin promoter. Eur. J. Biochem. 232, 627–632.

Clark, A.R., Wilson, N.E., London, N.J.M., James, R.F.L. and Docherty, K. (1995). Identification and characteristion of a thyroid hormone/retinoic acid response element upstream of the human insulin gene. Biochem. J. 309, 863–870.

Clark, J.L., and Steiner, D.F. (1969). Insulin biosynthesis in the rat: demonstration of two proinsulins. Proc. Natl. Acad. Sci. USA 62, 278–285.

Cohen, P. (1997). The search for physiological substrates of MAP and SAP kinases in mammalian cells. Trends Cell Biol. 7, 353–361.

Dandoy-Dron, F., Monthioux, E., Jami, J., and Bucchini, D. (1991). Regulatory regions of rat insulin I gene necessary for expression in transgenic mice. Nucleic Acids Res. 18, 4925–4930.

Davies, J.L., Kawaguchi, Y., Bennett, S.T., Copeman, J.B., Cordell, H.J., Pritchard, L.E., Reed, P.W., Gough, S.C.L., Jenkins, S.C., Palmer, S.M., Balfour, K.M., Rowe, B.R., Farrall, M., Barnett, A.H., Bain, S.C. and Todd, J.A. (1994). A genome-wide search for human type 1 diabetes susceptibility genes. Nature 371, 130–136.

DeVargas, L.M., Sobolewski, J. and Moss, L.G. (1997). Individual beta cells within the intact islet differentially respond to glucose. J. Biol. Chem. 272, 26573–26577.

Devaskar, S.U., Giddings, S.J., Rajakumar, P.A., Carnaghi, L.R., Menon, R.K. and Zahm, D.S. (1994). Insulin gene expression and insulin synthesis in mammalian neuronal cell. J. Biol. Chem. 269, 8445–8454.

Docherty, K. and Clark, A.R. (1994). Nutrient regulation of insulin gene expression. FASEB J. 8, 20–27.

Docherty, K. and Steiner, D.F. (1997). Molecular and cellular biology of the beta cell. In: *Ellenberg and Rifkin's Diabetes Mellitus* (5th ed.). (Porte, D. and Sherwin, R.S., Eds.), pp.29–48. Prentice-Hall, Englewood Cliffs, NJ.

Dutta, S., Bonner-Weir, S., Montminy, M. and Wright, C. (1998). Regulatory factor linked to late-onset diabetes? Nature 392, 560.

Edlund, T., Walker, M.D., Barr, P.J. and Rutter, W.J. (1985). Cell specific expression of the rat insulin gene: Evidence for the role of two distinct 5′ flanking sequences. Science 230, 912–916.

Efrat, S., Surana, M. and Fleischer N. (1991). Glucose induces insulin gene transcription in a murine pancreatic β-cell line. J. Biol. Chem. 266, 11141–11143.

Emens L.A., Landers, D.W. and Moss, L.G. (1992). Hepatocyte nuclear factor 1α is expressed in a hamster insulinoma line and transcativates the rat insulin I gene. Proc. Natl. Acad. Sci. USA 89, 7300–7304.

Fiedorek F.T. Jr., Carnaghi, L.R. and Giddings, S.J. (1990). Selective expression of the insulin I gene in rat insulinoma-derived cell lines. Mol. Endocrinol. 4, 990–999.

Froguel, P., Zouali, H., Vionnet, N., Velho, G., Vaxillaire, M., Sun, F., Lesage, S., Stoffel, M., Takeda, J. and Passa, P. (1993). Familial hyperglycaemia due to mutations in glucokinase: definition of a subtype of diabetes mellitus. N. Engl. J. Med. 328, 697–702.

Galsgaard, E.D., Gouilleux, F., Groner, B., Serup, P. Nielse, J.H. and Billestrup, N. (1996). Identification of a growth hormone responsive STAT-5 binding element in the rat insulin I gene. Mol. Endocrinol. 10, 652–660.

German, M., Ashcroft, S., Docherty, K., Edlund, H., Edlund, T., Goodison, S., Imura, H., Kennedy, G., Madsen, O., Melloul, D., Moss, L., Olson, K., Permutt, MA., Philippe, J., Robertson, RP., Rutter, WJ., Serup, P., Stein, R., Steiner, D., Tsai, M-J. and Walker, M. (1995). The insulin gene promoter: A simplified nomenclature. Diabetes 44, 1002–1004.

German, M. and Wang, J. (1994). The insulin gene contains multiple transcriptional elements that respond to glucose. Mol. Cell. Biol. 14, 4067–4075.

German, M.S., Moss, L.G., and Rutter, W.J. (1990). Regulation of insulin gene expression by glucose and calcium in transfected primary islet cultures. J. Biol. Chem. 265, 22063–22066.

German, M.S., Wang, J., Chadwick, R.B., and Rutter, W.J. (1992). Synergistic activation of the insulin gene by a LIM-homeodomain protein: Building a functional insulin minienhancer complex. Genes Dev. 6, 2165–2176.

Giddings, S.J., and Carnaghi, L. (1989) Rat insulin II gene expression by extraplacental membranes. J. Biol. Chem. 264, 9462–9469.

Giddings, S.J., Chirgwin, J. and Permutt, M.A. (1981). The effect of fasting and feeding on preproinsulin messenger RNA in rats. J. Clin. Invest. 67, 952–960.

Giddings, S.J., King, C.D., Harman, K.W., Flood, J.F. and Carnaghi, L.R. (1994). Allele-specific inactivation of insulin-I and insulin-2 in the mouse yolk sac indicates imprinting. Nat. Genet. 6, 310–313.

Gilligan, M., Welsh, G., Flynn, A., Bujalska, I., Diggle, T.A., Denton, R.M., Proud, C.G. and Docherty, K. (1996). Glucose stimulates the activity of the guanine exchange factor eIF-2B in isolated rat islets of Langerhans. J. Biol. Chem. 271, 2121–2125.

Goodison, S., Kenna, S. and Ashcroft, S.J.H. (1992). Control of insulin gene expression by glucose. Biochem J. 285, 563–568.

Goodman, P.A., Medina-Martinez, O. and Fernandez-Mejia, C. (1996). Identification of the human insulin negative regulatory element as a negative glucocorticoid response element. Mol. Cell. Endocrinol. 120, 139–146.

Grimaldi, K.A., Siddle, K. and Hutton, J.C. (1987). Biosynthesis of insulin secretory granule membrane proteins. Biochem. J. 245, 567–573.

Guest, P.C., Bailyes, E.M., Rutherford, N.G. and Hutton, J.C. (1991). Insulin secretory granule biogenesis. Co-ordinate regulation of the biosynthesis of the majority of constituent proteins. Biochem. J. 274, 73–78.

Guest, P.C., Rhodes, C.J. and Hutton, J.C. (1989). Regulation of the biosynthesis of insulin-secretory-granule proteins. Biochem. J. 257, 431–437.

Guz, Y., Montminy, M.R., Leonard, J., Garmer, L.W., Wright, C.V.E. and Teitelman, G. (1995). Expression of murine STF-1, a putative insulin gene transcription factor, in β cells of pancreas, duodenum epithelium and pancreatic exocrine and endocrine progenitors during ontogeny. Development 121, 11–18.

Hammond-Kosack, M.C.U., Dobrinski, B., Lurz, R., Docherty, K. and Kilpatrick, M.W. (1992a). The human insulin linked polymorphic region exhibits an altered DNA structure. Nucleic Acids Res. 20, 231–236.

Hammond-Kosack, M.C.U. and Docherty, K. (1992). A consensus repeat sequence from the human insulin gene linked polymorphic region adopts multiple quadriplex structures in vitro. FEBS Lett. 301, 79–82.

Hammond-Kosack, M.C.U., Kilpatrick, M.W. and Docherty, K. (1992b). Analysis of the human insulin gene linked polymorphic region *in vivo*. J. Mol. Endocrinol. 9, 221–225.

Hammond-Kosack, M.C.U., Kilpatrick, M.W. and Docherty, K. (1993). The human insulin gene linked polymorphic region adopts a quadriplex structure in chromatin assembled in vitro. J. Mol. Endocrinol. 10, 121–126.

Hattersley, A.T. (1998). Maturity-onset diabetes of the young: Clinical heterogeneity explained by genetic heterogeneity. Diabet. Med. 15, 15–24.

Inoue, H., Riggs, A.C., Tanizawa, Y., Ueda, K., Kuwano, A., Liu, L., Doniskeller, H. and Permutt, M.A. (1996). Isolation, characterisation and chromosomal mapping of the human insulin promoter factor-1 (IPF-1) gene. Diabetes 45, 789–794.

Itier, J.M., Douchet, P., Desbois, P., Joshi, R.L., Dandroy-Dron, F., Jami, J. and Bucchini, D. (1996). Human insulin gene expression in transgenic mice—mutational analysis of the regulatory region. Differentiation 60, 309–316.

Itkin-Ansari, P., Bain, G., Beattie, G.M., Murre, C., Hayek, A and Levine, F. (1996). E2A gene products are not required for insulin gene expression. Endocrinology 137, 3540–3543.

Itoh, N. and Okamoto, H. (1980). Translational control of proinsulin synthesis by glucose. Nature 283, 100–102.

Jensen, J., Serup, P., Karlsen, C., Nielsen T.F. and Madsen, O.D. (1996). mRNA profiling of rat islet tumours reveals Nkx 6.1 as a β cell specific homeodomain transcription factor. J. Biol. Chem. 271, 18749–18758.

Jonsson, J., Carlsson, _., Edlund, T. and Edlund H. (1994). Insulin promoter factor 1 is required for pancreas development in mice. Nature 371, 606–609.

Jonsson, J., Ahlgren, U., Edlund, T. and Edlund, H. (1995). IPF1, a homeodomain protein with a dual function in pancreas development. Int. J. Dev. Biol. 39, 789–798.

Kajimoto, Y., Watada, H., Matsuoka, T., Kaneto, H., Fujiani, Y., Miyasaki, J. and Yamasaki, Y. (1997). Suppression of transcription factor PDX1/IPF1/STF1/IDX1 causes no decrease in insulin mRNA in MIN6 cells. J. Clin. Invest. 100, 1840–1846.

Karlsson, O., Edlund, T., Moss, L.B., Rutter, W.J. and Walker, M.D. (1987). A mutational analysis of the insulin gene transcription control region: Expression in beta cells is dependent on two related sequences within the enhancer. Proc. Natl. Acad. Sci. USA 84, 8819–8823.

Karlsson, O., Walker, M.D., Rutter, W.J. and Edlund, T. (1989). Individual protein-binding domains of the insulin gene enhancer positively activate 8 cell specific transcription. Mol. Cell. Biol. 9, 823–827.

Kennedy, G.C., German, M.S. and Rutter, W.J. (1995). The minisatellite in the diabetes susceptibility locus IDDM2 regulates insulin transcription. Nat. Genet. 9, 293–298.

Kennedy, G.C. and Rutter, W.J. (1992). Pur-1, a zinc-finger protein that binds to purine-rich sequences, transactivates an insulin promoter in heterologous cells. Proc. Natl. Acad. Sci. USA 89, 11498–11502.

Knight, S.W. and Docherty, K. (1992). RNA-protein interactions in the 5′ untranslated region of preproinsulin mRNA. J. Mol. Endocrinol. 8, 225–234.

Lageux, M., Lwoff, L., Meister, M., Golten, J.M. and Hofmann, J.A. (1990). cDNAs from neurosecretory cells of brains of *Locusta migratorica* (Insecta, Orthoptera) encoding a novel member of the superfamily of insulins. Eur. J. Biochem. 187, 249–254.

Leonard, J., Peers, B., Johnson, T., Ferreri, K., Lee, S. and Montminy, M. (1993). Characterisation of somatostatin transactivating factor-1, a novel homeobox factor that stimulates somatostatin expression in pancreatic islet cells. Mol. Endocrinol. 7, 1275–1283.

Lu, M., Miller, C.P., and Habener, J.F. (1996). Functional regions of the homeodomain protein IDX-1 required for transactivation of the rat somatostatin gene. Endocrinol. 137, 2959–2967.

MacFarlane, W.M., Chapman, J.C., Shepherd, R.M., Hashmi, M.N., Kamimura, N., Cosgrove, K.E., Hart, A.W., Docherty, H.M., Lindley, K.L., Aynsley-Green, A., James, R.F.L., Docherty, K., and Dunne, M.J. (1999b). Engineering a glucose-responsive human insulin-secreting cell line from islets of Langerhans isolated from a patient with Persistent Hyperinsulinaemic Hypoglycaemia of Infancy (PHHI). J. Biol. Chem. (submitted).

Macfarlane, W.M., Cragg, H., Docherty, H.M., Read, M.L., James, R.F.L., Aynsley-Green, A. and Docherty, K. (1997b). Impaired expression of transcription factor IUF1in a cell line derived from a patient with persistent hyperinsulinaemic hypoglycaemia of infancy (nesidioblastosis). FEBS Lett. 413, 304–308.

Macfarlane, W.M., McKinnon, C.M., Felton-Edkins, Z.M., Cragg, H., James, R.F.L. and Docherty, K. (1999a). Glucose stimulates translocation of the homeodomain transcription factor PDX1 from the cytoplasm to the nucleus in pancreatic β cells. J. Biol. Chem. 274, 1011–1016.

MacFarlane, W.M., Read, M.L., Gilligan, M., Bujalska, I. and Docherty, K. (1994). Glucose modulates the binding activity of the β cell transcription factor, IUF1, in a phosphorylation-dependent manner. Biochem. J. 303, 625–631.

Macfarlane, W.M., Smith, S.B., James, R.F.L., Clifton, A.D., Doza, Y.N., Cohen, P. and Docherty, K. (1997a). The p38/reactivating kinase mitogen-activated protein kinase cascade mediates the activation of the transcription factor insulin upstream factor 1 and insulin gene transcription by glucose in pancreatic β cells. J. Biol. Chem. 272, 20936–20944.

Marshak, S., Totary, H., Cerasi, E. and Melloul, D. (1996). Purification of the β-cell glucose-sensitive factor that transactivates the insulin gene differentially in normal and transformed islet cells. Proc. Natl. Acad. Sci. USA 93, 15057–15062.

Melloul, D., Ben-Neriah, Y., and Cerasi, E. (1993). Glucose modulates the binding of an islet-specific factor to a conserved sequence within the rat I and human insulin promoters. Proc. Natl. Acad. Sci. USA 90, 3865–3869.

Miller, C.P., McGehee, R.E. Jr. and Habener, J.F. (1994). IDX-1: A new homeodomain transcription factor expressed in rat pancreatic islets and duodenum that transactivates the somatostatin gene. EMBO J. 13, 1145–1156.

Moss, L.G., Moss, J.B., and Rutter, W.J. (1988). Systematic binding analysis of the insulin gene transcriptional control region: insulin and immunoglobulin enhancers utilize similar transactivators. Mol. Cell. Biol. 8, 2620–2627.

Nagasawa, H., Kataoka, H., Ispgai, A., Tamura, S., Suzuki, A., Mizogichi, A., Fujiwara, Y., Suzuki, A., Takahashi, S.Y. and Ishizaki, H. (1986). Amino acid sequence of a prothoraciotrpic hormone of the silkworm *Bombyx mori*. Proc. Natl. Acad. Sci. USA 83, 5840–5843.

Naya, F.J., Huang, H-P., Qiu, Y., DeMayo, F.J., Leiter, A.B., and Tsai, M-J. (1997). Diabetes defective pancreatic morphogenesis, and abnormal enteroendocrine differentiation in BETA2/NeuroD-deficient mice. Genes and Development 11, 2323–2334.

Naya, F.J., Stellrecht, C.M., and Tsai, M.J. (1995). Tissue-specific regulation of the insulin gene by a novel basic helix-loop-helix transcription factor. Genes and Dev. 9, 1009–1019.

Nielsen, D.A., Welsh, M., Casadaban, M.J. and Steiner, D.F. (1985). Control of insulin gene expression in pancreatic β cells and in an insulin-secreting cell line, RIN-5F cells. I. Effects of glucose and cyclic AMP on the transcription of insulin mRNA. J. Biol. Chem. 260, 13585–13589.

Odagiri H., Wang, J. and German, M.S. (1996). Function of the human insulin promoter in primary cultured islet cells. J. Biol. Chem. 271, 1909–1915.

Ohlsson, H., Karlsson, K. and Edlund, T. (1993). IPF1, a homeodomain-containing transactivator of the insulin gene. EMBO J. 12, 4251–4259.

Olson, L.K., Redmon, J.B., Towle, H.C. and Robertson, R.P. (1993). Chronic exposure of HIT cells to high glucose concentrations paradoxically decreases insulin gene transcription and alters binding of insulin gene regulatory protein. J. Clin. Invest. 92, 514–519.

Olson, K.L., Sharma, A., Peshavaria, M., Wright, C,V,E., Towle, H.C., Robertson, R.P and Stein, R. (1995). Reduction of insulin gene transcription in HIT-T15 β cells chronically exposed to a supraphysiological glucose concentration is associated with a loss of STF-1 transcription factor expression. Proc. Natl. Acad. Sci. USA 92, 9127–9131.

Owerbach, D. and Gabbay, K.H. (1996). The search for IDDM susceptibility genes. The next generation. Diabetes 45, 544–551.

Peers, B., Leonard, J., Sharma, S., Teitelman, G. and Montminy, M.R. (1994). Insulin expression in pancreatic islet cells relies on coperative interactions between the helix loop helix factor E47 and the homeobox factor STF-1. Mol. Endocrinol. 8 1798–1806.

Peers, B., Sharma, S., Johnson, T., Kamps, M., and Montminy, M. (1995). The pancreatic islet factor STF-1 binds cooperatively with Pbx to a regulatory element in the somatostatin promoter: Importance of the FPWMK motif and of the homeodomain. Mol. Cell. Biol. 15, 7091–7097.

Permutt, M.A. and Kipnis, D.M. (1972). Insulin biosynthesis. 1. On the mechanism of glucose stimulation. J. Biol. Chem. 247, 1194–1199.

Peshavaria, M., Gamer, L., Henderson, E., Teitelman, G., Wright, C.V.E. and Stein, R. (1994). XlHbox 8, an endoderm-specific Xenopus homeodomain protein, is closely related to a mammalian insulin gene transcription factor. Mol. Endocrinol. 8, 806–816.

Peshavaria, M., Henderson, E., Sharma, A., Wright, C.V.E. and Stein, R. (1997). Functional characterisation of the transactivation properties of the PDX-1 homeodomain protein. Mol. Cell. Biol. 17, 3987–3996.

Petersen, H.V., Serup, P., Leonard, J., Michelsen, B.K. and Madsen, O.D. (1994). Transcriptional regulation of the human insulin gene is dependent on the homeodomain protein STF1/IPF1 acting through the CT boxes. Proc. Natl. Acad. Sci. USA 91, 10465–10469.

Polychronakos, C., Kukuvitis, A., Giannoukakis, N. and Colle, E. (1995). Parental imprinting effect at the INS-IGF2 diabetes susceptibility locus. Diabetologia 38, 715–719.

Pugliese, A., Zeller, M., Fernandez Jr. A., Zalcberg, L.J., Bartlett, R.J., Ricordi, C., Pietropaolo, M., Eisenbarth, G.S., Bennett, S.T. and Patel, D.D. (1997). The insulin geme is transcribed in the human thymus and transcription levels correlate with allelic variation at the INS VNTR susceptibility locus for type 1 diabetes. Nat. Genet. 15, 293–297.

Read, M.L., Clark, A.R., and Docherty, K. (1993). The helix loop helix transcription factor USF binds to a regulatory sequence in the human insulin gene enhancer. Biochem. J. 295, 233–237.

Read, M.L., Masson, M.R. and Docherty, K. (1997). A RIPE3b1-like factor binds to a novel site in the human insulin promoter in a redox-dependent manner. FEBS Lett. 418, 68–72.

Read, M.L., Smith, S.B. and Docherty, K. (1995). The insulin enhancer binding site 2 (IEB2; FAR) box of the insulin gene regulatory region binds at least three factors that can be distinguished by their DNA binding characteristics. Biochem. J. 309, 231–236.

Robertson, R.P., Olson, L.K. and Zhang, H-J. (1994). Differentiating glucose toxicity from glucose densensitisation: a new message from the insulin gene. Diabetes 34, 1085–1089.

Roeder, R.G. (1996).The role of general initiation factors in transcription by RNA polymerase II. Trends Biochem. Sci. 21, 327–335.

Schuit, R.C., In'tVeld, P.A. and Pipeleers, D.G. (1988). Glucose stimulates proinsulin biosynthesis by a dose-dependent recruitment of pancreatic beta cells. Proc. Natl. Acad. Sci. USA 85, 3865–3869.

Sharma, A., Olson, L.K., Robertson, R.P. and Stein, R. (1995). The reduction of insulin gene transcription in HIT-T15 β cells chronically exposed to high glucose concetration is associated with the loss of RIP3b1 and STF-1 transcription factor expression. Mol. Endocrinol. 9, 1127–1134.

Sharma, A. and Stein, R. (1994). Glucose-induced transcription of the insulin gene is mediated by factors required for β cell type specific expression. Mol. Cell. Biol. 14, 871–879.

Sharma, S., Leonard, J., Lee, S., Chapman, H.D., Leiter, E.H. and Montminy, M.R. (1996). Pancreatic islet expression of the homeobox factor STF-1 relies on an E-box motif that binds USF. J. Biol. Chem. 271, 2294–2299.

Shieh, S-Y. and Tsai, M-J. (1991). Cell-specific and ubiquitous factors are responsible for the enhancer activity of the rat insulin II gene. J. Biol. Chem. 266, 16708–16714.

Skelly, R.H., Bollheimer, L.C., Wicksteed, B.L., Corkey, B.E. and Rhodes, C.J. (1998). A distinct difference in the metabolic stimulus-response coupling pathways for regulating proinsulin biosynthesis and insulin secretion that lies at the level of a requirement for fatty acyl moities. Biochem. J. 331, 553–561.

Smit, A.B., Vreugdenhill, E., Ebberink, R.H.M., Garaerts, W.P.M., Klootwijk, J. and Joose, J. (1988). Growth-promoting mulloscan neurons produce the precursor of an isulin related peptide. Nature 331, 535–538.

Soares, M.B., Schon, E., Henderson, A., Karathanasis, S.K., Cate, R., Zeitlin, S., Chrigwin, J. and Efstradiatis, A. (1985). RNA-mediated gene duplication: the rat preproinsulin I gene is a functional retroposon. Mol. Cell. Biol. 5, 2090–2103.

Steiner, D.F., Chan, S.J., Welsh, J.M. and Kwok, S.C.M. (1985). Structure and evolution of the insulin gene. Ann. Rev. Genet. 19, 463–468.

Stoffers, D.A., Ferrer, J., Clarke, W.L. and Habener, J.L. (1997c) Early-onset type-II diabetes mellitus (MODY4) linked to *IPF1*. Nat. Genet. 17, 138–139.

Stoffers, D.A., Thomas, M.K. and Habener, J.F. (1997a). Homeodomain protein IDX-1. A master regulator of pancreas development and insulin gene expression. Trends Endocrinol. Metab. 8, 145–151.

Stoffers, D.A., Zinkin, N.T., Stanojevic, V., Clarke, W.L and Habener, J.L. (1997b). Pancreatic agenesis attributable to a single nucleotide deletion in the human *IPF1* gene coding sequence. Nat. Genet. 15, 106–110.

Vafiadis, P., Bennet, S.T., Todd, J.A., Nadeau, J., Grabs, R., Goodyer, C.G., Wickramasinghe, Colle, E. and Polychronakos, C. (1997). Insulin expression in human thymus is modulated by INS VNTR alleles at the IDDM2 locus. Nat. Genet. 15, 289–292.

Walker, M.D., Edlund, T., Boulet, A.M. and Rutter, W.J. (1998). Cell-specific expression controlled by the 5′ flanking region of insulin and chymotrypsin genes. Nature 306, 557–561.

Wang, J.H., Shen, L.P., Najafi, H., Kolberg, J., Matschinsky, F.M., Urdea, M. and German, M. (1997). Regulation of insulin preRNA splicing by glucose. Proc. Natl. Acad. Sci. USA 94, 4360–4365.

Weintraub, H. (1993). The MyoD family and myogenesis: redundancy, networks and thresholds. Cell 75, 1241–1244.

Welsh, M., Nielsen, D.A., MacKrell, A.J. and Steiner, D.F. (1985). Control of insulin gene expression in pancreatic β cells and in an insulin-secreting cell line, RIN-5F cells. II. Regulation of insulin mRNA stability. J. Biol. Chem. 260, 13590–13594.

Wu, K.L., Gannon, M., Peshavaria, M., Offield, M.F., Henderson, E., Ray, M., Marks, A., Gamer, L.W., Wright, C.V.E. and Stein, R. (1997). Hepatocyte nuclear factor 3β is involved in pancreatic $cell specific expression of the *pdx1* gene. Mol. Cell. Biol. 17, 6002–6013.

Xu, G., Marshall, C.A., Lin, T-A., Kwon, G., Munivenkatappa, R.B., Hill, J.R., Lawrence, J.C. and McDaniel, M.L. (1998). Insulin mediates glucose-stimulated phosphorylation of PHAS-1 by pancreatic Beta cells. J. Biol. Chem. 273, 4485–4491.

PROINSULIN PROCESSING AND INSULIN STORAGE

Christina Wasmeier, John W. M. Creemers, Robert S. Jackson, and John C. Hutton

I. CELL BIOLOGY OF THE REGULATED SECRETORY PATHWAY

Insulin, like most secreted proteins, enters the secretory pathway at the level of the endoplasmic reticulum (ER), being directed there by the hydrophobic leader (signal) sequence corresponding to the first 16 amino acids of the preproinsulin struc-

Advances in Molecular and Cell Biology
Volume 29, pages 125-150.

ISBN: 0-7623-0547-9

ture. Once in the lumen of the ER, the three disulphide bonds are formed and the leader sequence is removed by signal peptidases to form proinsulin. The molecule then moves to the Golgi apparatus by a vesicle or tubo-vesicular–mediated process and passes sequentially through stacks of the Golgi to reach the trans-Golgi network (TGN). Here it is sorted away from other proteins and targeted to the storage granules of the cell, where it can remain for several days until its release is stimulated. Nascent secretory granules are characteristically membrane-bounded spherical organelles enveloping an electron dense core. These transform in time to the mature secretory granules which typically have a condensed core, often crystalline in appearance, surrounded by a mantle of more electronlucent material.

Proteins travel between compartments via vesicular intermediates that bud from the donor compartment and fuse with the acceptor compartment. Formation of vesicles is controlled by a complex molecular machinery, many aspects of which are common to numerous intracellular transport steps. Major advances in identifying and characterizing the components involved have come from two complementary approaches; the development of cell-free transport assays, on the one hand, and yeast genetics, on the other. The finding that a number of genes defining secretory (sec) mutants in yeast represent proteins involved in constitutive and regulated secretion in mammalian cells illustrates the high degree of conservation of many underlying principles of vesicular transport.

The TGN gives rise to several distinct types of vesicles. A well-documented pathway is the sorting of lysosomal proteins into clathrin-coated vesicles. Coat proteins are thought to be essential for vesicle formation, and probably both have a mechanical role and serve to ensure the correct sequence of events during vesicle budding and fusion. The coat on TGN-derived clathrin-coated vesicles is composed of clathrin and the heterotrimeric adaptor complex AP-1. Sorting is mediated via binding of γ-adaptin, one of the subunits, to the cytoplasmic tail of the mannose-6-phosphate receptor, which specifically recognizes lysosomal proteins via a mannose-6-phosphate modification received in the Golgi (Pearse and Robinson, 1990).

Constitutive secretory vesicles also originate from the TGN. Three-dimensional reconstructions from electron micrographs of the TGN show a highly tubulated structure. Individual tubules produce either clathrin-coated vesicles or vesicles with a novel lacelike coat structure, which are thought to be constitutive secretory vesicles (Ladinsky et al., 1994). The proteins making up the lacelike coat have not been identified yet.

Specialized hormone-secreting cells such as the pancreatic β-cell possess an additional regulated secretory pathway, distinguished by the presence of dense-core granules. Immature secretory granules also form by budding from the TGN. Morphological and biochemical evidence suggests that the segregation of their contents largely occurs within the TGN. In the pancreatic β-cell, for example, immunoelectron microscopy showed that a marker protein for the constitutive transport route, the influenza virus protein hemagglutinin, could be detected in the

TGN and in electron-translucent vesicles but was largely excluded from proinsulin-rich budding and immature granules (Orci et al., 1987). Subcellular fractionation studies conducted with another neuroendocrine cell type, PC12 cells, supported this conclusion. Proteins were pulse-labelled in the TGN by sulphation, and chased to allow the formation of post-Golgi vesicles. Secretogranin II as a marker for the regulated pathway was found in a vesicle population distinct from that containing constitutively secreted heparin sulphate proteoglycans, without significant co-localization observed in post-TGN compartments (Tooze and Huttner, 1990).

A number of observations, however, suggest greater complexity in the case of the secretory granule. Electron micrographs reveal patchy clathrin coats on the surface of immature secretory granules (Orci, 1985; Tooze and Tooze, 1986). The adaptor complex mediating clathrin binding was identified as AP-1, the TGN-type adaptor (Dittié et al., 1996), but the function of the clathrin coat on immature granules is still unclear. There is no evidence of a coat encompassing the entire surface of nascent granules, arguing against a role for clathrin in driving secretory granule budding from the TGN. Rather, the coat may be involved in vesicular traffic originating from the immature granule.

Evidence for such pathways has been obtained in studies with pancreatic islet β-cells. Proinsulin is processed to mature insulin and the connecting C-peptide subsequent to its arrival in the secretory granule. C-peptide appears to be released in excess of insulin from unstimulated islets (Arvan et al., 1991). A model of "sorting by retention" was proposed, whereby proteins not destined for the granule may initially be included in the immature granule but are removed by vesicular traffic during the maturation process, while regulated secretory proteins are retained due to their ability to condense into an insoluble granule core (Kuliawat and Arvan, 1992, 1994). Similarly, lysosomal hydrolase precursors were found in the immature but not in the mature granule, suggesting that they were selectively retrieved during maturation. In PC12 cells, where sorting of regulated and constitutive secretory proteins occurs largely in the TGN (Tooze and Huttner, 1990), there is, nevertheless, evidence for some continued sorting after granule formation (Grimes and Kelly, 1992). There is some controversy as to the significance of this pathway in quantitative terms, but it is possible that the observed discrepancies are due to differences between cell types. A vesicular trafficking pathway originating from immature granules could serve a number of roles, for instance, an increase in the fidelity of TGN-based sorting, or the recycling of membrane proteins with a role in granule budding back to the TGN for another round of vesicle formation.

Comparatively little information is available about integral membrane proteins specific to the secretory granule. Recently, though, two such molecules have been cloned from β-cells—the closely related protein tyrosine phosphatase homologues phogrin and IA-2 (Lan et al., 1994; Solimena et al., 1996; Wasmeier and Hutton, 1996). While their structure marks them out as potential candidates for key roles in granule function, their physiological role is presently unknown. Their granule-specific localization, however, makes them valuable markers for cell biological studies of secretory granule membrane traffic during granule biogenesis and secretion.

What governs the delivery of proteins to their correct intracellular destination? Targeting signals have been described for proteins resident in other compartments, examples being the COOH-terminal KDEL sequence for lumenal ER proteins (Munro and Pelham, 1987) or the selective phosphorylation of mannose residues in proteins destined for the lysosome (Kornfeld and Mellman, 1989). The principles regulating the sorting of proteins to secretory granules are less well characterized. A number of studies introducing foreign secretory proteins into granulated cells allow the conclusion that these mechanisms are highly conserved. For example, AtT20 cells transfected with human preproinsulin (Moore et al., 1983), or human growth hormone (Moore and Kelly, 1985), and GH_3 cells expressing preprosomatostatin (Stoller and Shields, 1988) all efficiently transport the exogenous protein to the regulated secretory pathway. Similar principles seem to apply to the sorting of integral membrane proteins. P-selectin, a transmembrane protein of platelet and endothelial cell granules, is sorted to secretory granules both in transfected AtT20 and PC12 cells (Disdier et al., 1992; Koedam et al., 1992; Norcott et al., 1996).

The existence of sorting signals or domains has been demonstrated by studying the effects of deletions of candidate sequences, or their transfer to constitutively secreted proteins. Human growth hormone was able to divert a soluble form of the vesicular stomatitis virus glycoprotein to the regulated pathway (Moore and Kelly, 1986). The proregion of somatostatin mediated transport of the cytosolic protein α-globin to regulated secretory vesicles (Stoller and Shields, 1989). A series of chimeras between the soluble granule constituent PC2 and the related TGN resident furin identified two regions in PC2 that are important in its sorting (Creemers et al., 1996). The N-terminal 26 residues of the hormone precursor pro-opiomelanocortin (POMC) are sufficient for granule targeting (Tam et al., 1993), with the disulphide-bonded loop structure of this region proposed to be the crucial element (Cool et al., 1995). Similar evidence for the presence of granule sorting signals was obtained in studies with the integral membrane proteins peptidylglycine α-amidating monooxygenase (PAM) and P-selectin. However, while targeting of dense-core secretory granule proteins is clearly signal-dependent, no consensus primary sequence or structural similarities have been detected among the sorting domains identified.

Two basic models for the mechanism of signal-mediated sorting in the TGN have been proposed. The inclusion of soluble proteins into the forming granule may require their interaction with a transmembrane receptor. The proposed receptor(s) would need to be of broad specificity to explain the correct targeting of foreign proteins in granulated cells. A consideration of the membrane-to-content ratio of the secretory granule argues against the stoichiometric interaction of secretory proteins with granule membrane proteins. Alternatively, the primary sorting event may be a selective aggregation of secretory proteins, with certain proteins capable of spontaneously aggregating in the TGN promoting the co-aggregation of others (Halban and Irminger, 1994). This could be a result of the physicochemical properties of regulated secretory proteins. At pH 6.4 and 1 to 10 mM Ca^{2+}, the conditions thought to exist in the lumen of the TGN, several lumenal granule proteins could be induced to aggregate in vitro (Chanat and Huttner, 1991) while constitutive secretory proteins remained soluble (Colomer et al., 1996). Thus, acidic pH and high Ca^{2+} concentrations may be critical factors in segregating regulated from constitutive secretory proteins.

A recent study investigating possible sorting receptor(s) proposed the membrane-associated form of carboxypeptidase H (CPH or CPE) as the receptor for POMC and other hormones (Cool et al., 1997). However, membrane-associated CPH is a peripheral membrane protein and, as such, is not able to mediate typical receptor functions such as the assembly of coat proteins and granule budding. Moreover, a COOH-terminally truncated molecule that is entirely soluble is still sorted efficiently to the regulated pathway of secretion, indicating that its targeting is not dependent on an interaction with the membrane (Varlamov and Fricker, 1996). In addition, a prohormone-binding sorting receptor would be required at high concentrations, in particular in actively secreting cells, whereas the biosynthesis of CPH in the β-cell is not stimulated by secretagogues in parallel with that of insulin (Guest et al., 1989b). Clearly, the involvement of CPH in prohormone sorting requires further examination.

The formation of nascent secretory granules from the TGN is a complex process. Several classes of guanosine triphosphate (GTP)-binding proteins have been implicated. The nonhydrolysable GTP analogue GTPγS inhibits the formation of constitutive as well as regulated secretory vesicles in PC12 cells, demonstrating a requirement for GTP hydrolysis (Tooze et al., 1990). The drug brefeldin A inhibits a number of intracellular transport steps by acting on the small GTPase adenosine diphosphate ribosylation factor (ARF), required for coat protein recruitment and vesicle budding (Donaldson and Klausner, 1994). Brefeldin A, in intact PC12 cells and in a cell-free system, inhibits the formation of secretory granules (and constitutive secretory vesicles) from the TGN (Rosa et al., 1992), implicating an ARF or ARF-like protein, as well as the involvment of an as yet unidentified coat protein in the budding step. For Golgi-derived vesicles, ARF has been proposed to activate phospholipase D, leading to an increase in phosphatidic acid and, via this, increased recruitment of coat proteins. Whether such a mechanism also plays a role in the budding of secretory vesicles from the TGN is being investigated.

An involvement of intracellular pools of heterotrimeric G-proteins in membrane traffic has been recognized in recent years (for review, see Bomsel and Mostov, 1992). Multiple trimeric G-proteins are present on the TGN, where they exert both inhibitory and stimulatory effects on the formation of secretory vesicles (Leyte et al., 1992). It is not known which steps in the formation of secretory vesicles are controlled by heterotrimeric G-proteins, but it has been hypothesized that the G-proteins at the TGN membrane act as regulators of membrane traffic in response to, for instance, external signals or events related to sorting and vesicle budding.

Protein phosphorylation plays a central role in the regulation of many cellular events. Protein serine/threonine kinases or phosphatases are apparently not required for the formation of secretory vesicles per se, but studies in a PC12-derived cell-free system suggested that, by acting on a cytosolic phosphoprotein, they modulate the regulation of this process by TGN-associated trimeric G-proteins (Ohashi and Huttner, 1994). Recently, a role for tyrosine phosphorylation in secretory granule formation was suggested. Both tyrosine kinases and phosphatases appear to be required for budding of nascent secretory vesicles from permeabilized GH_3 cells, involving the increased tyrosine phosphorylation of several proteins on the TGN membrane (Austin and Shields, 1996). It is not clear, though, whether the vesicles in question are bona fide secretory granules, or whether they resemble constitutive secretory vesicles.

The use of cell-free systems has also demonstrated a dependence of secretory vesicle formation on cytosolic factors. One such factor is the phosphatidylinositol transfer protein. Functionally, it can be substituted for by yeast Sec14p, which is not homologous in its sequence but exhibits the same phospholipid transfer activity (Ohashi et al., 1995). Transiently increased levels of highly charged phosphoinositides may directly promote membrane budding through structural changes in the membrane, or mediate the recruitment of other key cytosolic proteins to the donor membrane. Phosphatidylinositol is a substrate for certain types of phosphoinositide 3-kinases, enzymes which have recently been implicated in a number of membrane traffic steps. They include Vps34p, a protein important for sorting to the yeast vacuole. Inhibitor studies in mammalian cells, using wortmannin, indicated that phosphoinositide 3-kinase activity is essential for transport between TGN and lysosomal compartments and within the endocytic system (Shepherd et al., 1996). An involvement of the phosphatidylinositol transfer protein in secretory granule formation may indicate a role for lipid kinases in this transport step, too.

II. ENZYMOLOGY OF PROINSULIN CONVERSION

Proinsulin, like most polypeptide hormones and neuropeptides which are secreted in a regulated manner, is produced from a larger biologically inactive precursor through intracellular post-translational proteolytic cleavage at sites marked by pairs of basic amino acid residues. Constitutively secreted peptides, viral membrane glycoproteins and precursor forms of resident membrane proteins are processed in a similar way though usually at sites marked by more extended basic motifs (Hutton, 1990; Steiner et al., 1992; Halban and Irminger, 1994; Seidah et al., 1994). Most enzymes responsible for processing belong to the family of serine proteases characterized by catalytic domains with structural homology to bacterial subtilisins. The seven mammalian enzymes so far identified (furin, PACE4,

PC1/3, PC2, PC4, PC5/6, PC7/8 [LPC]) have similar modular structures which precede variable COOH-termini (Figure 1). Most members of the family are capable of cleaving generic substrates with a P1 Arg, a P2 Arg or Lys, a P4 Arg or Lys, and a P6 Arg or Lys. A bulky hydrophobic residue is precluded in the P1′ position, whereas the P2′ is usually hydrophobic in character. Such specificity is quite distinct from bacterial subtilisin, which is more characteristic of a digestive enzyme involved in limited endoproteolytic cleavage. Where the individual members vary most significantly, however, is in their capacity to accommodate nonbasic amino acids in the P4 and P6 position and their accommodation of Lys in the P1 position. Where two enzymes hydrolyze the same substrate, there are obvious kinetic differences that need to be documented further. The other major determinant of their biological function is their cellular localization and that of their substrates. Thus PC1 and PC2 need to be coordinately synthesized with their substrates and co-packaged into dense-cored secretory granules where the enzyme activity can be expressed at the acidic pH and high calcium concentration prevailing in the granule (Davidson et al., 1988; Bailyes and Hutton, 1992; Bailyes et al., 1992; Guest et al., 1992; Lindberg et al., 1994; Bailyes et al., 1995; Malide et al., 1995). Furin, on the other hand, is localized to the TGN through an active retention and retrieval mechanism (Bosshart et al., 1994; Molloy et al., 1994; Schäfer et al., 1995) and only sees its substrates *en passant*. The tissue distribution of the eukaryote subtilisin-like enzymes is variable. Furin, PACE 4, and PC7/8 are expressed widely, PC1 and PC2 are confined to tissues of neuroendocrine lineage, and PC4 is restricted to germ cells of the testis (Bennett and Hutton, 1994).

This chapter focuses on recent advances in our understanding of the regulation of the catalytic function and cell biology of these enzymes gained from manipulation of their structure by mutagenesis or construction of chimeric molecules. Emphasis is directed towards PC1, PC2, and furin, which are the best characterized enzymes, and their function in cells with a regulated secretory pathway. Although furin is largely responsible for processing constitutively secreted proteins, it has an indirect role in the activation of PC1 and PC2 and may itself process secretory proteins in the TGN before their sorting to dense-core granules (Farhadi et al., 1997).

III. STRUCTURE OF PC1, PC2, AND CPH AND THEIR POST-TRANSLATIONAL PROCESSING

PC1, PC2, and CPH have many characteristic features of secretory proteins and are subjected to many of the same post-translational modifications as the secreted cargo. They also need to be sorted into the regulated pathway to exert their action; however, the mechanism of sorting may be fundamentally different from that of proinsulin. PC1 and PC2 (Figure 1) possess typical signal sequences that precede a moderately conserved (30% to 50%) prodomain of 70 to 90 amino acids that ter-

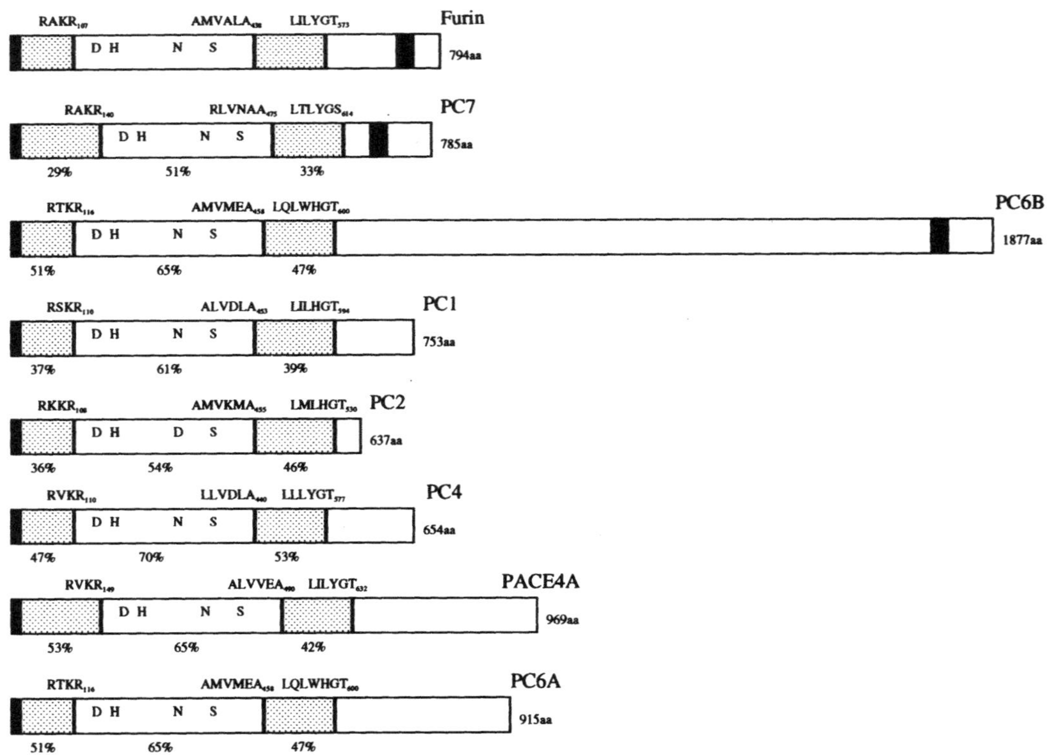

Figure 1. The mammalian subtilisinlike processing enzymes. The presence of an NH_2-terminal signal peptide, the prodomain that terminates in the sequence RXKR, the catalytic domain bearing the catalytically important D, H, and S residues, and the middle domain which incorporates the (H/Y) G (S/T) sequence is shown along with the variable COOH-terminal domain that, in the case of furin, PC7, and PC6B, incorporates a transmembrane segment (shaded). The percentage amino acid sequence identity of each domain relative to furin is also indicated.

minates in a cluster of basic residues. This borders on the highly conserved (50% to 70%) catalytic domain of 320 to 340 amino acids that terminates with the topological equivalent of the COOH-terminus of the prokaryotic serine protease subtilisin. Like all the eukaryote subtilisinlike convertases, they include a well conserved (35% to 50%) middle domain (also called P or Homo B domain) of 110 to 130 amino-acids followed by a variable COOH-terminus of 160 and 50 amino acids, respectively. CPH, although structurally unrelated, also has a signal peptide, a short prodomain terminating in a cluster of basic amino acids, an extended catalytic domain with only weak homology to the digestive enzymes carboxypeptidases A and B, and a short COOH-terminal tail, which has some similarities in secondary though not primary structure to that of PC1 and PC2 (see the discussion that follows).

The prodomains of PC1, PC2, and CPH, as in other secretory proteins, serve several functions: correct temporal and spatial activation of catalytic activity, promotion of folding and oligomerization, intracellular transport, stability, and sorting. PC1 is cleaved rapidly via an intramolecular process (Goodman and Gorman, 1994), whereas pro-PC2 cleavage is slow and can occur *in trans* (Matthews et al., 1994; Shennan et al., 1995). CPH prodomain cleavage is also rapid and presumably occurs in the ER; however, the nature of the enzyme responsible remains obscure. There is evidence, in each case, that removal of the prodomain affects the solubility characteristics of the proteins and their association with membranes. It has been postulated that this may be part of a sorting mechanism; however, it is not clear how and where this may operate in the cell (Shennan et al., 1994; Shennan et al., 1995; Guest et al., 1995). The proforms of PC1 and PC2 appear to be catalytically inactive towards exogenous substrate, a feature that may be part of the mechanism by which their activities are confined to the dense core secretory granule. Pro-CPH in contrast is active, a feature that may be of consequence in the trimming of the COOH-termini of proteins that are processed in the Golgi but not sorted to the regulated pathway.

Although they differ significantly in size and have little sequence homology, the COOH-terminal domains of PC1, PC2, and CPH have some similar characteristics, notably the presence of a putative amphipathic α-helix at the COOH-terminus preceded by a stretch of charged amino acids (Smeekens et al., 1991; Mitra et al., 1994). The membrane association or aggregation of the proteins is promoted by calcium and acidic pH (Shennan et al., 1994; Hill et al., 1995), a phenomenon which is dependent upon the COOH-terminal sequence being intact. The COOH-termini of PC1 (100 aa), PC2 (< 20 aa), and CPH (20 aa) are truncated in a post-trans–Golgi compartment, usually with quite slow kinetics (Guest et al., 1989a; Zhou and Mains, 1994; Bailyes et al., 1995). The truncated form of PC1 is reported to be more active, less stable, and to exhibit changes in its pH and calcium requirements towards synthetic substrates (Zhou et al., 1995b), though CPH and PC2 do not appear to be altered in their kinetic properties toward soluble substrates in dilute solutions. One can imagine that under physiological conditions, increased mobility of the molecules in the confines of the secretory granule may enhance their access to condensed prohormone substrates and enhance rates of prohormone conversion in vivo. Although the truncated proteins are subsequently secreted, this appears inconsequential since PC1, PC2, and CPH are virtually inactive at extracellular pH values.

Recent evidence suggests that the function of the COOH-termini of these molecules may relate more to the mechanism by which the proteins are sorted into the regulated pathway of secretion than to any intragranular function (Zhou et al., 1995a; Creemers et al., 1996). Thus, deletion of the tails perturbs their proper sorting. Furthermore, chimeric proteins, bearing the tails of CPH or PC2, can sort protein domains into the regulated pathway which would normally be committed to constitutive release. Studies on the overexpression of the truncated form of PC1 and of the COOH-terminal fragment alone in CHO cells transfected with prorenin or POMC (Jutras et al., 1997) have pointed to an additional function of the carboxyl-terminus as an endogenous inhibitor of PC1. The evidence suggests that full length PC1 binds prorenin at its dibasic cleavage site but fails to cleave due to the inhibitory effect of the COOH-terminus of PC1. As a result, the PC1/prorenin complex is targeted to the secretory granules, the COOH-terminal tail of PC1 is removed by proteolysis, which in turn renders PC1 capable of processing prorenin. In many ways, the model is similar to that formulated for CPH as a "sorting receptor" in neuro 2A cells transfected with POMC (Cool et al., 1997). As in this case, it is not clear whether the "sorting" occurs with a 1:1 stoichiometry or whether the mechanism has a more significant role in sorting the bulk of the granular cargo. The vast molar excess of the major secretory cargo to any of the processing enzymes in a typical neuroendocrine cell (~200 to 1,000-fold) precludes sorting through a 1:1 stoichiometry of cargo to processing enzyme. However, the processing enzymes could act as "sortases" if they functioned as nucleation centers for aggregation of cargo or docking sites for preformed aggregates of cargo. Such a mechanism is attractive as it would guarantee that each nascent granule would contain a complement of processing enzymes as opposed to a model where the enzymes were sorted by a self-association mechanism similar to that postulated for the major cargo (Chanat and Huttner, 1991).

PC2, unlike PC1 and CPH, is associated with another secretory granule protein, 7B2 (Braks and Martens, 1994), at least during the proximal segment of the secretory pathway. 7B2 shares a weak homology with the chaperonin 60 class of molecules in its amino-terminus and is postulated to either facilitate the folding of pro-PC2 to an active conformation (Martens et al., 1994; Zhu et al., 1996a) or to stabilize it during its transport to the TGN (I. Lindberg, personal communication). The carboxyl-terminus of 7B2, in addition, has a short segment of homology to the serpin class of serine proteinase inhibitors. Mutagenesis studies suggest that inhibitory activity results from the ability of the enzyme to recognize the dibasic motif in the sequence ^{169}VAKKSV but not to cleave it readily (Horssen et al., 1995). The 27kD proform of 7B2 is a substrate for PC1 and furin which cleave the sequence

^{150}RRKRR producing an NH_2-terminal 21kD and COOH-terminal 6kD fragment (Horssen et al., 1995). The cleaved molecule is inactive as a "chaperonin" and cleavage may serve to disengage 7B2 and pro-PC2 once they have reached the TGN or granule compartment. The COOH-terminal fragment is still inhibitory in the nanomolar range, which indicates that PC2 activity would never be expressed since the 7B2 and PC2 are present in granules at micromolar concentrations. Nevertheless, it is reported that the inhibitor domain is ultimately inactivated by cleavage of the internal ^{171}KK site by PC2 and removal of the COOH-terminal amino acids by CPH in the secretory granules (Zhu et al., 1996b). Such a mechanism could act to delay or dampen PC2 activity; however, the need for such a mechanism is not obvious since it is generally held that the protein arrives in the granule as the inactive proform. That said, there is conflicting evidence that pro-PC2 is processed before or soon after leaving the ER (Guest et al., 1992; Creemers et al., 1996), not least that it can cleave prosomatostatin in a pregranule compartment (Stoller and Shields, 1988). It thus remains unclear as to how the various interactions of 7B2 and PC2 control the passage of PC2 through the secretory pathway and its activity. A central issue is whether the interaction of 7B2 and PC2 is stoichiometric rather than catalytic in nature, which in turn raises the question of coordination of the synthesis of the two molecules and their relative molar ratio under different physiological conditions.

IV. REGULATION OF PROHORMONE PROCESSING

The processing of proinsulin in the pancreatic β-cell by PC1, PC2, and CPH in terms of its regulation needs to be considered in the broader context of the role that these enzymes play in processing other propeptide substrates both within the β-cell and in other neuroendocrine tissues. Many prohormones contain within their sequences more than a single bioactive peptide, each flanked by pairs of basic amino acids that are potential sites of cleavage for PC1 and PC2. POMC, for example, incorporates adrenocorticotrophic hormone (ACTH), three versions of melanocyte-stimulating hormone (MSH), an opiate peptide β-endorphin, and a lipotrophic activity; proglucagon contains not only glucagon but also glucagon-like peptide-1 (GLP-1). Expression of the bioactivity of any of these peptides depends upon their excision from the precursor by post-translational proteolysis, which often occurs in a cell-specific manner and may change under different physiological states. Such regulation of the proteolytic events in the cell permits a pleiotrophic and plastic response while being essentially conservative in terms of regulation of the gene encoding the precursor protein. For proinsulin processing to occur, it is sufficient only to ensure that all the genes are expressed in the tissue as there appears little variability in proinsulin conversion except perhaps the extent of conversion. The insulin granule, however, contains a wide range of other bioactive peptides, including peptides derived from chromogranin A (Hutton et al.,

1988), chromogranin B (Hoflehner et al., 1995), islet associated polypeptide (amylin) (Betsholtz et al., 1989), POMC (Rhodes et al., 1993), and thyrotropin-releasing hormone (TRH) (Wu and Jackson, 1988). Although their concentrations are low relative to insulin, they may have important functions as autocrine, paracrine, or vasoregulatory factors.

The level of endopeptidase activity in the regulated pathway of secretion appears to be the rate-limiting step in the overall rate of prohormone conversion in the cell. The controls on gene expression at the transcriptional and translational level are consequently important determinants of the rates of cleavage of proprotein substrates. Both the PC1 and PC2 gene are expressed in the β-cell in contrast to the α-cell where the predominant enzyme is PC2 (Itoh et al., 1996). The synthesis of PC1 and proinsulin increases in response to secretory stimuli (Alarcon et al., 1993), whereas PC2 and CPH appear to be synthesized constitutively (Guest et al., 1989b). This means that the relative composition of newly forming granules will likely change as a function of the stimulus received by the β-cell and its duration. Thus, the rate of conversion of proinsulin to insulin in newly formed granules may be variable and more or less unprocessed proinsulin could be released depending on the average maturation state of the granule pool in the cell. Changes in the relative proportion of proinsulin to CPH have a more hypothetical consequence arising from the suggestion that CPH acts as a sorting receptor (Cool et al., 1997).

PC1 and PC2 in in vitro assays are dependent on Ca^{2+} for activity and are active only in the acidic pH range (pH 4.5 to 6.5). Both pH and Ca^{2+} are thought to play a role in restricting the action of these enzymes to the granule compartment and preventing their activity in earlier vesicular compartments of the secretory pathway. Cellular depletion of Ca^{2+} prevents proinsulin processing, although it does so at several levels, including the inhibition of the proteolytic activation of the enzymes themselves (Guest et al., 1997). It remains an open question as to whether more subtle changes in intragranular Ca^{2+} concentration and pH, which are thought to accompany stimulation of the β-cell, would have any impact on the overall rates of conversion and maturation of enzymes and their catalytic activity.

V. MEASUREMENT OF PROINSULIN AS A DIAGNOSTIC TOOL

Sensitive assays for circulating levels of proinsulin have been developed by several groups based either on "proinsulin-specific antibodies," which do not cross-react with the insulin molecule, or by two-site assay procedures based on the antibodies that see various epitopes on proinsulin, C-peptide, and insulin. These have largely supplanted the more cumbersome high-performance liquid chromatography (HPLC) or chromatography-based assays that may require up to 20 ml of serum to execute. The two-site assays, like the combined immunoassay/HPLC

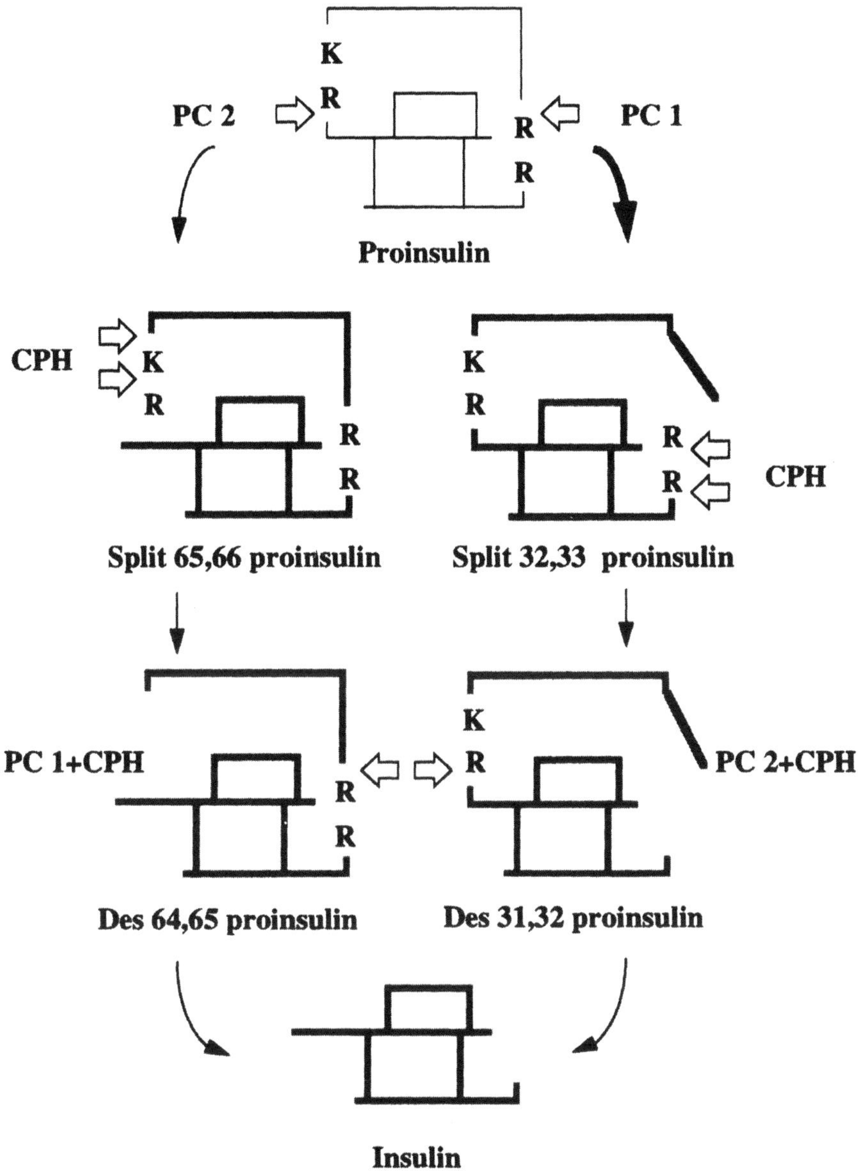

Figure 2. The pathway of conversion of human proinsulin. Proinsulin can be cleaved both by PC1 and PC2; however, each enzyme has a preferred site of action. Moreover, des 31,32 proinsulin is a much better substrate than proinsulin for PC2, which effectively means that the enzymes act sequentially and that the predominant intermediate found in the circulation is des 31,32 proinsulin

techniques, provide determinations of proinsulin, des 64,65 proinsulin and des 31,32 proinsulin (Figure 2), although correction for minor cross-reactivity with insulin is required. They do not distinguish between the common des intermediates and basic amino acid extended forms (65,66 split proinsulin and 32,33 split proinsulin); however, the relatively high levels of CPH compared to PC1 and PC2 in islets ensures that the concentration of extended forms is negligible. It is now possible with time-resolved fluorescence assays based on Europium-labeled antibodies to obtain a complete profile of the insulin-related molecules from 500 μl of serum. Studies of the secretion and circulating levels of proinsulin-related molecules in experimental animals, however, have been impeded by the lack of monoclonal antibodies of the required specificity for two site assays. HPLC/immunoassay-based analyses of rat and mouse samples are further complicated by the presence of two insulin genes in rodents and the need to resolve a minimum of eight molecular forms to complete the analysis. These limitations with rodent experiments, combined with the difficulty of obtaining human islets, has meant that little information is available on the cellular physiology of proinsulin secretion in either humans or experimental animals.

VI. GENETIC CAUSES OF PROINSULINEMIA

The most common causes of frank proinsulinemia of genetic origin arise from mutations of the proinsulin gene altering either one of the basic amino acids at a proteolytic cleavage site (Shibasaki et al., 1985; Yano et al., 1987), or more distally at the Zn^{2+}-binding His_{B10} residue (>Asp) (Chan et al., 1987). Such cases are mostly heterozygotes, and the impact on glucose metabolism is relatively small as normal insulin is produced from the other allele. In the former case, the defect arises from the limited ability of PC1 or PC2 to cleave at a site marked by a single basic amino acid. Although single basic sites do occur in a number of neuropeptides and prohormones such as dynorphin (Berman et al., 1995), other members of the eukaryote subtilisin enzyme family are thought to be responsible for the cleavage. The $_{B10}$His>Asp mutation is less readily explained. Initial transfection experiments in AtT20 cells suggested that the lack of a critical Zn^{2+} binding site impairs condensation of the prohormone in the TGN, resulting in missorting and constiutive release. However, further studies implicate a kinetic defect in processing, possibly brought about by alteration in secondary structural requirements for substrate recognition by the endopeptidases (Docherty et al., 1989).

Genetic deficiencies in the processing enzymes themselves have only recently been reported. The first of these, the *fat/fat* mouse (Naggert et al., 1995), was traced to a point mutation (Ser_{202}>Pro) of the CPH gene. The enzyme is synthesized at a relatively normal rate but fails to fold and is degraded rapidly in the ER. In addition, it appears to be catalytically inactive. Homozygous animals show hyperproinsulinemia, slowly developing obesity, infertility, and glucose intolerance. This is in associated with defects in proinsulin (Naggert et al., 1995), proenkephalin (Fricker et al., 1996), POMC (Cool et al., 1997), and CCK8 (Cain et al., 1997) processing, most notably the accumulation of intermediate forms extended by COOH-terminal basic amino acids. It is remarkable, given the substantial proportion of bioactive polypeptides that require a COOH-terminal amide, that the *fat/fat* mutation is not embryonically lethal. Carboxypeptidase D, an enzyme that probably acts to remove basic amino acid terminal residues from constitutively secreted proteins, can catalyze the reaction, albeit inefficiently (Song and Fricker, 1995) and rescues the mutant to some degree. A further feature of the *fat/fat* mutant is the finding that the proforms as well as the dibasic extended forms of hormones have accumulated in these animals. This is postulated as arising from failure in sorting of the propeptides into the regulated pathway caused by the deficiency of CPH and its proposed "sorting receptor" function (Cool et al., 1997). However, it could equally be accounted for by product inhibition or allosteric effects on the PC enzymes (Naggert et al., 1995). Secretion of proinsulin in the *fat/fat* animal is regulated by glucose, which suggests that sorting occurs normally. Moreover, diarginyl insulin, which accumulates and is secreted in these animals, could only have been generated in a granule compartment, which is again inconsistent with the "sorting receptor" defect hypothesis.

An isolated case of a PC1 gene mutation in humans has recently been identified (Jackson et al., 1997). This individual is a Caucasian female who presented at age 43 years with symptoms of postprandial hypoglycemia and a history of severe childhood obesity, idiopathic hypogonadotrophic hypogonadism, and gestational diabetes melllitus (O'Rahilly et al., 1995). Genomic sequencing revealed a missense mutation in the middle domain (Gly>Arg^{483}) of PC1 combined with an intron 5 donor splice site transversion (A>C+4) that results in truncation of the catalytic domain in the second allele. Inheritance was Mendelian and her four heterozygous children, at present teenagers, are phenotypically normal. Circulating insulin was undetectable in the proband but levels of intact proinsulin and des 64,65 proinsulin were elevated several hundredfold, consistent with the expression of nonfunctional PC1 in the pancreatic β-cell. POMC to adrenocorticotropic hormone (ACTH) ratios were also elevated in accordance with a PC1 defect; however, this is relatively mild suggesting that the mutant enzyme may still be active or perhaps compensated by another endopeptidase in this tissue (Azaryan et al., 1993; Hook et al., 1997). Her obesity is conceivably related to deficient processing of hypothalamic satiety factors such as α-melanocyte–stimulating hormone (MSH), cholecystokinin (CCK), and GLP-1. It has been postulated that a processing defect of pro-luteinizing hormone releasing factor (LHRH) or another hypothalamic peptide might be involved in her reproductive problems. However, previous studies indicate that pro-LHRH is processed by PC2 (W.C. Wetsel, personal communication). Likewise her thyroid status is normal, yet PC1 is thought to be crucial for pro-TRH conversion (Nillni et al., 1996). The question arises of whether the Gly>Arg^{483} allele in this individual retains some activity or whether the mutation affects the sorting of the enzyme to the regulated secretory pathway rather than its catalytic function. The site of the mutation is two amino acids from the end of the middle domain of PC1 and might either affect the contribution of the middle domain to catalytic activity or the inhibitory function of the COOH-terminus as discussed above. The particular Gly residue is conserved in all seven members of the eukaryote subtilisin family, and in all species so far studied, thus arguing for a critical structural function. Experiments with the PC1 Gly>Arg^{483} mutant in CHO cells suggest that the protein is synthesized normally and translocated into the lumen of the endoplasmic reticulum, but is retained there and not secreted (Jackson et al., 1997). It will be important to examine the activity and sorting of the mutant in a cell type with a regulated secretory pathway before drawing any firm conclusions about its biological function.

A PC2 variant, which lacks a substantial component of the prodomain and several NH_2-terminal amino acids of the catalytic domain, has recently been generated by a knock-out strategy in mice (Furuta et al., 1997). The mutant protein is synthesized in islets; however, it does not undergo post-translational processing and is not secreted. Homozygotes appear normal at birth, but grow slowly, exhibit chronic fasting hypoglycemia and a reduced rise in blood glucose levels during an intraperitoneal glucose tolerance test; features all of which are consistent with a deficiency of circulating glucagon. The processing of proglucagon, prosomatostatin, and proinsulin are severely impaired and their pancreatic islets show marked hyperplasia of α- and δ-cells and a relative diminution of β-cells. The phenotype is strikingly different from that of the *fat/fat* mouse and human PC1 mutant in not having diabetes or being associated with obesity. As in the case of PC1 Gly>Arg^{483}, there is some question of whether the mutant PC2 might have any biological activity. Nevertheless, it illustrates the importance of PC2 in the production of glucagon in the α-cell and the fact that, at least in rodents, proinsulin conversion can largely be achieved with PC1 alone.

The consequences of the loss of PC1, PC2, or CPH function on the whole organism are less than might be predicted from in vitro studies using cultured cells and isolated enzymes (Rouille et al., 1995). This fact may be due to compensation by negative feedback regulation in the whole organism and redundancy of enzyme function, such that the loss of one convertase does not cause complete loss of function (Fricker et al., 1996; Rothenberg et al., 1996). A practical diagnostic aspect of this finding is that, to investigate a putative case of convertase deficiency, it is wise to measure not only the concentration of product but also its precursor. PC1, PC2, and CPH deficiency grossly elevate proinsulin concentration and PC1 defects also increase POMC levels. A screen of non-Keleman syndrome cases of hypogonadotrophic hypogonadism has failed to reveal any further cases of proinsulinemia associated with infertility, and it is thus concluded that the indivdual with PC1 deficiency was a rare case. Likewise, no links have been found between a series of common PC2 gene polymorphisms with diabetes (Yoshida et al., 1995), which is again indicative that defects in these genes do not contribute greatly to the common multigene disorders of type I and type II diabetes. Nevertheless, the fact that gene ablation of the processing enzymes of the regulated pathway produces relatively mild phenotypes gives great encouragement for future studies in the area. It indicates that tissue specific antisense or gene knock-in technologies might be applied to answer questions regarding the in vivo as opposed to in vitro substrate specificities of these enzymes and to probe further their regulation, which is both biochemical and cellular in origin. Such models will be essential to follow up the hypothesis that PC1 plays an important role in the production of neuropeptides in the hypothalamus and gut that mediate satiety, since it is impossible at the moment to fully investigate the changes in hypothalamic neuropeptides noninvasively.

VII. PROINSULINEMIA OF NONGENETIC ORIGIN

Circulating "proinsulin" (proinsulin and the intermediates des 31,32 proinsulin and des 64,65 proinsulin) concentrations have been determined in a number of common clinical situations, and increases have been observed in association with the prediabetic phase of type I diabetes, type II diabetes, impaired glucose tolerance, insulin resistance syndromes, hepatic cirrhosis, pancreatitis cardiovascular disorders and neoplasia (Heaton et al., 1987; Ward et al., 1987; Temple et al., 1989; Bennett and Hutton, 1994; Rhodes and Alarcon, 1994; Gadot et al., 1995).

In the case of the endocrine pancreatic tumors, circulating levels one or more orders of magnitude greater than that in normal individuals are seen (Rubenstein et al., 1972). Moreover, the higher the ratio of (pro)insulin to intact insulin, the worse the prognosis in terms of the metastatic potential of the tumor. The proinsulin levels in this instance often do not respond to raising the blood glucose level, which could be a result of desensitization of the cell to the stimulus or could be due to the release of the prohormone by a constitutive mechanism. In the former case, release can be invoked by oral hypoglycemic agents such as tolbutamide. There are other tumors in which glucose can induce proinsulin secretion, suggesting some form of processing defect. In most cases, tumor resection is accompanied by a reduction of the level to the normal range, and the follow-up measurement of proinsulin provides a useful prognostic index.

The changes that occur in circulating proinsulin in the case of diabetes or prediabetes are far less dramatic than that associated with pancreatic tumors and for the most part appear to be attributable to the regulated release of proportionately more proinsulin and the intermediate des 31,32 proinsulin relative to insulin. The two- to threefold increase in circulating levels commonly observed corresponds to a difference in the secretion of approximately 95% processed material from the pancreas versus 98% in the normal individual. Changes of a similar magnitude can be produced in experimental animals by partial pancreatectomy or by chronic infusion of glucose, suggesting that the most probable origin of these changes is the decrease in functional β-cell mass and the proportionately greater release of immature secretory granules from the β-cell. The degree of functional impairment in an individual classified as having impaired glucose tolerance corresponds to the equivalent of a 50% pancreatectomy in such a situation. The possibility of a more subtle defect in either prohormone conversion or in pathways of sorting proteins to the secretory granule, however, should not be discounted. Unfortunately, the changes seen in the human are not readily reproduced in experimental models of type II diabetes. The GK rat, for example, which exhibits a mild form of type II diabetes characterized by impairment in the secretory responsiveness of islets to glucose but without insulin resistance or obesity, does not show marked changes in the circulating level of proinsulin (J.C. Hutton, C-G. Ostenson, and D. Gross, unpublished findings). No marked impairment in the rates of biosynthesis in response to glucose is seen for insulin or other secretory granule proteins in isolated islets from these animals (chromogranin A, PC2, PC1, and CPH). Nevertheless, chromogranin A, PC2, and CPH did not achieve the same degree of maturation in GK rats versus controls. The question remains as to whether other components of the diabetic syndrome bring about the changes in proinsulin seen in type II diabetic patients. The changes in circulating lipids, hypertension, and the intra-islet amyloidosis that occur in the human disease are not prominent features of the rodent type II diabetic models and could impact upon the prohormone processing defect seen in humans (Mohamed-Ali et al., 1995; Yudkin et al., 1997).

VIII. AMYLOID FORMATION IN THE ISLET

Type II diabetes is accompanied by the extracellular deposition of the amyloidogenic peptide IAPP (amylin) (Betsholtz et al., 1989), a process that occurs over a long period preceding the clinical emergence of disease (Clark et al., 1995). There is some evidence of lysosomal accumulation of the protein and the formation of immunoreactive amyloid fibrils in the cytoplasm, both signs of missorting or abnormal processing (de Koning et al., 1994a). IAPP has a signal sequence to direct it to the lumen of the ER and is processed at sites marked by basic amino acids by the enzyme PC2 (Badman et al., 1996) and thus behaves as a classical secretory pathway protein. Its secretion usually parallels that of insulin (Kahn et al., 1991), however, in vitro conditions have been defined in which the release of these granule constituents is dissociated. Whether this occurs by an alternative vesicular pathway or by a change in the "sorting by retention" model proposed by Arvan and colleagues is unclear (Kuliawat and Arvan, 1994). Mice with a high copy number of human IAPP transgenes do not typically develop diabetes or accumulate amyloid in their islets. Amyloid formation may, however, be induced in vitro in islets by prolonged culture in high glucose or in vivo by administration of high-fat diets, which suggests that the rate of biosynthesis of the protein can be regulated or that its colligative properties can be modulated by intracelular or extracellular factors (D'Alessio et al., 1994; de Koning et al., 1994b). The finding that amyloid deposits purified from the pancreas of type II diabetic individuals and those found in pancreatic tumors correspond to the processed form of the molecule argues that it is a substrate of normal endoproteolytic cleavage but does not indicate whether the rate of processing or intracellular site of processing is defective.

IX. CONCLUSION

Our understanding of the enzymology of proinsulin processing has expanded dramatically in the past decade following a search for almost 25 years for the enzymes responsible for catalyzing these reactions. Some of the fundamentals of the regulation of the expression of these proteins have been elucidated, both in the case of proinsulin conversion and for other prohormones that may undergo more complex differential processing. The recent lessons learned from null mutations in PC1, PC2, and CPH have been somewhat sobering in that they serve to illustrate that many of the in vitro findings cannot be readily extrapolated to the physiological state or that there is further redundancy that remains to be accounted for. Our understanding of the molecular events that are responsible for the sorting of these enzymes and other cargo proteins to the regulated secretory pathway has likewise increased dramatically thanks to genetic models in yeast and ciliates and to development of cell-free models of vesicular trafficking and secretory granule biogenesis. We remain largely ignorant about the molecular machinery that drives the formation of the insulin granules and regulates exocytosis and endocytic events coupled to exocytosis though some clues are emerging from the study of neurosecretory vesicles (Mizuta et al., 1994; Regazzi et al., 1995 and next chapter; Südhof, 1995; Regazzi et al., 1996). It is only in the last two years that granule-specific transmembrane proteins have been identified which might be used to track the movement of the membrane in the cell and determine the cytoplasmic and cytoskeletal elements with which the granule interacts (Solimena et al., 1996; Wasmeier and Hutton, 1996).

ACKNOWLEDGMENTS

The work from the authors' laboratories was supported by the Juvenile Diabetes Foundation International, Wellcome Trust, British Diabetic Association, and Fonds voor Wetenschappelijk Onderzoek Vlaanderen. Carrie John is thanked for help in preparing the manuscript.

REFERENCES

Alarcon, C., Lincoln, B. & Rhodes, C. J. (1993). The biosynthesis of the subtilisin-related proprotein convertase PC3, butno that of the PC2 convertase, is regulated by glucose in parallel to proinsulin biosynthesis in rat pancreatic islets. J. Biol. Chem. 268, 4276–4280.

Arvan, P., Kuliawat, R., Prabakaran, D., Zavacki, A. M., Elahi, D., Wang, S. & Pilkey, D. (1991). Protein discharge from immature secretory granules displays both regulated and constitutive characteristics. J. Biol. Chem. 266, 14171–14174.

Austin, C. D. & Shields, D. (1996). Formation of nascent secretory vesicles from the trans-Golgi network of endocrine cells is inhibited by tyrosine kinase and phosphatase inhibitors. J. Cell Biol. 135, 1471–1483.

Azaryan, A. V., Wong, M., Friedman, T. C., Cawley, N. X., Estivariz, F. E., Chen, H. C. & Loh, Y. P. (1993). Purification and characterization of a paired basic residue-specific yeast aspartic protease encoded by the YAP3 gene. Similarity to the mammalian pro-opiomelanocortin-converting enzyme. J. Biol. Chem. 268, 11968–11975.

Badman, M. K., Shennan, K. I., Jermany, J. L., Docherty, K. & Clark, A. (1996). Processing of pro-islet amyloid polypeptide (proIAPP) by the prohormone convertase PC2. FEBS Lett. 378, 227–31.

Bailyes, E. M. & Hutton, J. C. (1992). Kinetic analysis of the type 1 proinsulin endopeptidase using a monoclonal antibody-based immunoabsorbent assay. Biochem. J. 286, 223–229.

Bailyes, E. M., Shennan, K. I. J., Smeekens, S. P., Steiner, D. F., Hutton, J. C. & Docherty, K. D. (1992). A member of the eukaryotic subtilisin family (PC3) has the enzymic properties of the type 1 proinsulin-converting endopeptidase. Biochem. J. 285, 391–394.

Bailyes, E. M., Shennan, K. I. J., Usac, E. F., Arden, S. D., Gues, P. C., Docherty, K. & Hutton, J. C. (1995). Differences between the catalytic properties of recombinant human PC2 and endogenous rat PC2. Biochem. J. 309, 587–594.

Bennett, D. & Hutton, J. C. (1994). Proinsulin-processing endopeptidases and diabetes. In: *Molecular Biology of Diabetes, Autoimmunity and Genetics; Insulin Synthesis and Secretion.* (Draznin, B. & Leroith, D., eds.), Vol. 1, pp. 213–248. Humana Press, New York.

Berman, Y. L., Juliano, L. & Devi, L. A. (1995). Purification and characterization of a dynorphin-processing endopeptidase. J Biol Chem. 270, 23845–50.

Betsholtz, C., Svensson, V., Rorsman, F., Engstrom, U., Westermark, G. T., Wilander, E., Johnson, K. & Westermark, P. (1989). Islet amyloid polypeptide (IAPP):cDNA cloning and identification of an amyloidogenic region associated with the species-specific occurrence of age-related diabetes mellitus. Exp Cell Res. 183, 484–93.

Bomsel, M. & Mostov, K. (1992). Role of heterotrimeric G proteins in membrane traffic. Mol. Biol. Cell. 3, 1317–1328.

Bosshart, H., Humphrey, J., Deignan, E., Davidson, J., Drazba, J., Yuan, L., Oorschot, V., Peters, P. J. & Bonifacino, J. S. (1994). The cytoplasmic domain mediates localization of furin to the *trans*-Golgi Network en route to the endosomal/lysosomal system. J. Cell Biol. 126, 1157–1172.

Braks, J. A. & Martens, G. J. (1994). 7B2 is a neuroendocrine chaperone that transiently interacts with prohormone convertase PC2 in the secretory pathway. Cell. 78, 263–73.

Cain, B. M., Wang, W. & Beinfeld, M. C. (1997). Cholecystokinin (CCK) levels are greatly reduced in the brains but not the duodenums of Cpe(fat)/Cpe(fat) mice: a regional difference in the involvement of carboxypeptidase E (Cpe) in pro-CCK processing. Endocrinology. 138, 4034–7.

Chan, S. J., Seino, S., Gruppuso, P. A., Schwartz, R. & Steiner, D. F. (1987). A mutation in the B-chain coding region is associated with impaired proinsulin conversion in a family with hyperproinsulinaemia. Proc. Natnl. Acad. Sci. USA. 84, 2194–2197.

Chanat, E. & Huttner, W. B. (1991). Milieu-induced, selective aggregation of regulated secretory proteins in the trans-Golgi network. J. Cell Biol. 115, 1505–1519.

Clark, A., de Koning, E. J., Hattersley, A. T., Hansen, B. C., Yajnik, C. S. & Poulton, J. (1995). Pancreatic pathology in non-insulin dependent diabetes (NIDDM). Diabetes Res Clin Pract. 28, S39–47.

Colomer, V., Kicska, G. A. & Rindler, M. J. (1996). Secretory granule content proteins and the luminal domains of granule membrane proteins aggregate *in vitro* at mildly acidic pH. J. Biol. Chem. 271, 48–55.

Cool, D. R., Fenger, M., Snell, C. R. & Loh, Y. P. (1995). Identification of the sorting signal motif within proopiomelanocortin for the regulated secretory pathway. J. Biol. Chem. 270, 8723–8729.

Cool, D. R., Normant, E., Shen, F., Chen, H. C., Pannell, L., Zhang, Y. & Loh, Y. P. (1997). Carboxypeptidase E is a regulated secretory pathway sorting receptor: genetic obliteration leads to endocrine disorders in Cpe(fat) mice. Cell. 88, 73–83.

Creemers, J. W., Usac, E. F., Bright, N. A., Van de Loo, J. W., Jansen, E., Van de Ven, W. J. M. & Hutton, J. C. (1996). Identification of a transferable sorting domain for the regulated pathway in the prohormone convertase PC2. J Biol Chem. 271, 25284–91.

D'Alessio, D. A., Verchere, C. B., Kahn, S. E., Hoagland, V., Baskin, D. G., Palmiter, R. D. & Ensinck, J. W. (1994). Pancreatic expression and secretion of human islet amyloid polypeptide in a transgenic mouse. Diabetes. 43, 1457–61.

Davidson, H. W., Rhodes, C. J. & Hutton, J. C. (1988). Intraorganellar Ca and pH control proinsulin cleavage in the pancreatic β cell via two distint site-specific endopeptidases. Nature. 333, 93–96.

de Koning, E. J., Hoppener, J. W., Verbeek, J. S., Oosterwijk, C., van Hulst, K. L., Baker, C. A., Lips, C. J., Morris, J. F. & Clark, A. (1994a). Human islet amyloid polypeptide accumulates at similar sites in islets of transgenic mice and humans. Diabetes. 43, 640–4.

de Koning, E. J., Morris, E. R., Hofhuis, F. M., Posthuma, G., Hoppener, J. W., Morris, J. F., Capel, P. J., Clark, A. & Verbeek, J. S. (1994b). Intra- and extracellular amyloid fibrils are formed in cultured pancreatic islets of transgenic mice expressing human islet amyloid polypeptide. Proc Natl Acad Sci U S A. 91, 8467–71.

Disdier, M., Morrissey, J. H., Fugate, R. D., Bainton, D. F. & McEver, R. P. (1992). Cytoplasmic domain of P-selectin (CD62) contains the signal for sorting into the regulated secretory pathway. Mol. Biol. Cell. 3, 309–321.

Dittié, A. S., Hajibagheri, N. & Tooze, S. A. (1996). The AP-1 adaptor complex binds to immature secretory granules from PC12 cells, and is regulated by ADP-ribosylation factor. J. Cell Biol. 132, 523–536.

Docherty, K. D., Rhodes, C. J., Shennan, K. I. J., Taylor, N. A. & Hutton, J. C. (1989). Proinsulin endopeptidase substrate specificities defined by site-directed mutagenesis of proinsulin. J. Biol. Chem. 264, 18335–18339.

Donaldson, J. G. & Klausner, R. D. (1994). ARF: a key regulatory switch in membrane traffic and organelle structure. Curr. Opin. Cell Biol. 6, 527–532.

Farhadi, H., Pareek, S., Day, R., Dong, W., Chretien, M., Bergeron, J. J., Seidah, N. G. & Murphy, R. A. (1997). Prohormone convertases in mouse submandibular gland: Co-localization of furin and nerve growth factor. J. Histochem. Cytochem. 45, 795–804.

Fricker, L. D., Berman, Y. L., Leiter, E. H. & Devi, L. A. (1996). Carboxypeptidase E activity is deficient in mice with the fat mutation. Effect on peptide processing. J. Biol. Chem. 271, 30619–24.

Furuta, M., Yano, H., Zhou, A., Rouille, Y., Holst, J. J., Carroll, R., Ravazzola, M., Orci, L., Furuta, H. & Steiner, D. F. (1997). Defective prohormone processing and altered pancreatic islet morphology in mice lacking active Spc2. Proc. Natl. Acad. of Sci. USA 94, 6646–6651.

Gadot, M., Ariav, Y., Cerasi, E., Kaiser, N. & Gross, D. J. (1995). Hyperproinsulinemia in the diabetic Psammomys obesus is a result of increased secretory demand on the beta-cell. Endocrinology. 136, 4218–4223.

Goodman, L. J. & Gorman, C. M. (1994). Autoproteolytic activation of the mouse prohormone convertase mPC1. Biochem. Biophys. Res. Comm. 201, 795–804.

Grimes, M. & Kelly, R. B. (1992). Intermediates in the constitutive and regulated secretory pathways released in vitro from semi-intact cells. J. Cell Biol. 117, 539–549.

Guest, P. C., Arden, S. D., Bennett, D. L., Clark, A., Rutherford, N. G. & Hutton, J. C. (1992). The post-translational processing and intracellular sorting of PC2 in the islets of Langerhans. J. Biol. Chem. 267, 22401–22406.

Guest, P. C., Arden, S. D., Rutherford, N. G. & Hutton, J. C. (1995). The post-translational processing and intracellular sorting of carboxypeptidase H in the islets of Langerhans. Mol. Cell. Endocrinol. 113, 99–108.

Guest, P. C., Bailyes, E. M. & Hutton, J. C. (1997). Endoplasmic reticulum Ca2+ is important for the proteolytic processing and intracellular transport of proinsulin in the pancreatic beta-cell. Biochem J. 323, 445–450.

Guest, P. C., Pipeleers, D., Rossier, J., Rhodes, C. J. & Hutton, J. C. (1989a). Co-secretion of carboxypeptidase H and insulin from isolated rat islets of Langerhans. Biochem. J. 264, 503–508.

Guest, P. C., Rhodes, C. J. & Hutton, J. C. (1989b). Regulation of the biosynthesis of insulin-secretory-granule proteins: Co-ordinate translational control is exerted on some, but not all, granule matrix constituents. Biochem. J. 257, 431–437.

Halban, P. A. & Irminger, J.-C. (1994). Sorting and processing of secretory proteins. Biochem. J. 299, 1–18.

Heaton, D. A., Millward, B. A., Gray, P., Tun, Y., Hales, C. N., Pyke, D. A. & Leslie, R. D. G. (1987). Evidence of β cell dysfunction which does not lead to diabetes: a study of identical twins of insulin dependent diabetes. Brit. Med. J. 294, 145–146.

Hill, R. M., Ledgerwood, E. C., Brennan, S. O., Pu, L. P., Loh, Y. P., Christie, D. L. & Birch, N. P. (1995). Comparison of the molecular forms of the Kex2/subtilisin-like serine proteases SPC2, SPC3, and furin in neuroendocrine secretory vesicles reveals differences in carboxyl-terminus truncation and membrane association. Journal of Neurochemistry. 65, 2318–26.

Hoflehner, J., Eder, U., Laslop, A., Seidah, N. G., Fischer-Colbrie, R. & Winkler, H. (1995). Processing of secretogranin II by prohormone convertases: importance of PC1 in generation of secretoneurin. FEBS Lett. 360, 294–298.

Hook, V. Y., Moran, K., Kannan, R., Kohn, A., Lively, M. O., Azaryan, A., Schiller, M. & Miller, K. (1997). High-level expression of the prohormones proenkephalin, pro-neuropeptide Y, proopiomelanocortin, and beta-protachykinin for in vitro prohormone processing. Protein Expr Purif. 10, 80–8.

Horssen, A. M. V., Hurk, W. H. V. d., Bailyes, E. M., Hutton, J. C., Martens, G. J. M. & Lindberg, I. (1995). Identification of the region within the neuroendocrine polypeptide 7B2 responsible for the inhibition of prohormone convertase PC2. J. Biol. Chem. 270, 14292–14296.

Hutton, J. C. (1990). Subtilisin-like proteinases involved in the activation of proproteins of the eukaryotic secretory pathway. Current Opinion in Cell Biology. 2, 1131–1142.

Hutton, J. C., Nielsen, E. & Kastern, W. (1988). Molecular cloning of the chromogranin A-like precursor to β-granin and pancreastatin from the endocrine pancreas. FEBS Lett. 236, 269–274.

Itoh, Y., Tanaka, S., Takekoshi, S., Itoh, J. & Osamura, R. Y. (1996). Prohormone convertases (PC1/3 and PC2) in rat and human pancreas and islet cell tumors: subcellular immunohistochemical analysis. Pathol Int. 46, 726–37.

Jackson, R. S., Creemers, J. W., Ohagi, S., Raffin-Sanson, M. L., Sanders, L., Montague, C. T., Hutton, J. C. & O'Rahilly, S. (1997). Obesity and impaired prohormone processing associated with mutations in the human prohormone convertase 1 gene [see comments]. Nat Genet. 16, 303–6.

Jutras, I., Seidah, N. G., Reudelhuber, T. L. & Brechler, V. (1997). Two activation states of the prohormone convertase PC1 in the secretory pathway. Journal of Biological Chemistry. 272, 15184–8.

Kahn, S. E., Fujimoto, W. Y., D'Alessio, D. A., Ensinck, J. W. & Porte, D., Jr. (1991). Glucose stimulates and potentiates islet amyloid polypeptide secretion by the B-cell. Horm Metab Res. 23, 577–80.

Koedam, J. A., Kramer, E. M., Briend, E., Furie, B., Furie, B. C. & Wagner, D. D. (1992). P-selectin, a granule membrane protein of platelets and endothelial cells, follows the regulated pathway in AtT20 cells. J. Cell Biol. 116, 617–625.

Kornfeld, S. & Mellman, I. (1989). The biogenesis of lysosomes. Ann. Rev. Cell Biol. 5, 483–525.

Kuliawat, R. & Arvan, P. (1992). Protein targeting via the 'constitutive-like' secretory pathway in isolated pancreatic islets: Passive sorting in the immature granule compartment. J. Cell Biol. 118, 521–529.

Kuliawat, R. & Arvan, P. (1994). Distinct molecular mechanisms for protein sorting within immature secretory granules of pancreatic beta-cells. J. Cell Biol. 126, 77–86.

Ladinsky, M. S., Kremer, J. R., Furcinitti, P. S., McIntosh, J. R. & Howell, K. E. (1994). HVEM tomography of the trans-Golgi network: structural insights and identification of a lace-like vesicle coat. J. Cell Biol. 127, 29–38.

Lan, M. S., Lu, J., Goto, Y. & Notkins, A. J. (1994). Molecular cloning and identification of a receptor-type protein tyrosine phosphatase, IA-2, from human insulinoma. DNA Cell Biol. 13, 505–514.

Leyte, A., Barr, F. A., Kehlenbach, R. H. & Huttner, W. B. (1992). Multiple trimeric G-proteins on the trans-Golgi network excert stimulatory and inhibitory effects on secretory vesicle formation. EMBO J. 11, 4795–4804.

Lindberg, I., Ahn, S. C. & Breslin, M. B. (1994). Cellular distributions of the prohormone processing enzymes PC1 and PC2. Mol. Cell. Neurosci. 5, 614–622.

Malide, D., Seidah, N. G., Chrétien, M. & Bendayan, M. (1995). Electron microscopic immunocytochemical evidence for the involvement of the convertases PC1 and PC2 in the processing of proinsulin in pancreatic β-cells. Histochem. Cytochem. 43, 11–19.

Martens, G. J., Braks, J. A., Eib, D. W., Zhou, Y. & Lindberg, I. (1994). The neuroendocrine polypeptide 7B2 is an endogenous inhibitor of prohormone convertase PC2. Proceedings of the National Academy of Sciences of the United States of America. 91, 5784–7.

Matthews, G., Shennan, K. I., Seal, A. J., Taylor, N. A., Colman, A. & Docherty, K. (1994). Autocatalytic maturation of the prohormone convertase PC2. Journal of Biological Chemistry. 269, 588–92.

Mitra, A., Song, L. & Fricker, L. D. (1994). The C-terminal region of carboxypeptidase E is involved in membrane binding and intracellular routing in AtT-20 cells. J. Biol. Chem. 269, 19876–19881.

Mizuta, M., Inagaki, N., Nemoto, Y., Matsukura, S., Takahashi, M. & Seino, S. (1994). Synaptotagmin III is a novel isoform of rat synaptotagmin expressed in endocrine and neuronal cells. J. Biol. Chem. 269, 11675–11678.

Mohamed-Ali, V., Gould, M. M., Gillies, S., Goubet, S., Yudkin, J. S. & Haines, A. P. (1995). Association of proinsulin-like molecules with lipids and fibrinogen in non-diabetic subjects—evidence against a modulating role for insulin. Diabetologia. 38, 1110–6.

Molloy, S. S., Thomas, L., Slyke, J. K. V., Sternberg, P. E. & Thomas, G. (1994). Intracellular traffiking and activation of the furin proprotein convertase: localisation to the TGN and recycling from the cell surface. EMBO J. 13, 18–33.

Moore, H.-P. H. & Kelly, R. B. (1985). Secretory protein targeting in a pituitary cell line: Differential transport of foreign secretory proteins to distinct secretory pathways. J. Cell Biol. 101, 1773–1781.

Moore, H.-P. H. & Kelly, R. B. (1986). Re-routing of a secretory protein by fusion with human growth hormone sequences. Nature. 321, 443–446.

Moore, H.-P. H., Walker, M. D., Lee, F. & Kelly, R. B. (1983). Expressing a human proinsulin cDNA in a mouse ACTH-secreting cell. Intracellular storage, proteolytic processing, and secretion on stimulation. Cell. 35, 531–538.

Munro, S. & Pelham, H. B. R. (1987). A C-terminal signal prevents secretion of luminal ER proteins. Cell. 48, 899–907.

Naggert, J. K., Fricker, L. D., Varlamov, O., Nishina, P. M., Rouille, Y., Steiner, D. F., Carroll, R. J., Paigen, B. J. & Leiter, E. H. (1995). Hyperproinsulinaemia in obese fat/fat mice associated with a carboxypeptidase E mutation which reduces enzyme activity. Nature Genetics. 10, 135–42.

Nillni, E. A., Luo, L. G., Jackson, I. M. & McMillan, P. (1996). Identification of the thyrotropin-releasing hormone precursor, its processing products, and its coexpression with convertase 1 in primary cultures of hypothalamic neurons: anatomic distribution of PC1 and PC2. Endocrinology. 137, 5651–61.

Norcott, J. P., Solari, R. & Cutler, D. F. (1996). Targeting of P-selectin to two regulated secretory organelles in PC12 cells. J. Cell Biol. 134, 1229–1240.

O'Rahilly, S., Gray, H., Humphreys, P. J., Krook, A., Polonsky, K. S., White, A., Gibson, S., Taylor, K. & Carr, C. (1995). Brief report: impaired processing of prohormones associated with abnormalities of glucose homeostasis and adrenal function. New England Journal of Medicine. 333, 1386–90.

Ohashi, M., deVries, K. J., Frank, R., Snoek, G., Bankaitis, V., Wirtz, K. & Huttner, W. B. (1995). A role for phosphatidylinositol transfer protein in secretory vesicle formation. Nature. 377, 544–547.

Ohashi, M. & Huttner, W. B. (1994). An elevation of cytosolic protein phosphorylation modulates trimeric G-protein regulation of secretory vesicle formation from the trans-Golgi network. J. Biol. Chem. 269, 24897–24905.

Orci, L. (1985). The insulin factory: A tour of the plant surroundings and a visit to the assembly line. Diabetologia. 28, 528–546.

Orci, L., Ravazzola, M., Amherdt, M., Perrelet, A., Powell, S. K., Quinn, D. L. & Moore, H.-P. H. (1987). The trans-most cisternae of the Golgi complex: A compartment for sorting of secretory and plasma membrane proteins. Cell. 51, 1039–1051.

Pearse, B. M. F. & Robinson, M. S. (1990). Clathrin, adaptors and sorting. Ann. Rev. Cell Biol. 6, 151–171.

Regazzi, R., Ravazzola, M., Iezzi, M., Lang, J., Zahraoui, A., Andereggen, E., Morel, P., Takai, Y. & Wollheim, C. B. (1996). Expression, localisation and functional role of small GTPases of the Rab3 family in insulin-secreting cells. J. Cell Sci. 109, 2265–2273.

Regazzi, R., Wollheim, C. B., Lang, J., Theler, J. M., Rossetto, O., Montecucco, C., Sadoul, K., Weller, U., Palmer, M. & Thorens, B. (1995). VAMP 2 and cellubrevin are expressed in pancreatic beta-cells and are essential for Ca2+—but not for GTPgammaS-induced insulin secretion. EMBO J. 14, 2723–2730.

Rhodes, C. J. & Alarcon, C. (1994). What beta-cell defect could lead to hyperproinsulinemia in NIDDM? Some clues from recent advances made in understanding the proinsulin-processing. Diabetes. 43, 511–517.

Rhodes, C. J., Thorne, B. A., Lincoln, B., Thomas, G. & Hutton, J. C. (1993). Processing of proopiomelanocortin by the insulin secretory granule proinsulin processing endopeptidases. J. Biol. Chem. 268, 4267–4275.

Rosa, P., Barr, F. A., Stinchcombe, J. C., Binacchi, C. & Huttner, W. B. (1992). Brefeldin A inhibits the formation of constitutive secretory vesicles and immature secretory granules from the trans-Golgi network. Eur. J. Cell Biol. 59, 265–274.

Rothenberg, M. E., Eilertson, C. D., Klein, K., Mackin, R. B. & Noe, B. D. (1996). Evidence for redundancy in propeptide/prohormone convertase activities in processing proglucagon: an antisense study. Molecular Endocrinology. 10, 331–41.

Rouille, Y., Duguay, S. J., Lund, K., Furuta, M., Gong, Q., Lipkind, G., Oliva, A. A., Jr., Chan, S. J. & Steiner, D. F. (1995). Proteolytic processing mechanisms in the biosynthesis of neuroendocrine peptides: the subtilisin-like proprotein convertases. Frontiers in Neuroendocrinology. 16, 322–61.

Rubenstein, A. H., Block, M. B., Starr, J., Melani, F. & Steiner, D. F. (1972). Proinsulin and C-peptide in blood. Diabetes. 21, 661–672.

Schäfer, W., Stroh, A., Berghöfer, S., Seiler, J., Vey, M., Kruse, M.-L., Kern, H. F., Klenk, H.-D. & Garten, W. (1995). Two independent targeting signals in the cytoplasmic domain determine *trans-*

Golgi network localization and endosomal trafficking of the proprotein convertase furin. EMBO J. 14, 2424–2435.

Seidah, N. G., Chrétien, M. & Day, R. (1994). The family of subtilisin/kexin like pro-protein and pro-hormone convertases: Divergent or shared functions. Biochemie. 76, 197–209.

Shennan, K. I., Taylor, N. A. & Docherty, K. (1994). Calcium- and pH-dependent aggregation and membrane association of the precursor of the prohormone convertase PC2. Journal of Biological Chemistry. 269, 18646–50.

Shennan, K. I. J., Taylor, N. A., Jermany, J. L., Matthews, G. & Docherty, K. (1995). Differences in pH optima and calcium requirements for maturation of the prohormone convertases PC2 and PC3 indicates different intracellular locations for these events. J. Biol. Chem. 270, 1402–1497.

Shepherd, P. R., Reaves, B. J. & Davidson, H. W. (1996). Phosphoinositide 3-kinases and membrane traffic. Trends Cell Biol. 6, 92–97.

Shibasaki, Y., Kawakami, T., Kanazawa, Y., Akanuma, Y. & Takaku, F. (1985). Posttranslational cleavage of proinsulin is blocked by a point mutation in familial hyperproinsulinaemia. J. Clin. Invest. 89, 1902–1907.

Smeekens, S. P., Avruch, A. S., LaMendola, J., Chan, S. J. & Steiner, D. F. (1991). Identification of a cDNA encoding a second putative prohormone convertase related to PC2 in AtT-20 cells and islets of Langerhans. Proc. Natnl. Acad. Sci. USA. 88, 340–344.

Solimena, M., Dirkx, R., Hermel, J.-M., Pleasic-Williams, S., Shapiro, J. A., Caron, L. & Rabin, D. U. (1996). ICA 512, an autoantigen of type I diabetes, is an intrinsic membrane protein of neurosecretory granules. EMBO J. 9, 2102–2114.

Song, L. & Fricker, L. D. (1995). Purification and characterization of carboxypeptidase D, a novel carboxypeptidase E-like enzyme, from bovine pituitary. J. Biol. Chem. 270, 25007–25013.

Steiner, D. F., Smeekens, S. P., Ohagi, S. & Chan, S. J. (1992). The new enzymology of precursor processing endoproteases. J. Biol. Chem. 276, 23435–23438.

Stoller, T. J. & Shields, D. (1988). Retrovirus-mediated expression of preprosomatostatin: Posttranslational processing, intracellular storage, and secretion in GH3 pituitary cells. J. Cell Biol. 107, 2087–2095.

Stoller, T. J. & Shields, D. (1989). The propeptide of preprosomatostatin mediates intracellular transport and secretion of alpha-globin from mammalian cells. J. Cell Biol. 108, 1647–1655.

Südhof, T. C. (1995). The synaptic vesicle cycle: a cascade of protein-protein interactions. Nature. 375, 645–653.

Tam, W. W. H., Andreasson, K. I. & Loh, Y. P. (1993). The amino-terminal sequence of pro-opiomelanocortin directs intracellular targeting to the regulated secretory pathway. Eur. J. Cell Biol. 62, 294–306.

Temple, R. C., Carrington, C. A., Luzio, S. D., Owens, D. R., Schneider, A. E., Sobey, W. J. & Hales, C. N. (1989). Insulin deficiency in non-insulin dependent diabetes. Lancet. 1, 293–295.

Tooze, J. & Tooze, S. A. (1986). Clathrin-coated vesicular transport of secretory proteins during the formation of ACTH-containing secretory granules in AtT20 cells. J. Cell Biol. 103, 839–850.

Tooze, S. A. & Huttner, W. B. (1990). Cell-free protein sorting to the regulated and constitutive secretory pathways. Cell. 60, 837–847.

Tooze, S. A., Weiss, U. & Huttner, W. B. (1990). Requirement for GTP hydrolysis in the formation of secretory vesicles. Nature. 347, 207–208.

Varlamov, O. & Fricker, L. D. (1996). The C-terminal region of carboxypeptidase E involved in membrane binding is distinct from the region involved with intracellular sorting. J. Biol. Chem. 271, 6077–6083.

Ward, W. K., LaCava, E. C., Paquette, T. L., Beard, J. C., Wallum, B. J. & Porte, D., Jr. (1987). Disproportionate elevation of immunoreactive proinsulin in type 2 (non-insulin dependent) diabetes mellitus and in experimental insulin resistance. Diabetologia. 30, 698–702.

Wasmeier, C. & Hutton, J. C. (1996). Molecular cloning of phogrin, a protein tyrosine phosphatase homolgue localised to insulin secretory granule membranes. J. Biol. Chem. 271, 18161–18170.

Wu, P. & Jackson, I. M. (1988). Identification, characterization and localization of thyrotropin-releasing hormone precursor peptides in perinatal rat pancreas. Regul Pept. 22, 347–60.

Yano, H., Kitano, N., Morimoto, M., Polonsky, K. S., Imura, H. & Seino, Y. (1987). A novel point mutation in the human insulin gene giving rise to hyperproinsulinaemia. Proc. Natnl. Acad. Sci. USA. 84, 2194–2197.

Yoshida, H., Ohagi, S., Sanke, T., Furuta, H., Furuta, M. & Nanjo, K. (1995). Association of the prohormone convertase 2 gene (PCSK2) on chromosome 20 with NIDDM in Japanese subjects. Diabetes. 44, 389–93.

Yudkin, J. S., Denver, A. E., Mohamed-Ali, V., Ramaiya, K. L., Nagi, D. K., Goubet, S., McLarty, D. G. & Swai, A. (1997). The relationship of concentrations of insulin and proinsulin-like molecules with coronary heart disease prevalence and incidence. A study of two ethnic groups. Diabetes Care. 20, 1093–100.

Zhou, A. & Mains, R. E. (1994). Endoproteolytic processing of proopiomelanocortin and prohormone convertases 1 and 2 in neuroendocrine cells overexpressing prohormone convertases 1 or 2. J. Biol. Chem. 269, 17440–17447.

Zhou, A., Paquet, L. & Mains, R. E. (1995a). Structural elements that direct specific processing of different mammalian subtilisin-like prohormone convertases. J. Biol. Chem. 270, 21509–16.

Zhou, Y., Rovere, C., Kitabgi, P. & Lindberg, I. (1995b). Mutational analysis of PC1 (SPC3) in PC12 cells. 66-kDa PC1 is fully functional. J. Biol. Chem. 270, 24702–6.

Zhu, X., Lamango, N. S. & Lindberg, I. (1996a). Involvement of a polyproline helix-like structure in the interaction of 7B2 with prohormone convertase 2. J. Biol. Chem. 271, 23582–7.

Zhu, X., Rouille, Y., Lamango, N. S., Steiner, D. F. & Lindberg, I. (1996b). Internal cleavage of the inhibitory 7B2 carboxyl-terminal peptide by PC2: a potential mechanism for its inactivation. Proc. Natl. Acad. Sci. USA 93, 4919–24.

MECHANISM OF INSULIN EXOCYTOSIS

Romano Regazzi

Advances in Molecular and Cell Biology
Volume 29, pages 151-172.

ISBN: 0-7623-0547-9

I. INTRODUCTION

Pancreatic hormones released by the cells of the islets of Langerhans play a pivotal role in the regulation of nutrient disposal and metabolism. Indeed, the fine-tuning of insulin secretion from pancreatic β-cells is an essential requirement for the achievement of blood glucose homeostasis. Insulin is stored in secretory granules and is released by exocytosis, a process modulated by fluctuations in the concentration of nutrients (in particular glucose), hormones, and neurotransmitters (Wollheim et al., 1996). Both impaired insulin release from pancreatic β-cells and resistance to the insulin action of target tissues lead to various forms of noninsulin-dependent diabetes mellitus (Type 2 diabetes) (Porte, 1991; Taylor et al., 1994; Turner et al., 1995). Although the primary defect in Type 2 diabetes is still a matter of debate, a number of studies have suggested that perturbed β-cell secretory function precedes or coincides with the onset of the disease (Porte, 1991; Taylor et al., 1994; Turner et al., 1995). For this reason, the delineation of the molecular details of the process of insulin secretion and its regulation is essential for understanding the etiology of Type 2 diabetes. Despite decades of considerable efforts, our comprehension of the mechanism involved in the control of insulin release is still incomplete. However, in recent years several components of the cellular machinery that in eukaryotes governs the fusion of secretory vesicles with their target membrane have been identified. The aim of this chapter is to summarize the current hypothesis proposed to explain the secretory process in eukaryotes and to analyze how these models apply to insulin exocytosis.

II. STAGES IN THE SECRETORY PROCESS

In eukaryotes, the proteins to be released are synthesized on ribosomes attached to the endoplasmic reticulum. To reach the cell surface, they are conveyed in vesicles that bud off and fuse with different compartments (Rothman, 1994). Insulin is not an exception, and before being packaged in mature secretory granules ready to fuse with the plasma membrane, the hormone is transported to the Golgi apparatus. The release of granule content into the extracellular space is preceded by several reactions that prepare the fusion event. On the basis of electrophysiological, biochemical, and morphological observations, the complex phenomena occurring during the exocytotic process have been subdivided into three major steps (Bittner and Holz, 1992; Hay and Martin, 1992; Parsons et al., 1995; Steyer et al. 1997). The first step, usually defined as docking, involves the correct targeting of the secretory granule to the plasma membrane. Subsequently, during the priming step, the secretory granule undergoes rearrangements that prepare it for fusion. This process is reversible and requires submicromolar concentrations of calcium and adenosine triphosphate (ATP). The secretory granules that are primed proceed eventually to the third step, fusion with the plasma membrane, causing the release of their content into the extracellular space. This step occurs in the presence of micromolar calcium concentrations and does not need ATP hydrolysis.

This chapter focusses on the protein–protein interactions and on the modifications in the lipid composition of the vesicle membrane that control the docking, the priming, and the fusion steps.

III. THE COMPONENTS OF THE EXOCYTOTIC MACHINERY IN EUKARYOTES

In eukaryotes the transport of proteins between distinct cellular compartments occurs in vesicles that bud from a donor membrane and deliver their content to a receiving compartment (Rothman, 1994). This process is essential for the maintenance of cell integrity and for several cell functions, including protein secretion. Since each cellular compartment represents a destination for a specific set of proteins, each transport vesicle needs to be accurately targeted to the appropriate receiving membrane. The identification of the molecules ensuring the delivery of the vesicle cargo to the correct cellular location represents a major task for biologists. To this end, a variety of independent lines of research have been pursued. The most fruitful approaches have been the genetic dissection of the yeast secretory pathway, the reconstitution of membrane fusion in cell-free systems, and the study of the mode of action of clostridial neurotoxins.

A. Genetic Dissection of the Yeast Secretory Pathway

One of the best known vesicular transport pathways is the route taken by a protein destined for secretion. To elucidate this complex process, in the late 1970s Schekman and collaborators decided to use a genetic approach. They took advantage of the fact that yeast, a powerful system for genetic studies, secretes enzymes and mating factors into the extracellular space. Schekman's group was able to isolate a large number of mutant strains deficient in secretion (Novick et al., 1980). Each of these strains is defective in a gene whose protein product is required for at least one step in the secretory pathway. Thus, some of the yeast mutants have their secretory products blocked in the endoplasmic reticulum or in the Golgi apparatus, and others are unable to discharge the contents of their vesicles at the cell surface. The characterization of the mutant genes (called SEC genes) has contributed to clarifying the mechanism controlling the secretory pathway in yeast. Later on, the discovery of the homology between the SEC genes and proteins identified in mammalian cells (see the discussion that follows) extended the knowledge obtained in yeast to all eukaryotic cells.

B. Reconstitution of Membrane Fusion in Cell-free Systems

To shed light on the molecular mechanisms governing the movement of transport vesicles in mammalian cells, Rothman and colleagues followed a different line of investigation. Using purified organelles, they succeeded in reconstituting the process of vesicular transport in the test tube, and could thus observe the effects of adding or subtracting individual proteins. Using this approach they first identified a cytosolic protein called *N*EM-*s*ensitive *f*actor (NSF), able to reconstitute Golgi transport in cell extracts treated with the alkylating agent *N*-*e*thyl*m*aleimide (NEM) (Wilson et al., 1989). NSF turned out to be an ATPase essential for several intracellular fusion events. Membrane association and induction of fusion by NSF was found to require the presence of accessory proteins, termed α-, β-, and γ-SNAP (for *s*oluble *N*SF *a*ttachment *p*roteins) (Whiteheart et al., 1993). Cloning and sequencing of NSF and SNAPs revealed that these proteins represent the mammalian counterparts of the SEC18 and SEC17 genes identified by geneticists as being part of the secretory machinery in yeast. These results demonstrate that components of the molecular machinery for secretion are conserved in all eukaryotic cells.

However, NSF and SNAPs are cytosolic proteins and cannot alone account for the specificity with which vesicles are targeted to the acceptor compartment. It was clear that proteins embedded in the membrane of the vesicles and/or present in the acceptor compartment should be searched for. These proteins would constitute specific markers for the different cell compartments. Brain extracts immediately appeared to be the most abundant source for membrane proteins binding SNAPs and NSF.

The search for membrane *SNAP re*ceptors (abbreviated SNAREs) culminated in the isolation of three already known proteins: syntaxin, SNAP-25 (an unfortunate acronym for *syn*aptosomal *a*ssociated *p*rotein of *25*Kda), and VAMP (*V*esicle *A*ssociated *M*embrane *P*rotein) (Söllner et al., 1993). Syntaxin and SNAP-25 are proteins associated with cell membrane and were therefore defined as t-SNAREs (*t*arget SNAREs). VAMP, also known as synaptobrevin, is associated to synaptic vesicles and was defined as v-SNARE (vesicular SNARE).

The discovery that neuronal proteins implicated in neurotransmitter release are interacting with components of the general machinery for vesicle fusion was a major breakthrough. Indeed, these findings indicate that the highly regulated process of neurotransmitter release relies on the same proteins that participate in constitutive secretion in lower eukaryotes. This conclusion was strengthened by the subsequent isolation of yeast homologues for each of the SNARE proteins (Aalto et al., 1993; Protopopov et al., 1993; Brennwald et al., 1994).

C. Mode of Action of Clostridial Neurotoxins

Evidence confirming the importance of SNAREs in secretion came from the independent finding that each of these proteins is the target of at least one neurotoxin that causes the paralytic syndromes of tetanus and of botulism (Niemann et al., 1994; Rossetto et al., 1995). These extremely potent neurotoxins (lethal dose close to 0.1 ng/Kg) are produced by the bacteria *Clostridium tetanii* and *Clostridium botulinum*. Tetanus neurotoxin (TeTx) blocks neurotransmitter release from inhibitory interneurons, provoking spastic paralysis (Niemann et al., 1994; Rossetto et al., 1995). In contrast, the seven serotypes of botulinum neurotoxins (BoNT) block acetylcholine secretion at the neuromuscular junction, causing flaccid paralysis (Niemann et al., 1994; Rossetto et al., 1995). Clostridial neurotoxins were found to be zinc-dependent endopeptidases with very narrow substrate specificity. Thus, TeTx proteolyses specifically the v-SNARE VAMP (Schiavo et al., 1992). Despite opposing clinical symptoms of tetanus and botulism, BoNT types B, D, F, and G also attack VAMP proteins (Schiavo et al., 1992; Niemann et al. 1994; Rossetto et al., 1995) . In contrast, the other BoNT attack t-SNAREs. In fact, BoNT A and E cleave SNAP-25 and BoNT C acts on syntaxin (and partially also on SNAP-25) (Niemann et al., 1994; Rossetto et al., 1995). Therefore, the discovery of the mode of action of clostridial neurotoxins provides a beautiful confirmation of the involvement of SNARE proteins in the secretory process. In addition, the specificity of these toxins combined with the possibility of artificially introducing them in nonneuronal cells has been used as a tool to dissect the molecular mechanism of exocytosis in other cell systems, including pancreatic β-cells.

IV. THE SNARE HYPOTHESIS

The extraordinary convergence of results obtained in different experimental systems led Rothman and collaborators to propose a general model to explain how secretory vesicles are targeted to the appropriate acceptor membrane (Söllner et al., 1993; Rothman and Warren, 1994) (Figure 1).

According to this proposal, known as the SNARE hypothesis, each transport vesicle is endowed with a specific v-SNARE (or set of v-SNAREs) that joins with its cognate partner, a t-SNARE present on the intended target membrane. The association of v-SNAREs with t-SNAREs forms a scaffold that permits the recruitment of several cytosolic proteins, including NSF and SNAPs. The latter proteins assemble together with v- and t-SNAREs in a complex that ultimately leads to the fusion of the vesicle with the target membrane. According to the original proposal, the ATPase activity of NSF would generate the energy necessary for membrane fusion. As expected, the SNARE hypothesis has been the subject of intense investigation. The studies published since the first presentation of the hypothesis have slightly modified the model, but most of the original assumptions have been verified experimentally. Thus, as anticipated by the model, several v-SNAREs and t-SNAREs involved in vesicular trafficking within the Golgi apparatus in mammalian cells have been identified (Nagahama et al., 1996; Subramaniam et al., 1996). Recent findings have also confirmed the role of SNAPs and

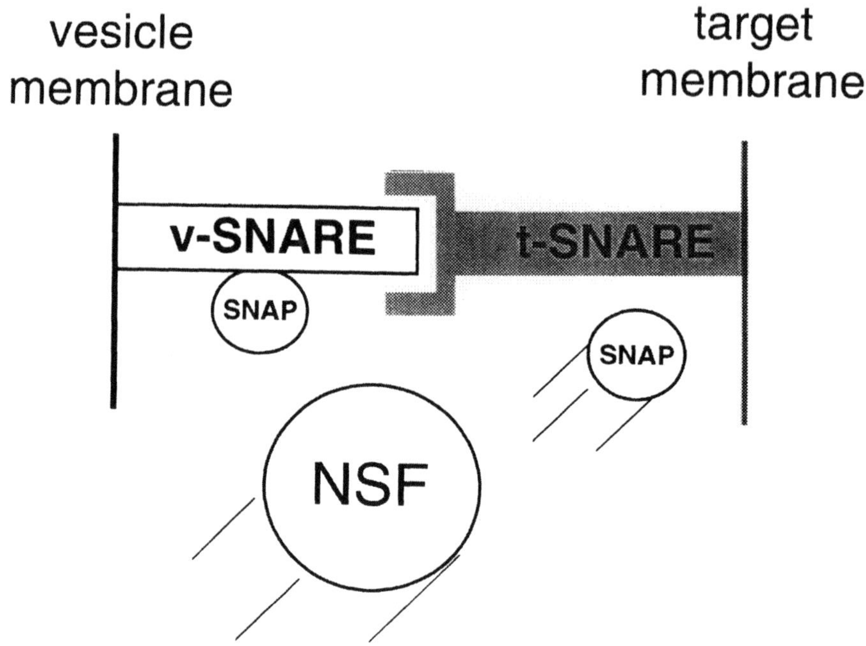

Figure 1. The SNARE hypothesis. Each secretory vesicle is tagged with one or more proteins related to VAMP and termed *v*esicular-*SNAP* *r*eceptors (v-SNARE). Each target membrane is tagged with a protein related to syntaxin termed *t*arget-*SNAP* *r*eceptor (t-SNARE). The specificity of this protein–protein interaction may underlie the specificity of vesicle targeting. The interaction of a v-SNARE with a t-SNARE results in the recruitment of soluble SNAPs (*s*oluble *N*SF *a*ttachment *p*roteins). In turn, the complex that is formed permits the anchoring of NSF (*N*EM-*s*ensitive *f*actor). The assembly of the SNAREs with the general cytosolic components is thought to lead, finally, to vesicle fusion.

NSF in vesicular transport, but ATP hydrolysis by NSF seems to be required for the priming step and not for membrane fusion (Morgan and Burgoyne, 1995; Nichols et al., 1997).

The SNARE hypothesis is a useful molecular description of the possible role of some components of the general machinery implicated in targeting and fusion of secretory vesicles. Obviously, the SNARE theory is an oversimplified view of the process and the mechanism regulating the assembly and the function of the fusion complex is still not understood. A number of proteins capable of interacting with the components of the SNARE complex have been identified (Figure 2). The list is continually increasing thanks to the input obtained from a variety of different research fields.

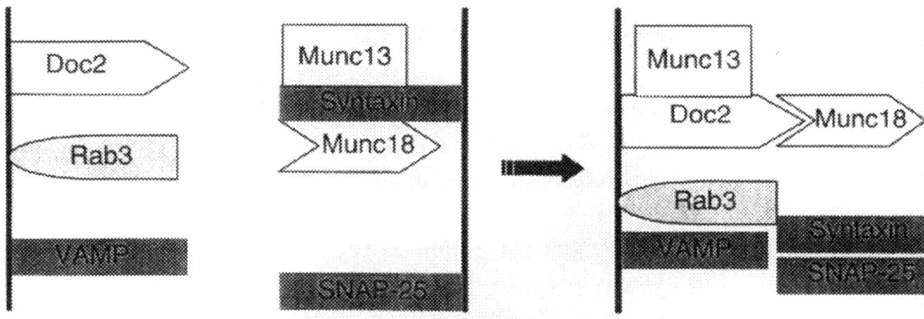

Figure 2. Model for the regulation of the assembly of the SNARE complex. Under resting conditions (left panel) Munc13 and Munc18 (also called sec-1) bind to syntaxin preventing its association with SNAP-25 and VAMP. During the process of targeting and fusion of a secretory vesicle (right panel), Rab3 interacts transiently with syntaxin and cause the displacement of Munc18. Doc2, which is also associated with the vesicle membrane, binds both to both Munc13 and Munc18, enabling syntaxin to form the SNARE complex.

Among these new components of the secretory machinery, Munc13 and Munc18, two proteins interacting with the t-SNARE syntaxin, are of particular interest. Munc13 represents the mammalian homologue of UNC13, a protein identified in the nematode *Caenorabditis elegans* (Maruyama and Brenner, 1991; Brose et al., 1995). Unc13 mutants in *C. elegans* are characterized by a severe deficit in neurotransmitter release (Maruyama and Brenner, 1991). Munc13 contains a phorbol ester-binding C1 domain and two C2 domains homologous to the calcium/phospholipid-binding region of protein kinase C (Maruyama and Brenner, 1991; Brose et al., 1995). This particular domain structure suggests that Munc13 may be involved in the calcium-diacylglycerol–dependent regulation of exocytosis. Munc18 is the mammalian homologue of UNC18 in *C. elegans*, Rop in *Drosophila*, and Sec1 in yeast (Hata et al., 1993; Pevsner et al., 1994; Garcia et al., 1994). Mutations in these genes result in the blockade of neurotransmitter release or in the accumulation of secretory vesicles, indicating an essential role for this protein in the secretory process. Both Munc13 and Munc18 exist in different isoforms, some of which appear to be brain specific while others are present in most cell types (Brose et al., 1995; Katagiri et al., 1995).

V. APPLICATION OF THE SNARE HYPOTHESIS TO INSULIN SECRETION

The SNARE hypothesis furnishes a molecular description of the process of targeting, docking and fusion in all eukaryotic cells. As predicted by the model, insulin-secreting cells, including primary β-cells, express a set of v-SNAREs and t-SNAREs that contribute to the control of insulin exocytosis.

A. v-SNAREs Expressed in Pancreatic β-Cells

In eukaryotes, three v-SNAREs belonging to the same protein family have been identified so far on post-Golgi secretory vesicles: VAMP-1, VAMP-2, and cellubrevin (sometimes referred to as VAMP-3) (Elferink et al., 1989; McMahon et al., 1993). My colleagues and I have shown by cDNA cloning that rat pancreatic β-cells express VAMP-2 and cellubrevin, but not VAMP-1, and that the amino acid sequence of these proteins is identical to that isolated from rat brain (Regazzi et al., 1995b). These findings are in agreement with the results obtained in other laboratories by in situ hybridization (Jacobsson et al, 1994; Rossetto et al., 1996) and by Western blotting (Wheeler et al., 1996). Morphological and biochemical studies demonstrate that VAMP-2 and cellubrevin are associated with both types of secretory vesicles present in pancreatic β-cells: dense core secretory granules containing insulin and synapticlike microvesicles carrying γ-aminobutyric acid (GABA) (Regazzi et al., 1995b). The insertion of VAMPs in the membrane of insulin-containing secretory granules is dictated by a short amino acid sequence located in the cytosolic portion of the molecule (amino acids 41 to 50) (Regazzi et al., 1996b). This corresponds to the domain of VAMP-2 responsible for the insertion of the protein in the membrane of synaptic vesicles in the neuroendocrine cell line PC12 (Grote et al., 1995). Introduction of TeTx or BoNT B in insulin-secreting cells in which the plasma membrane has been rendered permeable to macromolecules with streptolysin-O results in the selective cleavage of VAMP-2 and cellubrevin and abolishes Ca^{2+}-induced exocytosis (Regazzi et al., 1995b). My colleagues and I have shown that the blockade of Ca^{2+}-triggered secretion can be overcome by transfection of either a TeTx resistant mutant of VAMP-2 or of a corresponding mutant of cellubrevin, indicating that both VAMP isoforms are implicated in insulin exocytosis (Regazzi et al., 1996b). The restoration of exocytosis by the expression of a TeTx-resistant mutant of VAMP-2 has been useful for defining the domains of the protein that are essential for the exocytotic process. Using this approach, we could demonstrate that the central portion of VAMP-2, which is highly conserved between VAMP isoforms, is required for insulin secretion (Regazzi et al., 1996b). Moreover, it was possible to show that the presence of the two potential phosphorylation sites for Ca^{2+}-calmodulin-dependent protein kinase II (Thr^{35} and Ser^{61}) are not essential for exocytosis (Regazzi et al., 1996b).

B. t-SNAREs Expressed in Pancreatic β-Cells

The plasma membrane of pancreatic β-cells is equipped with t-SNAREs belonging to the SNAP-25 and syntaxin family. The involvement of these proteins in insulin exocytosis has been demonstrated using clostridial toxins or with inactivating antibodies. Thus, cleavage of SNAP-25 by BoNT A and E reduces Ca^{2+}-stimulated insulin release of permeabilized pancreatic β-cells (Sadoul et al., 1995). In contrast to TeTx, the inhibition of secretion caused by BoNT A and E was only partial, suggesting that other proteins with properties similar to SNAP-25 may be present in pancreatic β-cells. The implication of syntaxin-I in insulin secretion was demonstrated with inactivating antibodies against the protein (Martin et al., 1995) and with BoNT C (Lang et al., 1997b). Using both approaches, inactivation of syntaxin I resulted in the inhibition of insulin secretion indicating that this t-SNARE plays a positive role in the exocytotic process of β-cells. These findings are difficult to reconcile with the observation that the overexpression of syntaxin-Ia but not of syntaxin-Ib decreases insulin release (Nagamatsu et al., 1996). A possible explanation for the latter results is that a marked overexpression of syntaxin Ia reduces the availability of other important components for exocytosis. In line with this interpretation, marked overexpression of VAMPs impairs stimulated insulin secretion (Regazzi et al., 1996b).

C. NSF and SNAPs in Pancreatic β-Cells

The ATPase NSF and α-SNAP are general components of the machinery controlling vesicle fusion in eukaryotes. Therefore, it is not surprising that these proteins are also expressed in insulin-secreting cells (Kiraly-Borri et al., 1996). The involvement of NSF and α-SNAP was demonstrated using streptolysin-O permeabilized cells. In this experimental system, incubation of the cells in 10 μM Ca^{2+} immediately after permeabilization induces a marked increase in insulin release. If the incubation at stimulatory Ca^{2+} concentrations is delayed, the response of the cells is much less pronounced (Kiraly-Borri et al., 1996). The run-down of the response is due to the leakage of cytosolic components through the holes produced in the plasma membrane by streptolysin-O and can be counteracted by the addition of cytosol (Kiraly-Borri et al., 1996). Prior inactivation of cytosolic NSF with NEM abolished the reconstitution of Ca^{2+}-induced insulin secretion. Partial reconstitution of Ca^{2+}-triggered insulin secretion could also be achieved by the addition of recombinant α-SNAP (Kiraly-Borri et al., 1996). Taken together, these results indicate that NSF and α-SNAP participate in insulin exocytosis.

VI. INVOLVEMENT OF Ras-LIKE GTP-BINDING PROTEINS IN THE SECRETORY PATHWAY

SEC4 is one of the yeast strains with a defect in secretion identified using the genetic approach described earlier. This strain is characterized by the accumulation of post-Golgi vesicles that are unable to fuse with the plasma membrane. In 1988, Novick and collaborators identified the gene responsible for this phenotype. The product of the SEC4 gene was found to be a GTPase with structural homologies to the protein encoded by the Ras oncogene (Goud et al., 1988). The discovery that Ypt1, a protein mutated in another yeast strain deficient in secretion is also encoding a Ras-like GTPase (Segev et al., 1988) confirmed the important role played by this family of regulatory proteins in the secretory process. Today, about

40 GTPases with extensive homologies with Sec4 and Ypt1 have been identified in mammalian cells (for review, see Takai et al., 1992; Fischer von Mollard et al., 1994; Novick and Zerial, 1997). These proteins, grouped in the so-called Rab family, are associated with different cellular compartments and control vectorial movement of secretory vesicles to the appropriate acceptor membrane (Takai et al., 1992; Fischer von Mollard et al., 1994; Novick and Zerial, 1997).

Rabs are monomeric proteins with a molecular mass comprised between 20 and 30 kDa and are often referred to as small GTPases to distinguish them from heterotrimeric GTPases coupled to membrane receptors such as Gs or Gi. The small GTPases can reversibly associate with secretory vesicles thanks to isoprenoid groups that are post-translationally attached to the carboxyl-terminus of the proteins (Takai et al., 1992; Fischer von Mollard et al., 1994; Novick and Zerial, 1997). Rabs act as molecular switches that flip between two conformational states, the active GTP-bound and an inactive GDP-bound form. Guanine nucleotide binding and GTPase activity of Rab proteins, as well as their reversible association with membranes, are under the control of a variety of regulatory proteins. Rab GTPases in their GDP-bound state are cytosolic and associated with Rab*G*DP-*d*issociation *i*nhibitors (RabGDIs) (Regazzi et al., 1992a; Ullrich et al., 1993). RabGDIs are capable of chaperoning the GTPases in the cytosol and delivering them to the appropriate membrane (Soldati et al., 1994). RabGDIs control the interaction of small GTPases with phospholipids by binding to the isoprenylated carboxy terminus of Rabs in a guanine nucleotide-dependent manner (Takai et al., 1992; Fischer von Mollard et al., 1994; Novick and Zerial, 1997).

Once the Rabs have been delivered to the vesicle membrane, their interaction with another regulatory protein, the guanine nucleotide exchange factor (GEF), favors the exchange of guanosine diphosphate (GDP) for guanosine triphosphate (GTP). The signal induced by Rab GTPases is switched off thanks to the stimulation of the intrinsic GTPase activity of the proteins by isoform specific *G*TPase *a*ctivating *p*roteins (GAP) (Takai et al., 1992; Fischer von Mollard et al., 1994; Novick and Zerial, 1997). The small GTPases, again in GDP-liganded state, reform a complex with RabGDIs, dissociate from the membranes, and are, thereby, ready for another cycle.

Although the important role played by Rab proteins in the secretory process has been recognized for almost 10 years, the precise sequence of events permitting these proteins to control vesicular trafficking remains to be determined.

Different models have been proposed to account for the action of Rab GTPases on biological membrane fusion. It has been suggested that they ensure the docking of secretory vesicles with the appropriate target membranes by favoring the assembly of the v-SNARE/t-SNARE complex (see Figure 2). This model is supported by the observation made in yeast that the Rab-like protein Ypt1 transiently interacts with the t-SNARE Sed5 (Lupashin and Waters, 1997). Before the arrival of the secretory vesicle, Sed5 cannot engage in the SNARE complex because it is associated with Sly1, a Sec1-like protein that functions as a negative regulator.

The interaction between the small GTPase (Ypt1) associated with the vesicle and the t-SNARE (Sed5) causes the displacement of the negative regulator (Sly1), allowing the subsequent formation of the v-SNARE/t-SNARE complex.

This sequence of events would explain how Rab GTPases determine the specificity of docking of transport vesicles to the target membrane. However, it is not certain that this model can be directly applied to all Rab GTPases. In particular, it cannot fully account for the mode of action of Rab3 proteins, the GTPases controlling exocytosis. In fact, in contrast to what is observed with other Rab proteins, the activated form of Rab3 inhibits exocytosis. In addition, in Rab3A-deficient mice a nerve impulse was found to cause the fusion of a greater-than-usual number of synaptic vesicles (Geppert et al., 1997), suggesting that Rab3A might negatively regulate vesicle fusion. In contrast to the other steps in the secretory pathway, the fusion of the vesicles with the plasma membrane does not proceed at a constant rate but is modulated by fluctuations in the cellular concentration of Ca^{2+} and other second messengers. More information is required to assess whether Rab3 proteins function by interacting with the t-SNAREs located at the plasma membrane or in a manner different from the other proteins of the family.

VII. ROLE OF RAB GTPASES IN INSULIN EXOCYTOSIS

Pancreatic β-cells contain several GTPases of the Rab family associated with different cellular compartments, including insulin-containing dense core secretory granules (Regazzi et al., 1991; Regazzi et al., 1992b; Lankat-Buttgereit et al., 1992; Regazzi et al., 1996a; Iezzi et al., 1999). As in other cell systems, the members of the Rab3 family are the best candidates for the regulation of the final steps in the exocytotic process. Initial evidence for the possible involvement of Rab3 proteins in insulin secretion came from studies performed in permeabilized cells. In this system, the introduction of peptides mimicking the putative effector-binding domain of Rab3 proteins stimulates insulin release (Li et al., 1993; Olszewski et al., 1994; Regazzi et al., 1995a). A combination of techniques, including electron microscopy, demonstrates that Rab3A, the most abundant Rab3 isoform in neuronal cells, is associated with the membrane of secretory granules of rat pancreatic β-cells (Regazzi et al., 1996a). The functional implication of Rab3A in insulin exocytosis was shown by transfecting insulin-secreting cells with constructs of this GTPase bearing point mutations in key amino acids (Regazzi et al., 1996a). Overexpression of Rab3A mutants deficient in GTP hydrolysis decreases nutrient-stimulated insulin secretion, indicating that the hydrolysis of GTP by Rab3A is an essential requirement for insulin release (Regazzi et al., 1996a). Rat pancreatic β-cells also express other members of the Rab3 family, namely Rab3B, C, and D. Except in a small domain at the N-terminus and at the C-terminus, Rab3 proteins display highly homologous amino acid sequences (~75% amino acid identities). These isoforms are also implicated in the regulation of insulin exocytosis (Iezzi et al., 1999). Surprisingly, human pancreatic β-cells were found to express Rab3B and Rab3C, but not Rab3A and Rab3D (Lankat-Buttgereit et al., 1994; Regazzi et al., 1996a). The fine-tuning of insulin secretion is likely to result from the interplay between the different Rab3 isoforms present in the cells. Nevertheless, it is possible that in human β-cells, Rab3B and -C are sufficient for the control of insulin exocytosis.

VIII. ROLE OF CYTOSKELETON IN INSULIN SECRETION

Insulin is stored in more than 10,000 secretory granules that are assembled at the trans side of the Golgi complex and are then transported to the plasma membrane. The first phase of insulin release observed in response to glucose is thought to involve granules that are close to the plasma membrane, while the second phase, more sustained, probably needs the transport of granules over a certain distance. The transport of secretory vesicles toward the plasma membrane and their availability for the exocytotic process is thought to be controlled by cytoskeletal elements (Howell and Tyhurst, 1986; Li et al., 1994). Interruption of microtubules attenuates mainly the second phase of insulin secretion, indicating that the transport of granules requires microtubule-associated motor proteins (Howell and Tyhurst, 1986). Actin-containing microfilaments are also involved in the secretory process. The cortical actin network acts as a negative clamp, which blocks the fusion of secretory granules with the plasma membrane (Howell and Tyhurst, 1986; Li et al., 1994). Actin filaments also have a positive role in vesicular trafficking and exocytosis cannot occur without a minimal cytoskeletal structure (Howell and Tyhurst, 1986; Li et al., 1994).

A. Potential Role of Rho Proteins

Rho proteins constitute a family of Ras-like GTPases that regulate a variety of cellular functions including endo- and exocytosis (Ridley, 1996). Members of this family, including RhoA, Rac, and CDC42, are also expressed in β-cells and may participate in the exocytosis of insulin (Regazzi et al., 1992a; Regazzi et al., 1992b; Kowluru et al., 1996; 1997). Indeed, in pancreatic β-cells CDC42 is methylated at the carboxyl-terminus during stimulation of insulin release by glucose but not by depolarizing concentrations of K^+ (Kowluru et al., 1996).

The proteins of the Rho and Rab family have several properties in common. In fact, they are all monomeric and they interact with proteins controlling their subcellular distribution and their guanine nucleotide cycle such as GDIs, GEF, and GAPs (Takai et al., 1992). Although several potential downstream effectors of Rho proteins have been identified, the mechanism by which these GTPases regulate secretion is poorly understood. Increasing evidence has been accumulated indicating that members of the Rho family interact with phosphatidylinositol kinases (Tolias et al., 1995; Ren et al., 1996). This interaction may represent the missing link between activation of Rho GTPases, actin filament reorganization, and regulation of secretion. In fact, phosphorylation/dephosphorylation of the polar heads of phosphoinositides (PtdIns) in specific intracellular locations may cause the recruitment or the activation of proteins essential for cytoskeleton organization and vesicular transport (Liscovitch and Cantley, 1995; DeCamilli et al., 1996). On the one hand, several PtdIns have been reported to uncap F-actin filaments and promote actin polymerization (Hartwig et al., 1995). On the other hand, two cytosolic factors required for ATP-dependent priming and exocytosis from neuroendocrine cells have been identified as PtdIns transfer protein and PtdIns-5 kinase (Hay and Martin, 1993; Hay et al., 1995). Moreover, inhibitors of PtdIns-4 kinase or PtdIns-3 kinase affect exocytosis in different cell systems, including β-cells (Wiedenmann et al., 1996; Straub and Sharp, 1996). It is still unclear whether the effects of PtdIns on exocytosis are linked to the rearrangement of cytoskeletal elements or whether changes in the polyphosphorylated PdtIns content of secretory vesicles promote fusion by altering the physical properties of the membrane (Summers et al., 1996).

B. Involvement of Phospholipase D

Rho GTPases and PtdIns 4,5 bisphosphate are potent stimulators of phospholipase D (PLD) (Exton, 1990; Hammond et al., 1997). Alteration in lipid content by PLD could be an important determinant in vesicular dynamics (Liscovitch and Cantley, 1995). Indeed, PLD has been implicated in the regulation of secretion in different cell types (Stuchfield and Cockcroft, 1993), including rat pancreatic islets (Metz and Dunlop, 1990; see also the chapter by Metz later in this volume). The mechanism by which PLD affects the secretory process is still largely unknown. A model has been proposed whereby PdtIns 4,5 bisphosphate synthesis and PLD activation by small GTPases may coordinate vesicle fusion with acceptor membranes (Liscovitch and Cantley, 1995). According to this model, phosphatidic acid produced by PLD would stimulate PtdIns(4) 5-kinase resulting in massive synthesis of PtdIns 4,5 bisphosphate. This, in turn, would cause further stimulation of PLD. This positive feedback loop would profoundly alter lipid composition of vesicular membranes, leading to the generation of microdomains greatly enriched in phosphatidic acid and PtdIns 4,5 bisphosphate and with a higher propensity for fusion (Summers et al., 1996).

IX. FINE-TUNING OF THE EXOCYTOTIC PROCESS

The discoveries made in the past few years led to the identification of the general components of the machinery implicated in vesicle fusion. In eukaryotes, most fusion events proceed at a constant rate and are not influenced by short-term variations in the concentration of cellular second messengers. In contrast, in cells with regulated secretion such as neurons, endocrine cells, or exocrine cells the fusion of vesicles with the plasma membrane is tightly controlled. The clarification of the molecular mechanism by which secretagogues such as glucose stimulate exocytosis in pancreatic β-cells is essential for elucidating how such regulation becomes deranged in disease states including diabetes. Although alterations in the function of some of the general components of the fusion machinery in certain forms of diabetes cannot be excluded, such defects would be expected to affect several other cell systems. It seems more likely that the lesions leading to disturbed insulin secretion in pre-Type 1 and Type 2 diabetes concern regulatory components expressed selectively in β-cells.

A. Ca^{2+} Sensors

In most cell systems equipped for regulated secretion, an increase in cytosolic Ca^{2+} is the main trigger for exocytosis. Although the release of insulin is known to be modulated by fluctuations in several second messengers, Ca^{2+} is also a major stimulus for exocytosis in pancreatic β-cells (Wollheim et al., 1996). Insulin-secreting cells contain proteins capable of sensing variations in the intracellular concentration of Ca^{2+} and of transmitting the signal to the secretion machinery (Figure 3).

Synaptotagmins have all the characteristics required to be a Ca^{2+} sensor for exocytosis. They possess two so-called C2 domains which are homologous to the second conserved region of protein kinase C. These domains confer them Ca^{2+}- and phospholipid-binding properties (Südhof and Rizo, 1996). In addition, synaptotagmins localize to secretory vesicles and directly interact with components of the fusion machinery (Südhof and Rizo, 1996). Synaptotagmin I and II are present in different β-cell lines and co-localize mainly with insulin-containing secretory granules (Lang et al., 1997a). In permeabilized cells, antibodies directed against the first C2 domain of these proteins inhibited exocytosis triggered by Ca^{2+} (Lang et al., 1997a). Similar results were obtained after overexpression of synaptotagmin constructs in which the C2 domains were mutated (Lang et al., 1997a).

Doc2 is another protein associated with secretory vesicles and with two C2 domains. Two isoforms called α and β have been described. The α isoform is expressed selectively in neurons while the β isoform appears to be ubiquitous (Orita et al., 1995; Sakaguchi et al., 1996). At present it is not known which Doc2 isoform is expressed in β-cells. In neuroendocrine cells, overexpression of Doc2 potentiates catecholamine secretion while reduction of endogenous Doc2, achieved by transfection of antisense DNA, decreases Ca^{2+}-triggered exocytosis (Orita et al., 1996). Overexpression of Doc2 in insulin-secreting cells results in an increase in stimulus-induced secretion (R. Regazzi, unpublished observation). This indicates that Doc2 may participate in the regulation of insulin exocytosis. The precise mode of action of Doc2 is still unclear. Interestingly, Doc2 binds Munc13 and Munc18, two proteins interacting with syntaxin I. Since Munc13 and Munc18 are localized on the plasma membrane, their interaction with Doc2 may contribute to the docking of secretory vesicles. Doc2 may modulate the assembly of the SNARE complex by displacing Munc13 and Munc18 from syntaxin. In addition, since Doc2 and Munc13 bind Ca^{2+}, they could serve as Ca^{2+} sensors and

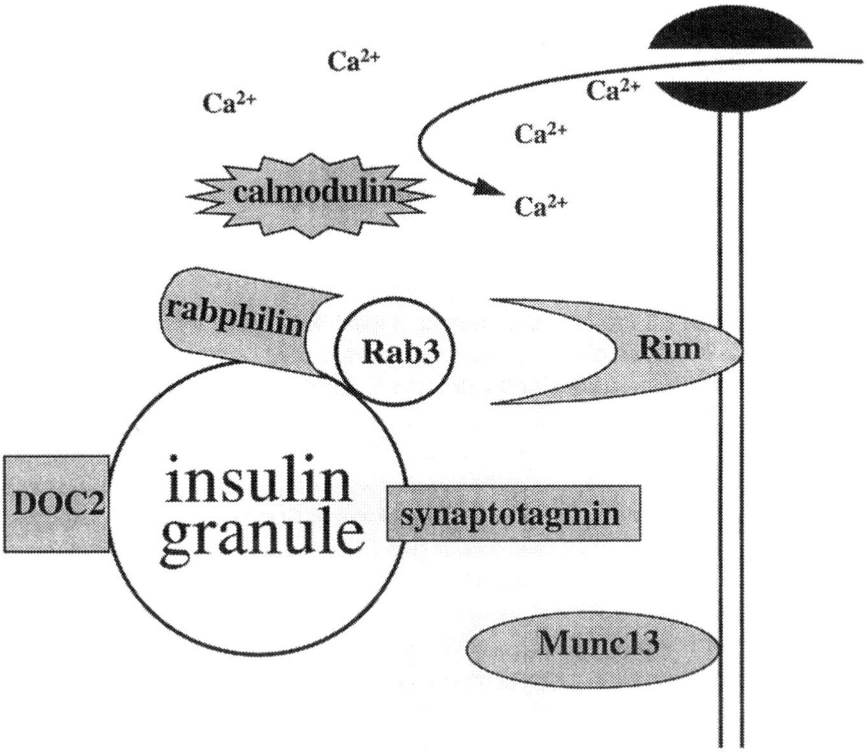

Figure 3. Putative Ca^{2+}-sensors involved in insulin secretion. Several proteins capable of binding calcium at concentrations encountered in cells exposed to secretagogues are associated with the membrane of secretory granules or with the plasma membrane. Doc2, Munc13, Synaptotagmin, Rim, and rabphilin all possess C2 domains homologous to the second conserved domain of protein kinase C that enable them to bind calcium. Rabphilin and Rim are also targets of Rab3 and can interact specifically with the guanosine triphosphate (GTP)–bound form of this GTPase. A direct involvement of these proteins in insulin exocytosis has been demonstrated only in the case of synaptotagmin (Lang et al., 1997a), rabphilin (Arribas et al., 1997), and Doc2 (Regazzi, unpublished observation). Recently, in vitro studies have shown that calmodulin can dissociate Rab3 from secretory vesicles in a Ca^{2+}-dependent manner, indicating that calmodulin is an additional potential Ca^{2+}-sensor in the process of exocytosis.

regulate fusion itself. Interestingly, Munc13 contains a so-called C1 homologous to the first conserved region of protein kinase C. This confers to the protein the ability to bind diacylglycerol and phorbol esters. In agreement with this finding, the interaction of Munc13 with Doc2 is stimulated by phorbol esters (Orita et al., 1997). Thus, it is possible that phorbol esters and secretagogues generating diacylglycerol, such as glucose or acetylcholine (Wollheim et al., 1996), modulate the exocytotic process by favoring the formation of the Munc13–Doc2 complex.

Rabphilin is a protein related to synaptotagmin and Doc2 that interacts with the GTP-bound form but not with the GDP-bound form of Rab3 proteins (Shirataki et al., 1992; Shirataki et al., 1993). Rabphilin is expressed exclusively in neuronal cells and in some insulin-secreting cell lines but is undetectable in rat and in human pancreatic β-cells (Inagaki et al., 1994; Regazzi et al., 1996a). The protein contains two C2 domains and binds phospholipids in a Ca^{2+}-dependent manner (Yamaguchi et al., 1993). Overexpression of rabphilin in neuroendocrine and in insulin-secreting cells leads to an increase in stimulus-induced secretion (Chung et al., 1995; Arribas et al., 1997). At present, since rabphilin is not detectable in pancreatic β-cells, the role of this protein in insulin exocytosis remains unclear. It is possible that in pancreatic β-cells, rabphilin is replaced by Noc2, a protein with high similarity to the Rab3-binding domain of rabphilin (Kotake et al., 1997).

Recently, a new putative effector for Rab3 called Rim has been identified using a yeast two-hybrid approach (Wang et al., 1997). Rim is localized to presynaptic active zones in synapses and could serve as a Rab3-dependent regulator of synaptic vesicle fusion (Wang et al., 1997). Rim possesses several interesting features, including the presence in its sequence of two C2 domains, suggesting that the protein could be a Ca^{2+}-sensor (Wang et al., 1997). No information is available concerning the presence of Rim in pancreatic β-cells.

In addition to all these proteins that could act as specific Ca^{2+}-sensors in the process of insulin exocytosis, a recent report has demonstrated that calmodulin is able to extract Rab3 from synaptosomal membranes in a Ca^{2+}-dependent manner (Park et al., 1997). However, the functional significance in vivo of the effect of calmodulin on Rab3 remains to be established.

B. Heterotrimeric GTPases

Most heterotrimeric GTPases are associated with the plasma membrane, where they couple hormone and neurotransmitter receptors to enzymes such as phospholipase C and adenylate cyclase (for review, see Neer, 1995). Some heterotrimeric GTPases have also been localized on intracellular membranes, suggesting a role of these proteins in vesicular trafficking (Burgoyne, 1992). In contrast to Ras-like GTPases, the members of this family of GTPases, are composed of three subunits. The α subunit is responsible for binding and hydrolysis of GTP (Neer, 1995). The βγ dimer reversibly associates with the α subunit, depending on the occupancy of the receptor, and modulates the interaction of the α subunit with downstream effectors (Neer, 1995). Pancreatic β-cells express a large number of heterotrimeric GTPases that are involved in the generation of second messengers such as cyclic adenosine monophosphate (cAMP), diacylglycerol, and inositol triphosphate (Wollheim et al., 1996). All these molecules play an essential role in the transduction of extracellular signals arriving to pancreatic β-cells and, in turn, in the regulation of insulin secretion. In addition to their involvement in the generation of second messengers (discussed elsewhere in this book), heterotrimeric GTPases control insulin exocytosis at a step situated beyond the regulation of cAMP and Ca^{2+}. Thus, epinephrine inhibits insulin secretion both in intact and in permeabilized cells (Ullrich et al., 1988; Lang et al., 1995). This effect is mediated by α_2-adrenergic receptors that are coupled to the heterotrimeric GTPases called G_i and G_o (Ullrich et al., 1988; Lang et al., 1995). Since these GTPases are negative modulators of the adenylate cyclase, the binding of epinephrine results in a decrease in the production of cAMP (Ullrich et al., 1988). However, this effect of epinephrine cannot explain the inhibition of insulin secretion in permeabilized cells because small molecules such as cAMP rapidly leak out through the holes in the plasma membrane. The inhibition of insulin secretion by epinephrine is mimicked by the addition of nanomolar amounts of preactivated G_i/G_o α subunits or by the overexpression of constitutively active mutants of different isoforms of G_i α subunits (Lang et al., 1995). These results indicate that G_i α subunits exert a direct control on the exocytotic process. The mechanism by which G_i α subunits interact with the fusion machinery remains to be elucidated.

X. CONCLUDING REMARKS

During the past few years, our understanding of the mechanism permitting targeting and fusion of secretory vesicles in eukaryotes has improved enormously. The evolutionary conservation of the basic components of the fusion machinery has attracted researchers from many different fields. This extraordinary convergence of interests is very fruitful and promising for the future. The studies carried out in vitro and in lower eukaryotes opened the way for the identification of the proteins participating in exocytosis in pancreatic β-cells. Although some of these components have now been identified, much remains to be done to elucidate how the targeting and fusion processes are regulated. The exocytotic machinery of β-cells is under the control of a variety of cellular messengers including not only Ca^{2+} and protein kinases, but also molecules coupling cellular metabolism to hormone release (Maeschler et al., 1997). Clearly, the characterization of the mechanism that ensures the coupling of metabolism to secretion and its relevance for diabetes constitutes a major task for future research in the field.

ACKNOWLEDGMENTS

I would like to thank Dr. P. Clarke for critically reading the manuscript and for helpful suggestions.

REFERENCES

Aalto, M.K., Ronne, H. and Keranen, S. (1993). Yeast syntaxins Sso1p and Sso2p belong to a family of related membrane proteins that function in vesicular transport. EMBO J. 12, 4095–4104.

Arribas, M., Regazzi, R., Garcia, E., Wollheim, C.B. and De Camilli, P. (1997). The stimulatory effect of Rabphilin3A on regulated exocytosis from insulin-secreting cells does not require an association-dissociation cycle with membranes mediated by Rab3. Eur. J. Cell Biol. 74, 203–216.

Bittner, M.A. and Holz, R.W. (1992). Kinetic analysis of secretion from permeabilized adrenal chromaffin cells reveals distinct components. J. Biol. Chem. 267, 16219–16225.

Brennwald, P., Kearns, B., Champion, K., Keränen, S., Bankaitis, V. and Novick P. (1994). Sec9 is a SNAP-25-like component of a yeast SNARE complex that may be the effector of Sec4 function in exocytosis. Cell 79, 245–258.

Brose, N., Hofmann, K., Hata, Y. and Südhof, T.C. (1995). Mammalian homologues of Caenorhabditis elegans unc-13 gene define novel family of C2 domain proteins. J. Biol. Chem. 270, 25273–25280.

Burgoyne, R.D. (1992) Trimeric G proteins in Golgi transport. Trends Biochem. Sci. 17, 87–88.

Chung, S.-H., Takai, Y. and Holz, R.W. (1995). Evidence that the Rab3a-binding protein, Rabphilin3a, enhances regulated secretion. J. Biol. Chem. 270, 16714–16718.

DeCamilli, P., Emr, S.D., McPherson, P.S. and Novick, P. (1996). Phosphoinositide as regulators in membrane traffic. Science 271, 1533–1539.

Elferink, L.A., Trimble, W.S. and Scheller, R.H. (1989). Two vesicle-associated membrane protein genes are differentially expressed in the rat central nervous system. J. Biol. Chem. 264, 11061–11064.

Exton, J.H. (1990). Signaling through phosphatidylcholine breakdown. J. Biol. Chem. 265, 1–4.

Fischer von Mollard, G., Stahl, B., Li, C., Südhof, T.C. and Jahn, R. (1994). Rab proteins in regulated exocytosis. Trends. Biochem. Sci. 19, 164–168.

Garcia, E.P, Gatti, E., Butler, M., Burton, J. and De Camilli, P. (1994). A rat brain sec1 homologue related to Rop and Unc18 interacts with syntaxin. Proc. Natl. Acad. Sci. USA 91, 2003–2007.

Geppert, M., Goda, Y., Stevens, C.F. and Südhof, T.C. (1997). The small GTP-binding protein Rab3A regulates a late step in synaptic vesicle fusion. Nature 387, 810–814.

Goud, B., Salminen, A., Walworth, N.C. and Novick, P.J. (1988). A GTP-binding protein required for secretion rapidly associates with secretory vesicles and the plasma membrane in yeast. Cell 53, 753–768.

Grote, E., Hao, J.C., Bennett, M.K. and Kelly, R.B. (1995). A targeting signal in VAMP regulating transport to synaptic vesicles. Cell 81, 581–589.

Hammond, S.M., Jenco, J.M., Nakashima, S., Cadwallader, K., Gu, Q.-M., Cook, S., Nozawa, Y., Prestwich, G.D., Frohman, M.A. and Morris, A.J. (1997). Characterization of two alternatively spliced forms of phospholipase D1. J. Biol. Chem. 272, 3860–3868.

Hartwig, J.H., Bokoch, G.M., Carpenter, C.L., Janmey, P.A., Taylor, L.A., Toker, A. and Stossel, T.P. (1995). Thrombin receptor ligation and activated Rac uncap actin filaments barbed ends through phosphoinositide synthesis in permeabilized human platelets, Cell 82, 643–653.

Hata, Y., Slaughter, C.A. and Südhof, T.C. (1993). Synaptic vesicle fusion complex contains unc-18 homologue bound to syntaxin. Nature 366, 347–351.

Hay, J.C., Fisette, P.L., Jenkins, G.H., Fukami, K., Takenawa, T., Anderson, R.A. and Martin T.F.J. (1995). ATP-dependent inositide phosphorylation required for Ca^{2+}-activated secretion. Nature 374, 173–177.

Hay, J.C. and Martin T.F.J. (1992). Resolution of regulated secretion into sequential MgATP-dependent and calcium-dependent stages mediated by distinct cytosolic proteins. J. Cell Biol. 119, 139–151

Hay, J.C. and Martin T.F.J. (1993). Phosphatidylinositol transfer protein required for ATP-dependent priming of Ca^{2+}-activated secretion. Nature 366, 572–575.

Howell, S.L. and Tyhurst, M. (1986). The cytoskeleton and insulin secretion. Diab. Metab. Rev. 2, 107–123.

Iezzi, M., Escher, G., Meda, M., Charollais, A., Baldini, G., Darchen, F., Wollheim, C.B., and Regazzi, R. (1999). Subcelullar distribution and function of Rab3A, B, C, and D isoforms in insulin-secreting cells. Mol. Endocrinol. 13, 202–212.

Inagaki, N., Mizuta, M. and Seino, S. (1994). Mouse rabphilin-3A is expressed in hormone-secreting cells. J. Biochem. 116, 239–242.

Jacobsson, G., Bean, A.J., Scheller, R.H., Juntti-Berggren, L., Deeney, J.T., Berggren, P.-O. and Meister, B. (1994). Identification of synaptic proteins and their isoform mRNAs in compartments of pancreatic endocrine cells. Proc. Natl. Acad. Sci. USA 91, 12487–12491.

Katagiri, H., Terasaki, J., Murata, T., Ishihara, H., Ogihara, T., Inukai, K., Fukushima, Y., Anai, M., Kikuchi, M., Miyazaki, J.I., Yazaki, Y. and Oka, Y. (1995). A novel isoform of syntaxin-binding protein homologous to yeast sec1 expressed ubiquitously in mammalian cells. J. Biol. Chem. 270, 4963–4966.

Kiraly-Borri, C., Morgan, A., Burgoyne, R., Weller, U., Wollheim, C.B., Lang, J. (1996). Soluble N-ethylmaleimide sensitive factor attachment protein and N-ethylmaleimide insensitive factors are required for Ca^{2+} stimulated exocytosis of insulin. Biochem. J. 314, 199–203.

Kotake, K., Ozaki, N., Mizuta, M., Sekiya, S., Ingaki, N., and Seino, S. (1997). Noc2, a putative zinc finger protein involved in exocytosis in endocrine cells. J. Biol. Chem. 272, 29407–29410.

Kowluru, A, Seavey, SE, Li, G, Sorenson, RL, Weinhaus, AJ, Nesher, R, Rabaglia, ME, Vadakekalam, J and Metz, SA (1996). Glucose- and GTP-dependent stimulation of the carboxyl methylation of CDC42 in rodent and human pancreatic islets and pure β-cells. J. Clin. Invest. 98, 540–555.

Kowluru, A., Li, G., Rabaglia, M.E., Segu, V.B., Hofmann, F., Aktories, K., and Metz, S.A. (1997). Evidence for differential roles of the Rho subfamily of GTP-binding proteins in glucose- and calcium-induced insulin secretion from pancreatic beta cells. Biochem. Pharmacol. 54, 1097–1108.

Lang, J., Fukuda, M., Zhang, H., Mikoshiba, K. and Wollheim, C.B. (1997a). The first C2 domain of synaptotagmin is required for exocytosis of insulin from pancreatic β-cells: Action of synaptotagmin at low micromolar calcium. EMBO J. 16, 5837–5846.

Lang, J., Niemann, H. and Wollheim, C.B. (1997b). Effect of botulinum neurotoxin C1 on exocytosis of insulin and SNARE-protein cleavage in pancreatic β-cells. Diabetologia 40, A43.

Lang, J., Nishimoto, I., Okamoto, T., Regazzi, R., Kiraly, C., Weller, U. and Wollheim, C.B. (1995). Direct control of exocytosis by receptor-mediated activation of the heterotrimeric GTPases Gi and Go or by the expression of their active Ga subunits. EMBO J. 14, 3635–3644.

Lankat-Buttgereit, B., Fehmann, H.C., Hering, B.J., Bretzel, R.G. and Göke, B. (1994). Expression of the ras-related rab3a gene in human insulinomas and normal human pancreatic islets. Pancreas 9, 434–438.

Lankat-Buttgereit, B., Göke, R., Fehmann, H.C., Niess, C. and Göke B. (1992). Expression of the ras-related rab3a gene in insulinoma-derived cell lines. FEBS Lett. 312, 183–186.

Li, G., Regazzi, R., Balch, W.E. and Wollheim, C.B. (1993). Stimulation of insulin release from permeabilized HIT-T15 cells by a synthetic peptide corresponding to the effector domain of the small GTP-binding protein rab3. FEBS Lett. 327, 145–149.

Li, G., Rungger-Brändle, E., Just, I., Jonas, J.-C. Aktories, K. and Wollheim, C.B. (1994). Effect of disruption of actin filaments by *Clostridium botulinum* C2 toxin on insulin secretion in HIT-T15 cells and pancreatic islets. Mol. Biol. Cell 5, 1199–1213.

Liscovitch, M. and Cantley, L.C. (1995). Signal transduction and membrane traffic: The PITP/phosphoinositide connection. Cell 81, 659–662.

Lupashin, V.V. and Waters, M.G. (1997). t-SNARE activation through transient interaction with a Rab-like guanosine triphosphatase. Science 276, 1255–1258.

Maeschler, P., Kennedy, E.D., Pozzan, T. and Wollheim, C.B. (1997). Mitochondrial activation directly triggers the exocytosis of insulin in permeabilized pancreatic β-cells. EMBO J. 16, 3833–3841.

Martin, F, Moya, F, Gutierrez, LM, Reig, JA and Soria B (1995). Role of syntaxin in mouse pancreatic beta cells. Diabetologia 38, 860–863.

Maruyama, I.N. and Brenner, S. (1991). A phorbol ester/diacylglycerol-binding protein encoded by the unc-13 gene of Caenorhabditis elegans. Proc. Natl. Acad. Sci. USA 88, 5729–5733.

McMahon, H.T., Ushkaryov, Y.A., Edelman, L., Link, E., Binz, T., Niemann, H., Jahn, R. and Südhof, T. C. (1993). Cellubrevin is a ubiquitous tetanus-toxin substrate homologous to a putative synaptic vesicle fusion protein. Nature 364, 346–349.

Metz, S.A. and Dunlop, M. (1990). Stimulation of insulin release by phospholipase D. Biochem. J. 270, 427–435.

Morgan, A. and Burgoyne, R.D. (1995). Is NSF a fusion protein? Trends Cell. Biol. 5, 335–339.

Nagahama, M., Orci, L., Ravazzola, M., Amherdt, M., Lacomis, L., Tempst, P., Rothman, J.E. and Söllner, T.H. (1996). A v-SNARE implicated in intra-Golgi transport. J. Cell Biol. 133, 507–516.

Nagamatsu, S., Fujiwara, T., Nakamichi, Y., Watanabe, T., Katahira, H., Sawa, H. and Akagawa, K. (1996). Expression and functional role of syntaxin 1/HPC-1 in pancreatic beta cells: syntaxin 1A, but not 1B, plays a negative role in regulatory insulin release pathway. J. Biol. Chem. 271, 1160–1165.

Neer, E.J. (1995). Heterotrimeric G proteins: organizers of transmembrane signals. Cell 80, 249–257.

Nichols, B.J., Ungermann, C., Pelham, H.R.B., Wickner, W.T. and Haas, A. (1997). Homotypic vacuolar fusion mediated by t- and v-SNAREs. Nature 387, 199–202.

Niemann, H., Blasi, J. and Jahn, R. (1994). Clostridial neurotoxins: New tools for dissecting exocytosis. Trends Cell Biol. 4, 179–185.

Novick, P., Field, C. and Schekman, R. (1980). Identification of 23 complementation groups required for post-translational events in the yeast secretory pathway. Cell 21, 205–215.

Novick, P. and Zerial, M. (1997) The diversity of Rab proteins in vesicle transport. Curr. Op. Cell Biol. 9, 496–504.

Olszewski, S., Deeney, J.T., Schuppin, G.T., Williams, K.P., Corkey, B.E. and Rhodes, C.J. (1994). Rab3A effector domain peptides induce insulin exocytosis via a specific interaction with a cytosolic protein doublet. J. Biol. Chem. 269, 27987–27991.

Orita, S., Naito, A., Sakaguchi, G., maeda, M., Iragarashi, H., Sasaki, T. and Takai, Y. (1997). Physical and functional interactions of DOC2 and Munc13 in Ca^{2+}-dependent exocytotic machinery. J. Biol. Chem. 272, 16081–16084.

Orita, S., Sasaki, T., Komuro, R., Sakaguchi, G., Maeda, M., Igarashi, H. and Takai, Y. (1996). DOC2 enhances Ca^{2+}-dependent exocytosis from PC12 cells. J. Biol. Chem. 217, 7257–7260.

Orita, S., Sasaki, T., Naito, A., Komuro, R., Ohtsuka, T., Maeda, M., Suzuki, H., Igarashi, H. and Takai Y. (1995). Doc2: A novel brain protein having two repeated C2-like domains. Biochem. Biophys. Res. Commun. 206, 439–448.

Park, J.B., Farnsworth, C.C. and Glomset, J.A. (1997) Ca^{2+}/calmodulin causes Rab3A to dissociate from synaptic membranes. J. Biol. Chem. 272, 20857–20865.

Parsons, T.D., Coorssen, J.R, Horstmann, H and Almers, W. (1995). Docked granules, the exocytotic burst, and the need for ATP hydrolysis in endocrine cells. Neuron 15, 1085–1096.

Pevsner, J., Hsu, S.C. and Scheller, R.H. (1994). n-Sec1: A neural-specific syntaxin-binding protein. Proc. Natl. Acad. Sci. USA 91, 1445–1449.

Porte, D. (1991). β-cells in type II diabetes mellitus. Banting lecture 1990. Diabetes 40, 166–180.

Protopopov, V., Govindan, B., Novick, P. and Gerst, J.E. (1993). Homologs of the synaptobrevin/VAMP family of synaptic vesicle proteins function on the late secretory pathway in S. cerevisiae. Cell 74, 855–861.

Regazzi, R., Kikuchi, A., Takai, Y. and Wollheim, C.B. (1992a). The small GTP-binding proteins in the cytosol of insulin-secreting cells are complexed to GDP dissociation inhibitor proteins. J. Biol. Chem. 267, 17512–17519.

Regazzi, R., Ravazzola, M., Iezzi, M., Lang, J., Zahraoui, A., Andereggen, E., Morel, P., Takai, Y. and Wollheim, C.B. (1996a) Expression, localization and functional role of small GTPases of the Rab3 family in insulin-secreting cells. J. Cell Sci. 109, 2265–2273.

Regazzi, R., Sadoul, K., Meda, P., Kelly, R.B., Halban, P.A. and Wollheim, C.B. (1996b). Mutational analysis of VAMP domains implicated in Ca^{2+}-induced insulin exocytosis. EMBO J. 15, 6951–6959.

Regazzi, R., Sasaki, T., Takahashi, K., Jonas, J.C., Volker, C., Stock, J.B., Takai, Y. and Wollheim, C.B. (1995a). Prenylcysteine analogs mimicking the C-terminus of GTP-binding proteins stimulate exocytosis from permeabilized HIT-T15 cells. Biochim. Biophys. Acta 1268, 268–278.

Regazzi, R., Ullrich, S., Kahn, R.A. and Wollheim, C.B. (1991). Redistribution of ADP-ribosylation factor during stimulation of permeabilized cells with GTP analogues. Biochem. J. 275, 639–644.

Regazzi, R., Vallar, L., Ullrich, S., Ravazzola, M., Kikuchi, A., Takai, Y. and Wollheim, C.B. (1992b). Characterization of small-molecular-mass guanine-nucleotide-binding regulatory proteins in insulin-secreting cells and PC12 cells. Eur. J. Biochem. 208, 729–737.

Regazzi, R., Wollheim, C.B., Lang, J., Theler, J.M., Rossetto, O., Montecucco, C., Sadoul, K., Weller, U., Palmer, M. and Thorens, B. (1995b). VAMP-2 and cellubrevin are expressed in pancreatic β-cells and are essential for Ca^{2+}—but not for GTPγS-induced insulin secretion. EMBO J. 14, 2723–2730.

Ren, X.-D., Bokoch, G.M., Traynor-Kaplan, A., Jenkins, G.H., Anderson, R.A. and Schwartz, M.A. (1996). Physical association of the small GTPase Rho with a 68-kDa phosphatidylinositol 4-phosphate 5-kinase in swiss 3T3 cells. Mol. Biol. Cell 7, 435–442.

Ridley, A.J. (1996). Rho: Theme and variations. Curr. Biol. 6, 1256–1264.

Rossetto, O., Deloye, F., Poulain, B., Pellizzari, R., Schiavo, G. and Montecucco, C. (1995). The metallo-proteinase activity of tetanus and botulism neurotoxins. J. Physiol. 89, 43–50.

Rossetto, O., Gorza, L., Schiavo, G., Schiavo, N., Scheller, R.H. and Montecucco, C. (1996). Vamp/synaptobrevin isoforms 1 and 2 are widely and differentially expressed in nonneuronal tissues. J. Cell Biol. 132, 167–179.

Rothman, J.E. (1994). Mechanisms of intracellular protein transport. Nature 372, 55–63.

Rothman, J.E. and Warren, G. (1994). Implications of the SNARE hypothesis for intracellular membrane topology and dynamics. Curr. Biol. 4, 220–233.

Sadoul, K., Lang, J., Montecucco, C., Weller, U., Regazzi, R., Catsicas, S., Wollheim, C.B. and Halban, P.A. (1995). SNAP-25 is expressed in islets of Langerhans and is involved in insulin release. J. Cell Biol. 128, 1019–1028.

Sakaguchi, G., Orita, S., Maeda, M., Igarashi, H. and Takai, Y. (1996). Molecular cloning of an isoform of DOC2 having two C2-like domains. Biochem. Biophys. Res. Commun. 217, 1053–1061.

Schiavo, G., Benfenati, F., Poulain, B., Rossetto, O., Polverino de Laureto, P., DasGupta, B.R. and Montecucco, C. (1992). Tetanus and botulinum-B neurotoxins block neurotransmitter release by proteolytic cleavage of synaptobrevin. Nature 359, 832–835.

Segev, N., Mulholland, J. and Botstein, D. (1988). The yeast GTP-binding YPT1 protein and a mammalian counterpart are associated with the secretion machinery. Cell 52, 915–924.

Shirataki, H., Kaibuchi, K., Sakoda, T., Kishida, S., Yamaguchi, T., Wada, K., Miyazaki, M. and Takai, Y. (1993). Rabphilin-3A, a putative target protein for smg p25/rab3A p25 small GTP-binding protein related to synaptotagmin. Mol. Cell. Biol. 13, 2061–2068.

Shirataki, H., Kaibuchi, K., Yamaguchi, T., Wada, K., Horiuchi, H. and Takai, Y. (1992). A possible target protein for smg-25A/rab3A small GTP-binding protein. J. Biol. Chem. 267, 10946–10949.

Soldati, T., Shapiro, A.D., Dirac-Svejstrup, B.A. and Pfeffer, S.R. (1994). Membrane targeting of the small GTPase Rab9 is accompanied by nucleotide exchange. Nature 369, 76–78.

Söllner, T., Whiteheart, S.W., Brunner, M., Erdjument-Bromage, H., Geromanos, S.,Tempst, P. and Rothman, J. E. (1993). SNAP receptors implicated in vesicle targeting and fusion. Nature 362, 318–324.

Steyer, J.A., Horstmann, H. and Almers, W. (1997). Transport, docking and exocytosis of single secretory granules in live chromaffin cells. Nature 388, 474–478.

Straub, S.G. and Sharp, G.W.G. (1996). A wortmannin-sensitive signal transduction pathway is involved in the stimulation of insulin release by vasoactive intestinal polypeptide and pituitary adenylate cyclase-activating polypeptide. J. Biol. Chem. 271, 1660–1668.

Stuchfield, J. and Cockcroft, S. (1993). Correlation between secretion and phospholipase D activation in differentiated HL60 cells. Biochem. J. 293, 649–655.

Subramaniam, V.N., Peter, F., Philp, R., Wong, S.H. and Hong, W. (1996). GS28, a 28-kilodalton Golgi SNARE that participates in ER-Golgi transport. Science 272, 1161–1163.

Südhof, T.C. and Rizo, J. (1996). Synaptotagmins: C2-domain proteins that regulate membrane traffic. Neuron 17, 379–388.

Summers, S.A., Guebert, B.A. and Shanahan, M.F. (1996). Polyphosphoinositide inclusion in artificial lipid bilayer vesicles promotes divalent cation-dependent membrane fusion. Biophys. J. 71, 3199–3206.

Takai, Y., Kaibuchi, K., Kikuchi, A. and Kawata, M. (1992). Small GTP-binding proteins. Int. Rev. Cytol. 133, 187–230,

Taylor, S.I., Accili, D. and Imai, Y. (1994). Insulin resistance or insulin deficiency: Which is the primary cause of NIDDM? Diabetes 43, 735–740.

Tolias, K.F., Cantley, L.C. and Carpenter, C.L. (1995). Rho family GTPases bind to phosphoinositide kinases. J. Biol. Chem. 270, 17656–17659.

Turner, R.C., Hattersley, A.T., Shaw, J.T.E. and Levy J.C. (1995). Type II diabetes: Clinical aspects of molecular biological studies. Diabetes 44, 1–10.

Ullrich, O., Stenmark, H., Alexandrov, K., Huber, L.A., Kaibuchi, K., Sasaki, T., Takai, Y. and Zerial, M. (1993). Rab GDP dissociation inhibitor as a general regulator for the membrane association of rab proteins. J. Biol. Chem. 268, 18143–18150.

Ullrich, S. and Wollheim, C.B. (1988). GTP-dependent inhibition of insulin secretion by epinephrine in permeabilized RINm5F cells. J. Biol. Chem. 263, 8615–8620.

Wang, Y., Okamoto, M., Schmitz, F., Hofmann, K. and Shdhof, T.C. (1997) Rim is a putative Rab3 effector in regulating synaptic-vesicle fusion. Nature 388, 593–598.

Wheeler, M.B., Sheu, L., Ghai, M., Bouquillon, A., Grondin, G., Weller, U., Beaudoin, A.R. Bennett, M.K., Trimble, W.S. and Gaisano, H.Y. (1996). Characterization of SNARE protein expression in 8 cell lines and pancreatic islets. Endocrinology 137, 1340–1348.

Whiteheart, S.W., Griff, I.C., Brunner, M., Clary, D.O., Mayer, T., Buhrow, S.A. and Rothman, J.E. (1993). SNAP family of NSF attachment proteins includes a brain-specific isoform. Nature 362, 353–355.

Wiedenmann, C., Schäfer, T. and Burger, M.M. (1996). Chromaffin granule-associated phosphatidylinositol 4-kinase activity is required for stimulated secretion. EMBO J. 15, 2094–2101.

Wilson, D.W., Wilcox, C.A., Flynn, G.C., Chen, E., Kuang, W., Henzel, W.J., Block, M.R., Ullrich, A. and Rothman, J.E. (1989). A fusion protein required for vesicle-mediated transport in both mammalian cells and yeast. Nature 339, 355–359.

Wollheim, C.B., Lang, J. and Regazzi, R. (1996). The exocytotic process of insulin secretion and its regulation by Ca^{2+} and G-proteins. Diabetes Rev. 4, 276–297.

Yamaguchi, T., Shiratachi, H., Kishida, S., Miyazaki, M., Nishikawa, J., Wada, K., Numata, S., Kaibuchi, K. and Takai, Y. (1993). Two functionally different domains of rabphilin-3A, Rab3A p25/smg p25A-binding and phospholipid- and Ca^{2+}-binding domains. J. Biol. Chem. 268, 27164–27170.

SECTION III

REGULATION OF INSULIN SECRETION

POTENTIATORS AND INHIBITORS OF INSULIN SECRETION

Bo Ahrén

Advances in Molecular and Cell Biology
Volume 29, pages 175-197.

ISBN: 0-7623-0547-9

I. INTRODUCTION

A. Glucose Is a Main Regulator of β-Cell Function

Insulin is the most important anabolic hormone and of most importance for the acute regulation of carbohydrate metabolism. It is therefore expected that its secretion from the β-cells is carefully regulated by a variety of factors. The main regulator of insulin secretion is glucose. The relationship between circulating glucose and insulin secretion is sigmoidal, and in healthy subjects, the linear portion of the sigmoidal curve is seen at glucose concentrations between 5 and 9 mmol/l (Karam et al., 1974; Byrne et al., 1995), which is the range of glucose occurring physiologically. For example, after an overnight fast, circulating glucose is approximately 3 to 5 mmol/l, which is below the threshold for insulin secretion; hence, fasting insulin levels are low. After meal intake, glucose levels increase to 6 to 8 mmol/l; this stimulates insulin secretion, resulting in increased circulating insulin in order to facilitate the peripheral glucose uptake to preserve normal glucose levels. When glucose levels subsequently decline, the stimulus for insulin secretion is reduced and therefore circulating insulin returns to baseline.

B. Glucose Is Not the Only Regulator of β-Cell Function

Although glucose is the main regulator of insulin secretion, other regulators exist as well, and these factors may be more important than glucose under certain conditions. For example, during fasting, circulating glucose levels decline toward 3 mmol/l, which is below the threshold for glucose to stimulate insulin secretion. Nevertheless, circulating insulin is rarely reduced to levels below 40 to 60 pmol/l, indicating that other factors than glucose replace the role of the sugar to stimulate insulin secretion under this condition. Simultaneously, other factors than hypoinsulinemia, like glucagon and catecholamines, prevent the evolvement of frank hypoglycemia under this condition (Boyle et al., 1995). Furthermore, following food intake, insulin secretion is stimulated more markedly than expected from the resulting hyperglycemia, which is due to the potentiating actions of the so-called incretin hormones released from the gut (Creutzfeldt, 1979; Fehmann et al., 1995). Moreover, during certain types of stress, such as severe diseases (Taylor and Majid, 1971) or physical exercise (Galbo et al., 1978), circulating insulin is inadequately increased when judged in relation to the concurrent hyperglycemia, which is due to inhibition of insulin secretion by stress-related factors, as for example the sympathoadrenal system.

Thus, although glucose is the main regulator of insulin secretion, nonglucose factors modulate the insulinotropic response to the sugar in a potentiating or inhibiting direction to allow the optimal insulin secretion yielding the required circulating insulin for each moment. Conversely, however, glucose is also permissive for these nonglucose factors to affect insulin secretion. For example, the insulinotropic action of cholecystokinen (CCK) and glucagonlike peptide-1 (GLP-1) is strictly dependent on the glucose level (Karlsson and Ahrén, 1992; Fehmann et al., 1995). Thus, when glucose levels fall, CCK and GLP-1 are no longer able to stimulate insulin secretion, and, conversely, when hyperglycemia exists, the insulinotropic action of these non-glucose stimuli are potentiated. This is of particular interest in Type 2 diabetes, when the insulinotropic actions of the nonglucose stimuli are potentiated by the concurrent hyperglycemia, which will compensate the initial defect in insulin secretion accompanying the disease (Porte, 1991). The integrated regulation of β-cells by glucose and nonglucose potentiators and inhibitors of glucose-stimulated insulin secretion is thus of major importance for normal carbohydrate metabolism, and its possible perturbations during physiological challenges and the disturbed metabolism in diabetes are important to delineate.

II. ISLET ANATOMY AND SIGNALING PATHWAYS

A. Anatomical Basis for Nonglucose Regulation of Insulin Secretion

The nonglucose factors that modulate glucose-stimulated insulin secretion in a potentiating or inhibiting direction are of endocrine, neurocrine, paracine, or autocrine nature (Table 1). The anatomical basis for the multiregulated integrated influence on insulin secretion by these factors is the microorganization of the pancreatic islets (Bonner-Weir, 1991). Figure 1 illustrates the components in an islet that are of importance for the modulation of insulin secretion. The islets consist of different endocrine cell types: β-cells occurring in the center of the islet surrounded by α-, δ-, and ζ-cells in the peripheral mantle zone. Each of these endocrine cells produces regulatory peptides with potential influence on insulin secretion. Furthermore, a rich neural network innervates the islets, being both parasympathetic, sympathetic, and sensory, and neurotransmitters released from their nerve terminals exert potential influence on insulin secretion. Finally, there is a rich blood supply to the islets, allowing circulating hormones together with glucose and other nutrient insulin secretagogues, like amino acids and fatty acids, to reach the β-cells and affect insulin secretion. All these impulses integrate on the level of the β-cells, enabling an integrated response in relation to the requirement

Table 1. Potentiators and Inhibitors of Insulin Secretion

Name	*Abbreviation*	*Source*	*Effect on Insulin Secretion*	*Mode of influence*
Islet amyloid polypeptide	IAPP	β-cells	Inhibitory	Autocrine
Pancreastatin	—	β-cells	Inhibitory	Autocrine
Glucagon	—	α-cells	Stimulatory	Paracrine
Peptide YY	PYY	α-cells	Inhibitory	Paracrine
Somatostatin		δ-cells	Inhibitory	Paracrine
Calcitonin gene-related polypeptide	CGRP	δ-cells and sensory nerves	Inhibitory	Paracrine and neurocrine
Diazepam binding inhibitor	DBI	δ-cells	Inhibitory	Paracrine
Pancreatic polypeptide	PP	ζ-cells	Inhibitory	Paracrine
Norepinephrine	NE	Sympathetic nerves	Inhibitory or stimulatory	Neurocrine
Galanin	—	Sympathetic nerves	Inhibitory	Neurocrine
Neuropeptide Y	NPY	Sympathetic nerves	Inhibitory	Neurocrine
Acetylcholine	ACh	Parasympathetic nerves	Stimulatory	Neurocrine
Gastrin releasing peptide	GRP	Parasympathetic nerves	Stimulatory	Neurocrine
Vasoactive intestinal polypeptide	VIP	Parasympathetic nerves	Stimulatory	Neurocrine
Pituitary adenylate cyclase–activating polypeptide	PACAP	Nerves	Stimulatory	Neurocrine
Cholecystokinin	CCK	Nerves ι-cells	Stimulatory	Neurocrine Endocrine
Glucagonlike peptide-1	GLP-1	λ-cells in the gut	Stimulatory	Endocrine
Gastric inhibitory polypeptide	GIP	κ-cells in the gut	Stimulatory	Endocrine
Epinephrine	—	Adrenal medulla	Stimulatory or inhibitory	Endocrine

as reflected by the level of circulating factors, neuronal activity, or secretory degree of the other islet endocrine cells.

B. Signaling Basis for Nonglucose Regulation of Insulin Secretion

The messages modulating insulin secretion are integrated at the level of the β-cells, enabling an optimal discharge of insulin from each individual cell. The modulating factors interact with signaling pathways in the β-cells which are of importance for the exocytosis of the secretory granules containing insulin (Prentki and Matschinsky, 1987; Wolf et al., 1994; Sharp, 1996). For simplicity, the mechanisms of insulin secretion may be divided in proximal and distal events. The proximal events are those initially responding to a stimulus and leading to production

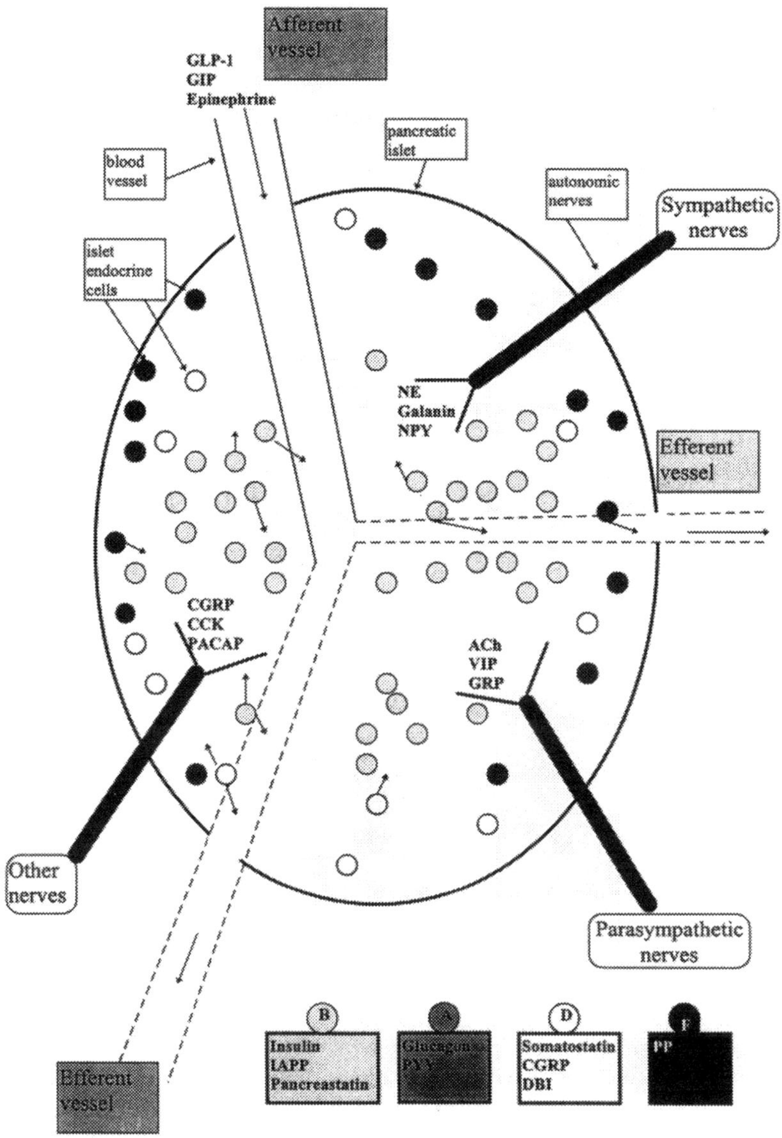

Figure 1. Anatomy of a typical pancreatic islet. The β-cells (B) are located in the center of the islet, surrounded by the α-, δ- and ζ-cells (A, D, and F). The regulatory peptides produced by these endocrine cells, shown at bottom in the figure, are released from the cells and may affect nearby cells or leave the islet through the efferent vessels. Furthermore, afferent vessels penetrate to the center of the islet, where permeable, fenestrated, efferent vessels are formed. This organization implies that nutrients and hormones reach the islet endocrine cells. Finally, different types of nerves innervate the islets and the islet endocrine cells, and some neurotransmitters are illustrated. For a key to the abbreviations of regulatory peptides and neurotransmitters used in this figure, see Table 1.

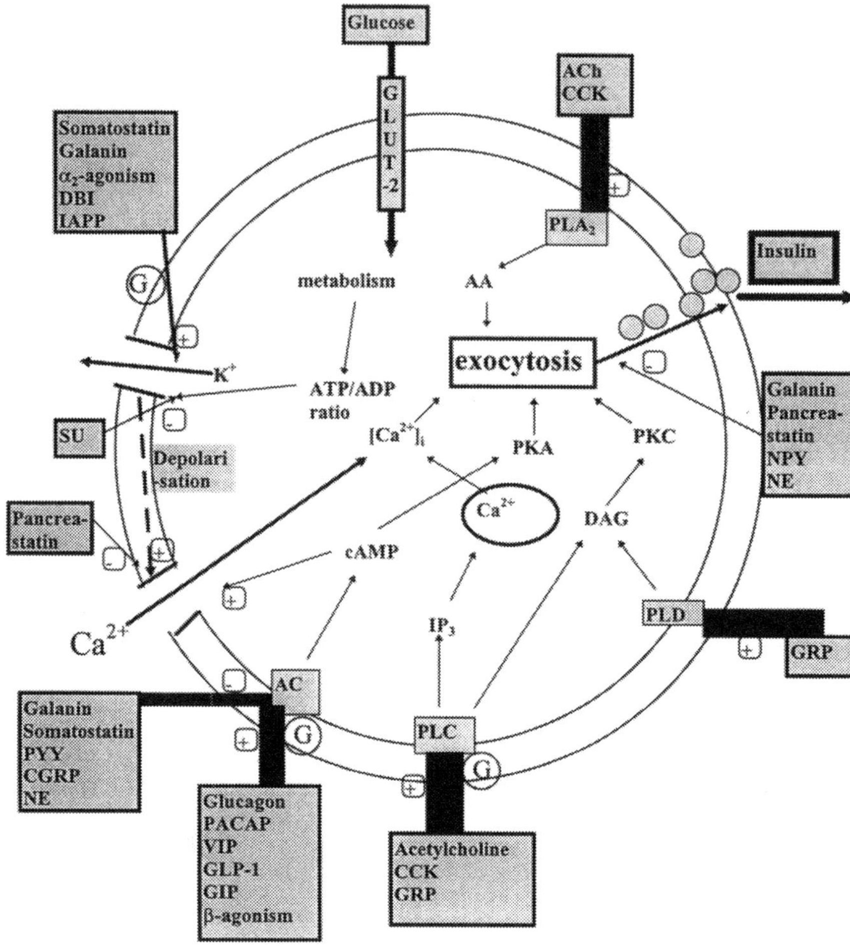

Figure 2. Overview of signaling pathways involved in the regulation of insulin secretion with emphasis on processes involved in modulation of exocytosis by inhibitors and potentiators. Several routes of effectuation of the exocytosis are shown. A first route is the uptake of glucose into the cell by a facilitating glucose transporter (GLUT-2 in rat islets). Inside the cell, glucose is metabolized and adenosine trisphosphate (ATP) is generated, which increases the ATP/adenosine diphosphate (ADP) ratio and closes ATP-regulated K^+ channels, which also are closed by exogenously added sulphonyureas (SU). This will, through depolarization, stimulate opening of voltage-sensitive Ca^{2+} channels, which will, in turn, stimulate Ca^{2+} uptake and thereby increase the cytosolic concentration of Ca^{2+} ($[Ca^{2+}]_i$). A second route is activation of adenylate cyclase, which will stimulate formation of cyclic AMP (cAMP). The cAMP then stimulates opening of voltage-sensitive Ca^{2+} channels as well as activation of protein kinase A (PKA). A third route is activation of phospholipase C (PLC) which, through hydrolysis of phosphoinositides, stimulates the formation of inositol-1,4,5-trisphosphate (IP_3) and diacyl glycerol (DAG). IP_3 stimulates the release

or activation of a second messenger which transduces the signal to the exocytosis process. Five different proximal events are of particular interest in view of potentiating insulin secretion (Figure 2).

1. One mechanism is depolarization of the plasma membrane due to reduction of K^+ permeability, after closure of adenosine trisphosphate (ATP)–dependent K^+ channels or other K^+ channels. This results in opening of voltage-sensitive Ca^{2+} channels, uptake of extracellular Ca^{2+}, and a raise of the cytosolic concentration of Ca^{2+}. This chain of events is induced by the glucose metabolism and its increase in the rate between ATP and adenosine diphosphate (ADP), which is initiated after uptake of glucose by means of specific glucose transporters in the plasma membrane (GLUT-2 in the rat) and phosphorylation of the glucose molecule by means of glucokinase.
2. A second mechanism is the activation of adenylate cyclase, which generates cyclic AMP, subsequently activating protein kinase A (PKA) and increasing the uptake of extracellular Ca^{2+}, raising the cytosolic concentration of Ca^{2+}.
3. A third mechanism is activation of phospholipase C (PLC), which hydrolyzes phosphoinositides, yielding production of inositol-1,4,5-trisphosphate (IP_3) and diacyl glycerol (DAG). IP_3 diffuses into the cytoplasm and promotes liberation of Ca^{2+} from Ca^{2+} storage sites, which increases the cytosolic Ca^{2+} concentration, whereas DAG activates protein kinase C (PKC).
4. A fourth mechanism is stimulation of phosphdipase D and *de novo* synthesis of DAG.
5. Finally, a fifth mechanism is the activation of phospholipase A_2 (PLA_2) which causes the production of arachidonic acid and subsequent production of arachidonic acid metabolites through the cyclooxygenase and lipoxygenase pathways, mainly prostaglandins and leukotrienes.

Figure 2. (continued)

of Ca^{2+} from intracellular storage sites into the cytoplasm, whereas DAG stimulates the activity of protein kinase C (PKC). A fourth route is the stimulation of phospholipase D (PLD) activity, which stimulates the de novo synthesis of DAG. Finally, a fifth route is the stimulation of phospholipase A_2 (PLA_2) activity, which induces formation of arachidonic acid (AA). All these routes will, through their end products (Ca^{2+}, PKA, PKC, and AA) or metabolites or events induced by them, activate the exocytotic discharge of secretory granules with extrusion of insulin to the extracellular space, which is a process involving several different proteins. Potentiators and inhibitors of insulin secretion modulate the activity of these different routes, as indicated in the figure. For a key to the abbreviations of regulatory peptides and neurotransmitters used in this figure, see Table 1.

These proximal events thus increase the cytosolic concentration of Ca^{2+}, stimulate the formation of cyclic AMP and arachidonic acid and its metabolites, as well as activate PKA and PKC, which all are stimulatory inducers of the distal events in the signal transductions of the β-cells. These distal events are mainly related to the translocation of the secretory granules to the plasma membrane, fusion of the granules and the cell membrane, and discharge of the contents of the secretory granules to the extracellular space. These mechanisms are not as well understood as the primary events of insulin secretion. It has been thought, however, that the distal events involve the proteins located in the membrane of secretory granules, synaptotagmin, vesicle-associated membrane protein, and synaptobrevin, as well as the proteins located at the plasma membrane, syntaxin and synaptosome-associated protein (Sharp, 1996). In any case, factors that modulate insulin secretion induce changes either in the proximal events or in the distal events, or both, and the receptors for the potentiators and inhibitors of insulin secretion characteristically induce signals through one or several of these mechanisms, often following transduction of the message through G-proteins.

III. LOCAL (AUTOCRINE/PARACRINE) REGULATION OF INSULIN SECRETION

The four main endocrine cells in the pancreatic islets produce a variety of regulatory peptides, which have stimulatory or inhibitory actions on insulin secretion. They, therefore, have the potential to affect cells located closely to their cells of origin, which is called autocrine or paracrine regulation.

A. β-Cell Regulation of Insulin Secretion

β-*cells* produce, besides insulin, the two regulatory peptides, *pancreastatin* and *islet amyloid polypeptide (IAPP)* (Smith and Bloom, 1995; Sanchez-Margalet et al., 1996; Ahrén, 1996a). Pancreastatin is a 49-amino acid peptide derived from proteolytic cleavage of chromogranin A and has been demonstrated to be produced in β-cells, as well as in δ- and α-cells (Sanchez-Margalet et al., 1996). The peptide is released from the pancreas in parallel with insulin (Ahrén et al., 1991b) and has been demonstrated to inhibit glucose-stimulated secretion under a variety of experimental conditions (Ahrén et al., 1988a; Lindskog et al., 1992; Ahrén et al., 1996a; Sanchez-Margalet et al., 1996), although no pancreastatin receptors have so far been demonstrated in islets. Mechanistically, pancreastatin has been shown to abolish the efflux of $^{45}Ca^{2+}$ from prelabeled islets without affecting K^{+} permeability, suggesting that the peptide impairs the uptake of extracellular Ca^{2+} through a direct action on the Ca^{2+} channels (Lindskog et al., 1992). This assumption is supported by results showing that the cytosolic concentration of Ca^{2+} is reduced by the peptide in isolated islet cells (Ma et al., 1996). On the other hand, pancreastatin has also been shown to inhibit insulin secretion by a distal action directly on the exocytotic mechanisms (Hertelendy et al., 1996). Therefore, different inhibitory mechanisms may be induced by the peptide. The physiological importance of the inhibitory action of the peptide on insulin secretion remains to be established. Being a product of the β-cells, the autocrine inhibition may be a means to protect the β-cells from exhaustion, or a mean to finely tune a specific stimulus. It has also been suggested that pancreastatin is a hormone of importance for the hyperglycemia that occurs during stress, as besides its inhibition of insulin secretion, the peptide also stimulates hepatic glucose production, thus creating hyperglycemia (Sanchez-Margalet et al., 1996).

IAPP (sometimes called amylin) is produced in the islet β-cells. IAPP was initially isolated from amyloids in the islets, which occur in long-standing Type 2 diabetes (Clark, 1992; Westermark et al., 1992). The peptide is processed from an 89 amino acid propeptide through proteolytic cleavage at dibasic sites by the same enzymes that process proinsulin; in its final form it consists of 37 amino acids (Cooper, 1994). It is stored in the β-cell secretory granules, in which the content of IAPP is approximately 1% to 2% of that of insulin, and under normal conditions, IAPP is secreted from the β-cells in parallel with insulin (Kahn et al., 1990). It has been studied whether the peptide affects insulin secretion. Most studies on its potential influence on insulin secretion have reported an inhibitory action, although high dose levels of the peptide are required for this action (Ahrén and Pettersson, 1990; Ar'Rajab and Ahrén, 1991; Wagoner et al., 1993; Cooper, 1994). These high dose levels may, however, be physiological, since a high local concentration of IAPP may occur in the close vicinity of the β-cells due to its combined release with insulin. Furthermore, it has been argued that exogenously applied IAPP has to break through the endogenous IAPP to execute any effect, which may require high doses (Cooper, 1994). That IAPP inhibits insulin secretion under physiological conditions is also supported by studies showing that an IAPP antagonist, $IAPP_{8-37}$, potentiates insulin secretion (Smith and Bloom, 1995) and that IAPP hyperpolarizes islet cells and thereby prevents glucose from depolarizing the cells, which might be induced by increasing the K^+ permeability (Wagoner et al., 1993). However, more studies are required to elucidate the mechanism of the inhibitory action of IAPP on insulin secretion, and it has to be emphasized that so far the IAPP receptor has not been identified or localized.

It has also been claimed that IAPP as a hormone exerts insulin antagonistic actions on the peripheral glucose metabolism. Thus, IAPP has been shown to inhibit glycogen synthesis in skeletal muscle (Leighton and Cooper, 1988) and to inhibit peripheral glucose uptake in rats (Ahrén and Pettersson, 1990), although it has been argued that these effects are seen only at supraphysiological doses of the peptide and therefore are probably not physiological (Smith and Bloom, 1995). However, it was recently shown that the glucose elimination rate following intravenous or oral glucose challenge is increased in mice targeted for a disruption of the IAPP gene in spite of exhibiting a normal insulin response, suggesting a physiological peripheral effect of IAPP, at least in rodents (Gebre-Medhin et al., 1997).

Whether this occurs due to indirect influences on other hormones or through altered direct insulinlike actions remains to be etablished. Thus, IAPP as a regulatory peptide may inhibit both insulin secretion and insulin action.

During recent years, much attention has also been given to the ability of IAPP to form amyloid in islets of subjects with Type 2 diabetes (Clark, 1992; Westermark et al., 1992; Cooper, 1994). Thus, patients with long-standing Type 2 diabetes often exhibit islet amyloid. Whether this is secondary to islet dysfunction or a primary event thus leading to islet dysfunction is a question of debate. Recent findings show that expression of IAPP is increased under conditions of increased demand of the β-cells, such as insulin resistance and experimental diabetes in rodents (Mulder et al., 1995). This could be of importance for the pathogenesis of Type 2 diabetes, since increased expression of the peptide could result in exaggerated release and, therefore, an accentuation of its inhibitory action on insulin secretion and action. Increased IAPP expression could also exaggerate the amyloid formation, which tentatively might be deleterious for islet function since IAPP fibrils have been claimed to be toxic for the β-cells (Lorenzo et al., 1994). However, more research is required to establish the physiological and pathophysiological implication of IAPP and its islet actions, as well as its potential mechanisms of action in the islets.

B. α-Cell Regulation of Insulin Secretion

The *α-cells* produce and secrete the two regulatory peptides: *glucagon* and *peptide YY* (PYY) (Ahrén, 1996a). Glucagon is the main α-cell product; it is a 29-amino acid peptide that is processed from the 160-amino acid, proglucagon (Fehmann et al., 1995). It should be emphasized that proglucagon is expressed both in the pancreatic α-cells and the intestinal λ-cells, but due to tissue-specific differences in the proglucagon processing, glucagon is produced only in the pancreatic α-cells. The other regulatory peptide of the α-cells, PYY, is a 36-amino acid peptide that shows structural stimilarities to neuropeptide Y (NPY) and pancreatic polypeptide (PP) (Tatemoto et al., 1982). The two peptides, glucagon and PYY, are stored together in the secretory granules of the α-cells (Böttcher et al., 1989).

When the α-cells are activated, which occurs during parasympathetic and sympathetic nerve activation and hypoglycemia, both glucagon and PYY are secreted; therefore, these two regulatory peptides may influence insulin secretion by a paracrine manner under these conditions. Glucagon has thereby been demonstrated to stimulate insulin secretion (Ahrén et al., 1987), which is an effect exerted through activation of high-affinity glucagon receptors on the β-cells and mediated by the formation of cyclic adenosine monophosphate (cAMP) (Goldfine et al., 1972). PYY, on the other hand, has been demonstrated to inhibit glucose-stimulated insulin secretion (Böttcher et al., 1989; Nieuwenhuizen et al., 1994). The mechanism of inhibition by PYY of glucose-stimulated insulin secretion has been thought to be dependent both on inhibition of the production and on effects of cAMP. Thus, in isolated islets, PYY inhibits both the concentration of cAMP and the cAMP-induced insulin secretion, whereas the peptide is without influence on the cytosolic concentration of Ca^{2+} in isolated islet cells (Nieuwenhuizen et al., 1994). Hence, the α-cell might by a local (paracrine) action influence the nearby β-cells both in a stimulatory (glucagon) and in an inhibitory (PYY) manner. Interestingly, a recent study showed that the inhibition of insulin secretion exerted by PYY is of physiological importance, since incubation of isolated islets in the presence of a specific PYY antiserum was found to increase insulin secretion (Karlsson and Ahrén, 1996b). The physiological implication of this dual action of the α-cells on the β-cells remains, however, to be established.

C. δ-Cell Regulation of Insulin Secretion

The *δ-cells* produce three different regulatory peptides: *somatostatin, calcitonin gene-related peptide (CGRP),* and *diazepam-binding inhibitor (DBI)* (Ahrén and Pettersson, 1990; Östenson et al., 1991; Ahrén, 1996a). Somatostatin occurs both as a 14-amino acid and as a 28-amino acid form, both processed from the 92-amino acid precursor, prosomatostatin (Holst et al., 1991). CGRP, which is also a constituent of pancretic sensory nerves and, in some species, is produced in the β-cells, is a 37-amino acid peptide showing structural similarities with IAPP (Ahrén and Pettersson, 1990). The third peptide localized to δ-cells in some species, notably also in humans, is DBI, which is an 86-amino acid peptide initially isolated from the small intestine (Östenson et al., 1991). These regulatory peptides are released from the δ-cells when the δ-cells are activated, which is in parallel with the β-cells. Thus, both glucose and parasympathetic activation stimulate the secretion of the δ-cells, whereas sympathetic activation inhibits δ-cell secretion.

Several studies have demonstrated that all three δ-cell regulatory peptides—somatostatin, CGRP, and DBI—inhibit insulin secretion (Ahrén and Pettersson, 1990; Östenson et al., 1991; Ahrén, 1996a). The mechanisms underlying the inhibitory action on insulin secretion by somatostatin and CGRP have been studied in detail. It has thereby been demonstrated that somatostatin inhibits β-cell secretion by inducing hyperpolarization of the plasma membrane through opening of the ATP-dependent K^+ channels, which will close the voltage-sensitive Ca^{2+} channels and therefore reduce the cytosolic concentration of Ca^{2+}; an action that seems to be mediated by G-proteins (Sharp, 1996; Ma et al., 1996). CGRP, on the other hand, seems to inhibit insulin secretion mainly by inhibiting the production of cAMP, and this effect also seems induced by a G-protein (Pettersson and Ahrén, 1990). In addition, CGRP slightly reduces the cytosolic Ca^{2+}, as evident from reduced $^{45}Ca^{2+}$-efflux from preloaded rat islets; but this effect is slight and probably of marginal importance in relation to the abolishment of glucose-stimulated increase in the islet concentration of cAMP (Pettersson and Ahrén, 1990). The mechanism of the inhibitory action of DBI on insulin secretion has not yet been established, although it has been suggested that the peptide interfers either with the closure of ATP-regulated K^+ channels or with the formation of cAMP induced by glucose (Östenson et al., 1991). Thus, somatostatin, CGRP, and DBI all inhibit insulin secretion, suggesting that the δ-cells, which are activated concomitantly with β-cells, as a main effect restrain insulin secretion. The δ-cells may therefore be of importance for protecting β-cells from exhaustion damage during continuous stimulation, and, furthermore, the δ-cells may, if overproducing their inhibitory peptides, be of pathophysiological importance in diabetes. However, the physiological and pathophysiological implications of the modulatory inhibition by δ-cell peptides on β-cells remain to be established.

D. ζ-Cell Regulation of Insulin Secretion

ζ-cells produce *pancreatic polypeptide (PP)* (Larsson et al., 1976), which is a peptide consisting of 36 amino acids and with structural similarities to NPY and PYY (Tatemoto et al., 1982). PP is stored in secretory granules of the ζ-cells (Larsson et al., 1976) and is released mainly during parasympathetic nerve activation (Ahrén et al., 1986). The peptide has been demonstrated to inhibit insulin secretion under experimental conditions (Ahrén, 1996a), and it may therefore be suggested that PP locally exerts a negative feedback influence on insulin secretion during conditions when the parasympathetic tonus is high. However, no clear evidence for a physiological role of PP in the regulation of insulin secretion exists, and, furthermore, the mechanisms of the proposed inhibitory influence on insulin secretion have not been studied.

E. Summary

Both morphological and functional studies, as reviewed thus far, indicate that the regulatory peptides produced by the islet endocrine cells affect insulin secretion through paracrine or local actions, as illustrated in Figure 1. The paracrine modulations are either stimulatory (glucagon) or inhibitory (PYY, somatostatin, CGRP, DBI, PP), to which autocrine inhibitory modulation may be added (IAPP, pancreastatin). The evidence for a local regulatory concept, the main influence thus being inhibitory, include, beside the functional studies, also the close anatomical apposition of the various endocrine cells to each other where gap or tight junctions in some places unite nearby δ-cells (Maeda, 1996). Furthermore, a local intra-islet blood flow enables secretory products from the β-cells to reach α- and δ-cells (Samols and Stagner, 1991). However, the paracrine concept of regulation of insulin secretion has also been questioned in light of studies using antibodies directed against the peptides in the perfused dog pancreas, as well as the argument that a local high concentration of a peptide would have downregulated its own receptor (Samols and Stagner, 1991). Nevertheless, the specific action of certain of the effects reported above by islet peptides on insulin secretion and the likelihood that these effects are exerted only under certain conditions make it likely that insulin secretion is regulated both in an autocrine and a paracrine manner. For future studies, it is important to establish the physiological conditions of these local regulations and their possible pathophysiological implication.

IV. NEUROCRINE REGULATION OF INSULIN SECRETION

A. Parasympathetic Regulation of Insulin Secretion

Preganglionic parasympathetic nerves traverse within the vagus and enter the pancreas together with the arteries. They terminate in intrapancreatic ganglia, from which postganglionic fibers emanate, which end in close proximity to vessels and islet endocrine cells (Ahrén et al., 1986). The nerve terminals of the postganglionic cholinergic nerves harbor not only the classical neurotransmitter, *acetylcholine*, but also the neuropeptides *vasoactive intestinal polypeptide (VIP)* and *gastrin-releasing polypeptide (GRP)* (Ahrén and Karlsson, 1994). This suggests that at least three different signals might mediate the message of parasympathetic nerve activation on islet function. It is thereby of importance that vagal nerve activation as well as acetylcholine, VIP, and GRP all stimulate insulin secretion (Ahrén et al., 1986; Ahrén and Karlsson, 1994). All three parasympathetic neurotransmitters may therefore contribute to the stimulation of insulin secretion obtained by vagal nerve activation. However, the relative contribution of the various neurotransmitters might differ between species. For example, in the dog, the insulinotropic action of parasympathetic nerve activation is largely inhibited by muscarinic receptor antagonism by atropine, indicating mediation by acetylcholine in this species (Ahrén and Taborsky, 1986), whereas in the pig, atropine does not prevent vagally induced insulin secretion, indicating mediation by the neuropeptides in the pig (Holst et al., 1981). Whether the relative contribution of these neurotransmitters also differs under physiological or pathophysiological conditions remains to be established.

The three parasympathetic neurotransmitters use different mechanisms to stimulate insulin secretion. The stimulation of insulin secretion by acetylcholine or its analogue, carbachol, has been shown to involve activation of muscarinic receptors of the m_3 subtype (Karlsson and Ahrén, 1993) with a subsequent activation of PLC and formation of IP_3 and DAG (Ahrén et al., 1990). Carbachol has also been shown to activate PLA_2, suggesting also that this signaling mechanism is involved in cholinergically induced insulin secretion (Konrad et al., 1992). In contrast, a main mechanism for VIP in its action to stimulate insulin secretion is the activation of adenylate cyclase and formation of cAMP, although also a cAMP-independent mechanism seems to be involved (Straub and Sharp, 1996). Finally, the mechanism of the insulinotropic action of GRP involves both activation of PLC (Karlsson and Ahrén, 1996a)—that is, a similar mechanism as acetylcholine—and a mechanism suggested to involve yet another phospholipase species, phospholipase D (PLD), which induces de novo synthesis of DAG (Gregersen and Ahrén, 1996).

The parasympathetic nervous system is of main importance for insulin secretion during food intake, primarily in the early, so-called cephalic, phase. Due to activation of the parasympathetic nervous system, insulin secretion is stimulated prior to any changes in the circulating level of glucose, and it has been suggested that this action is of importance for the preservation of normal glucose tolerance during meal intake. Thus, prevention of the early cephalic insulin secretion during food intake results in glucose intolerance, in spite of minimal reduction in absolute levels of insulin secretion during the entire period of the food intake (Teff and Engelman, 1996).

B. Sympathetic Regulation of Insulin Secretion

The islets are also innervated by sympathetic nerves. These fibers originate in the celiac ganglion, with the preganglionic fibers passing in the splanchnic nerves, and after entering the pancreas, the postganglionic sympathetic nerves terminate in close connection to vessels and islet endocrine cells (Ahrén et al., 1986). As demonstrated by immunocytochemistry, these nerves harbor in most species the classical neurotransmitter *norepinephrine*, as well as the neuropeptides *galanin* and *neuropeptide Y (NPY)* (Ahrén and Karlsson, 1994). Activation of the sympathetic nerves has been demonstrated to inhibit both basal and stimulated insulin secretion (Ahrén et al., 1986). The inhibition of glucose-stimulated insulin secretion has been shown to be reproduced by infusion of norepinephrine and since the selective α_2-adrenoceptor agonist, clonidine, exerts a similar action, it is thought that norepinephrine as a sympathetic neurotransmitter inhibits glucose-stimulated insulin secretion through activating α_2-adrenoceptors (Ahrén, 1996a). In contrast, although sympathetic nerve activation also inhibits basal insulin secretion, this is not reproduced by intrapancreatic infusion of norepinephrine or prevented by complete α- and β-adrenoceptor antagonism (Ahrén and Karlsson, 1994). Therefore, inhibition of basal insulin release by sympathetic nerve activation seems instead to be mediated by the two neuropeptides confined to the sympathetic nerve terminals in islets, galanin and NPY. Both these neuropeptides are released from the pancreas, as is norepinephrine, upon sympathetic nerve activation (Ahrén and Karlsson, 1994) and both inhibits basal insulin secretion in experimental animals (Ahrén and Lindskog, 1992; Skoglund et al., 1993; Smith and Bloom, 1995). Galanin has been of particular interest in this respect, since it has been found that the peptide is released from islet nerve terminals during sympathetic nerve stimulation in sufficient quantities to inhibit insulin secretion in the dog (Dunning and Taborsky, 1988) and that inhibition of insulin secretion during physical exercise in mice is prevented by immunoneutralization of galanin (Dunning et al., 1991). However, species differences seem to exist and, for example, although the peptide is present in human islet nerves and inhibits insulin secretion from isolated human islets at a high dose level (Ahrén et al., 1991a), it does not suppress insulin secretion in vivo in humans (Ahrén, 1990), and furthermore, in baboons, galanin seems to be a parasympathetic neurotransmitter in the islets without influencing insulin secretion (Verchere et al., 1996).

The mechanisms of the inhibitory influences by α_2-adrenoceptor activation and galanin of insulin secretion show similar characteristics. Thus, both norepinephrine (or clonidine) and galanin induce hyperpolarization of the β-cells, due to opening of K^+ channels, which through closure of the voltage-sensitive Ca^{2+} prevents the uptake of extracellular Ca^{2+}, which, in turn, will reduce the cytoplasmic concentration of Ca^{2+}. In addition, norepinephrine and galanin inhibit the production of cAMP, and, furthermore, both galanin and norepinephrine inhibit insulin secretion through a direct inhibitory action on the exocytosis (Ahrén and Lindskog, 1992; Sharp, 1996). In contrast, the mechanism underlying the inhibitory action of NPY on insulin secretion awaits to be established. Studies undertaken thus far have indicated that NPY does not influence the production of cAMP in islets; that NPY inhibits insulin secretion stimulated by forskolin, which activates adenylate cyclase; and that NPY and the Ca^{2+}-channel antagonist verapamil exert additive inhibitory actions on insulin secretion (Skoglund et al., 1993). These results suggest that NPY inhibits insulin secretion at a distal site in the signaling events, although no direct evidence for this exists as yet.

The sympathetic nerves are probably of main importance for the regulation of insulin secretion during stress, when the sympathetic activation is concurrent with stress-induced hyperglycemia and inhibited insulin secretion. The relative contribution of norepinephrine, galanin, and NPY to this influence remains to be established. The potential mediation by three neurotransmitters may, however, offer selectivity for the sympathetic nervous system.

C. Regulation of Insulin Secretion by Other Neuropeptides

Besides VIP and GRP as constituents of parasympathetic nerves, and galanin and NPY as constituents of sympathetic nerves, several other neuropeptides have also been demonstrated in islet nerves (Ahrén and Karlsson, 1994). These neuropeptides are neurotransmitters in sensory nerves, or in other, not yet completely identified, nerves. One of these neuropeptides is CGRP, which, as reported above also occurs in islet endocrine cells. CGRP has been localized to sensory nerves and inhibits insulin secretion (Ahrén and Karlsson, 1994). Another pancreatic neuropeptide is *pituitary adenylate cyclase–activating polypeptide (PACAP),* which has been demonstrated in nerve terminals in intrapancreatic ganglia as well as within the exocrine portion of the pancreas and in the islets (Ahrén and Karlsson, 1994). PACAP was initially localized to the hypothalamus, but later was demonstrated to have widespread localization in nerves throughout the body. At least three different receptors with affinity for

PACAP exist. The peptide is a potent stimulator of insulin secretion (Straub and Sharp, 1996; af Klinteberg et al., 1996) and, in one study, PACAP was shown to stimulate insulin secretion from isolated islets at an extremely low concentration, suggesting that it is more potent on a molar basis than other insulinotropic peptides (Yada et al., 1994). Mechanistic studies have shown that the peptide increases the formation of cAMP in insulin-producing tissues, which is a likely main signaling mechanism underlying its potent insulinotropic action (Straub and Sharp, 1996; af Klinteberg et al., 1996). The peptide also slightly increases the cytosolic concentration of Ca^{2+} and stimulates phosphoinositide hydrolysis which, however, most likely is secondary to the raised content of cAMP (af Klinteberg et al., 1996). The physiological function of this potent insulinotropic polypeptide remains to be established.

Another pancreatic neuropeptide with potent insulinotropic activity is *cholecystokinin (CCK)*. This peptide has been demonstrated in nerve terminals in islet nerves and stimulates insulin secretion (Karlsson and Ahrén, 1992). CCK is also a gastrointestinal hormone released into the bloodstream during food intake, and several lines of evidence exist suggesting that CCK might contribute, albeit not as the main factor, to the large insulin response to intake of a meal by acting as an incretin (Karlsson and Ahrén, 1992). CCK has been shown to stimulate insulin by activating the CCK-A-receptors in a glucose-dependent manner: at glucose levels below 3 mmol/l, CCK does not stimulate insulin secretion, whereas a clear insulinotropic action is seen at glucose levels exeeding 7 mmol/l (Karlsson et al., 1994). Its signaling mechanism involves activation of PLC with stimulated formation of IP_3, a subsequent release of Ca^{2+} from intracellular storage sites, and an increased cytoplasmic concentation of Ca^{2+}. The activation of PLC also causes formation of DAG which, together with Ca^{2+}, activates PKC, in turn stimulating the exocytosis process. A second mechanism underlying the insulinotropic action of CCK is therefore activation of PKC, which is evident by the impairment of CCK-stimulated insulin secretion following downregulation of PKC activity (Karlsson and Ahrén, 1992). However, this mechanism seems not to be the sole mechanism underlying CCK-stimulated insulin secretion, since downregulation of PKC only partially inhibits CCK-stimulated insulin secretion, suggesting that other mechanisms also are involved. Recently, activation by CCK of PLA_2 was suggested as such an additional mechanism of CCK in islets (Simonsson et al., 1996). Thus, CCK was found to increase the efflux of ^3H-labeled arachidonic acid from prelabeled islets, which is an indirect method for determination of PLA_2 activity, and CCK-stimulated insulin secretion from islets was inhibited by a specific PLA_2-inhibitor. The formed AA might be the final second messenger, rather than its cyclooxygenase- or lipoxygenase-derived metabolites, since inhibition of these enzymes did not affect CCK-stimulated insulin secretion. Hence, CCK may, as both a gut hormone and a pancreatic neuropeptide, stimulate insulin seretion in a glucose-dependent manner by activating PLC and perhaps also PLA_2.

V. ENDOCRINE REGULATION OF INSULIN SECRETION

A. Gastrointestinal Hormones and Insulin Secretion

Several different gastrointestinal hormones have been characterized; most of them are released into the circulation after food intake to regulate various homeostatic mechanisms after food intake (Ahrén, 1996b). Some of the gastrointestinal hormones have been found to affect insulin secretion, and since it has been demonstrated that gut hormones profoundly potentiate the postprandial insulin secretion, the hormones have been thought to allow a marked insulin secretion during food intake. It has been discussed which of the gut hormones are the most important *incretin* factors (Creutzfeldt, 1979). Today, *CCK*, which also is a pancreatic neuropeptide and as such discussed previously, *gastric inhibitory polypeptide (GIP),* and *glucagon-like peptide-1 (GLP-1)* are the most recognized of the gut hormones in this respect. All of these gut hormones are released into the circulation following food intake and stimulate insulin secretion. Their roles as incretins have been verified in rats, where inhibition of CCK, GIP, or GLP-1-receptors has been demonstrated to inhibit insulin secretion after food intake (Ahrén, 1996b). However, although this may be true in some species, like rodents, it has been questioned whether CCK and GIP are physiological incretins in humans, since inhibition of their release or action does not seem to inhibit postprandial insulin secretion in the human (Ahrén, 1996b). In contrast, GLP-1 is considered an important incretin hormone in humans (Kreyman et al., 1987). It should be emphasized, however, that synergistic actions of these hormones should not be overlooked, and the gut hormones may therefore be of importance for postprandial islet function by acting together. Nevertheless, most data today suggest that GLP-1 is the gut hormone of main importance in this respect.

The mechanism of the insulinotropic action of CCK has been discussed earlier, as CCK also is a pancreatic neuropeptide. GIP, however, is exclusively a hormone which is produced in the so-called κ-cells of the upper small intestine (Ahrén, 1996b); the peptide consists of 42 amino acids and shows structural similarities to glucagon and VIP. Its potentiating action on glucose-stimulated insulin secretion has been verified in human islets, where its receptors have also been cloned and the signaling of GIP to stimulate insulin secretion has been found to rely on a potentiated formation of cAMP (Gremlich et al., 1995). GLP-1, which consists of 30 amino acids, is processed from proglucagon in the λ-cells in the distal small intestine (Fehmann et al., 1995). GLP-1 potently stimulates insulin secretion both in experimental animals and in humans (Fehmann et al., 1995). The GLP-1 receptor has been cloned, and its signaling pathway involves formation of cAMP and increase of the cytoplasmic concentration of Ca^{2+}, the latter effect possibly being secondary to the adenylate cyclase–activating effect (Fehmann et al., 1995; Ahrén et al., 1996b). It is also of interest that

GLP-1 exerts antidiabetogenic action in subjects with Type 2 diabetes (Gutniak et al., 1992; Ahrén et al., 1997), and, therefore, the use of this gut hormone in the treatment of diabetes has been discussed as a possibility in the future (Gutniak et al., 1996).

B. Adrenal Medullary Hormone (Epinephrine) and Insulin Secretion

Epinephrine (adrenalin) has also been suggested to be involved in the regulation of insulin secretion. Epinephrine activates both the α- and the β-adrenoceptors, and the net effect in regard to insulin secretion may therefore be complex, since activation of α-adrenoceptors inhibits insulin secretion whereas activation of β-adrenoceptors stimulates insulin secretion (Ahrén et al., 1986). A rationale for these complex interactions might be that epinephrine exerts opposite action on insulin secretion under baseline versus stimulated conditions. For example, it has been demonstrated that epinephrine inhibits glucose-stimulated insulin secretion (Porte, 1967), whereas the hormone stimulates basal insulin secretion (Ahrén et al., 1988b). Thus, epinephrine might be part of the sympathoadrenal system inhibiting insulin seretion during stress, and at the same the hormone might participate in the stabilization of insulin secretion under fasting conditions, although this latter hypothesis remains to be verified experimentally. Another possibility is that the effect of epinephrine on insulin secretion depends on its concentration, and therefore perhaps on the degree of stress. Thus, it has been demonstrated in human islets that epinephrine produces a large inhibition of insulin secretion at a high concentration, whereas at a low concentration, a stimulated insulin secretion is observed (Lacey et al., 1993). Similarly, in dogs, epinephrine at a low dose stimulates insulin secretion whereas at a high dose, an inhibitory effect is executed (Ahrén et al., 1988b). On the other hand, it has also been suggested that epinephrine does not contribute to changes in insulin secretion during physiological conditions (Houwing et al., 1995). Therefore, more studies are required to solve the issue of whether epinephrine is of physiological importance for insulin secretion, and the relative contribution of α- versus β-adrenoceptors for such a potential influence needs to be established.

The mechanisms underlying influences on insulin secretion by α- and β-adrenoceptor activation, respectively, have been fairly well characterized. Thus, the inhibition of insulin secretion induced by activation of the α-adrenoceptors is caused by opening of ATP-regulated K^+ channels, hyperpolarization, and reduction in cytosolic concentration of Ca^{2+} (see earlier discussion), although specifically for epinephrine, opening of another K^+ channel which is distinct from the ATP-regulated one has also been suggested (Rorsman et al., 1991). In contrast, β-adrenoceptor activation has been demonstrated to stimulate the formation of cAMP in islets (Lacey et al., 1993).

VI. CONCLUSIONS

The pancreatic islet is a complexly organized sturcture composed of at least four different endocrine cells, each producing a variety of different regulatory peptides. Furthermore, the islet is richly innervated by autonomic nerves, harboring a variety of neuropeptides in addition to the classical neurotransmitters, norepinephrine and acetylcholine. Moreover, the islet is richly supplied with vessels, allowing nutrients and hormones to reach the islet endocrine cells. This structural organization enables a complex regulation of insulin secretion, since the islet β-cells are affected by regulatory peptides produced by the islet endocrine cells, by neuropeptides and other neurotransmitters from the islet nerves, and by circulating hormones. This creates local autocrine and paracrine as well as neurocrine and endocrine regulation of insulin secretion, with the inhibitors and potentiators of insulin secretion modulating the minute-to-minute insulin secretion. A conceptual issue at present is to understand the similarity both in action and mechanism of several of the regulatory peptides. For example, somatostatin and galanin on one hand, and VIP and PACAP on the other hand, exert similar actions through similar mechanisms, although they seem to have different sources and possible different physiological functions. One potential explanation for this is to enable modulations under a variety of different conditions, since the cellular sources of these seemingly similar regulatory peptides differ, enabling them to be of importance under different conditions. However, although the cellular localization, potential effects, and mechanisms are known for a number of the peptides, their contribution to physiological regulation remains in most cases to be established. The exploratory research during the past two decades has, therefore, created the basis for more integrative research, from both a physiological and a pathophysiological perspective.

ACKNOWLEDGMENTS

The studies by the author have been supported by the Swedish Medical Research Council (Grant No. 14X-6834), Ernhold Lundström, Albert Påhlsson, Crafoord and Novo Nordic Foundations, Swedish Diabetes Association, and the Faculty of Medicine, Lund University.

REFERENCES

af Klinteberg, K., Karlsson, S. and Ahrén, B. (1996). Signaling mechanisms underlying the insulinotyropic effect of pituitary adenylate cycliase-activating polypeptide in HIT-T15 cells. Endocrinology 137, 2791–2798.

Ahrén, B. (1990). Effects of galanin and calcitonin gene-related peptide on insulin and glucagon secretion in man. Acta Endocrinol. 123,591–597.

Ahrén, B. (1996a). In: *Essentials of Experimental Surgery: Gastroenterology and Endocrinology.* (Jensen, S.L., Gregersen, H., Moody, F. and Shokouh-Amiri, M.H., Eds.), pp. 43/1–43/18. Harwood Academic, London.

Ahrén, B. (1996b). In: *Essentials of Experimental Surgery: Gastroenterology and Endocrinology.* (Jensen, S.L., Gregersen, H., Moody, F. and Shokouh-Amiri, M.H., Eds.), pp. 12/1–12/24. Harwood Academic, London.

Ahrén, B., Ar'Rajab, A., Böttcher, G., Sundler, F. and Dunning, B.E. (1991a). Presence of galanin in human pancreatic nerves and inhibition of insulin secretion from isolated human islets. Cell Tissue Res. 264, 263–267

Ahrén, B., Bertrand, G., Roye, M. and Ribes, G. (1996a). Pancreastatin modulates glucose-stimulated insulin secretion from the perfused rat pancreas. Acta Physiol. Scand. 158, 63–70.

Ahrén, B. and Karlsson, S. (1994). In: *Insulin Secretion and Pancreatic β-cell Research.* (Flatt, R. and Lenzen, S., Eds.), pp. 313–318. Smith-Gordon, London.

Ahrén, B., Karlsson, S. and Lindskog, S. (1990). In: *Progress in Brain Research*, vol. 84: *Cholinergic Neurotransmission.* (Aquilonius, S. and Gillberg, P.G., Eds.), pp. 209–218. Elsevier, Amsterdam.

Ahrén, B. Larsson, H. and Holst, J.J. (1997). Effects of glucagon-like peptide-1 on islet function and insulin sensitivity in noninsulin-dependent diabetes mellitus. J. Clin. Endocrinol. Metab. 82, 473–478.

Ahrén, B. and Lindskog, S. (1992). Galanin and the regulation of islet hormone secretion. Int. J. Pancreatol. 11, 147–160.

Ahrén, B., Lindskog, S., Tatemoto, K., Efendic, S. (1988a). Pancreastatin inhibits insulin secretion and stimulates glucagon secretion in mice. Diabetes 37, 281–285.

Ahrén, B., Nobin, A. and Scherstén, B. (1987). Insulin and C-peptide secretory responses to glucagon in man: Studies on the dose-response relationships. Acta Med. Scand. 221, 185–190.

Ahrén, B., Östenson, C.G., Mårtensson, H. and Efendic, S. (1991b). Pancreatic release of pancreastatin in the pig. Pancreas 6, 324–329.

Ahrén, B. and Pettersson, M. (1990). Calcitonin gene-related peptide (CGRP) and amylin and the endocrine pancreas. Int. J. Pancreatol. 6, 1–15.

Ahrén, B., Simonsson, E. and Karlsson, S. (1996b). Glucagon-like peptide-$1_{(7\text{-}36)}$amide and cytoplasmic calcium in insulin producing cells. Acta Physiol. Scand. 157, 333–341.

Ahrén, B. and Taborsky, G.J. Jr. (1986). The mechanism of vagal nerve stimulation of glucagon and insulin secretion in the dog. Endocrinology 118, 1551–1557.

Ahrén, B., Taborsky, G.J. Jr. and Porte, D. Jr. (1986). Neuropeptidergic versus cholinergic and adrenergic regulation of islet hormone secretion. Diabetologia 28, 827–836.

Ahrén, B., Veith, R.C. and Taborsky, G.J. Jr. (1988b). The effects of epinephrine on islet hormone secretion in the dog. Int. J. Pancretol. 3, 375–388.

Ar'Rajab, A. and Ahrén, B. (1991). Effects of amidated rat islet amyloid polypeptide on glucose-stimulated insulin secretion *in vivo* and *in vitro* in rats. Eur. J. Pharmacol. 192, 443–445.

Bonner-Weir, S. (1991). In: *The Endocrine Pancreas.* (Samoles, E., Ed.), pp. 15–27. Raven Press, New York.E21–E27.

Böttcher, G., Ahrén, B., Lundquist, I. and Sundler, F. (1989). Peptide YY: Intrapancreatic localization and effects on insulin and glucagon secretion in the mouse. Pancreas 4, 282–288.

Boyle, P.J., Shah, S.D. and Cryer, P.E. (1989). Insulin, glucagon, and catecholamines in prevention of hypoglycemia during fasting. Am. J. Physiol. 256, E651–661.

Byrne, M.M., Sturis, J. and Polonsky, K.S. (1995). Insulin secretion and clearance during low-dose graded glucose infusion. Am. J. Physiol. 268, E21-E27.

Clark, A. (1992). Islet amyloid: An enigma of type 2 diabetes. Diabetes/Metabol. Rev. 8, 117–132.

Cooper, G.J.S. (1994). Amylin compared with calcitonin gene-related peptide: structure, biology, and relevance to metabolic disease. Endocr. Rev. 15, 163–201.

Creutzfeldt, W. (1979). The incretin concept today. Diabetologia 16, 75–85.

Dunning B.E., Karlsson, S. and Ahrén, B. (1991). Contribution of galanin to stress-induced impairment of insulin secretion in swimming mice. Acta Physiol. Scand. 143, 145–152.

Dunning, B.E. and Taborsky G.J. Jr. (1988). Galanin—sympathetic neurotransmitter in endocrine pancreas? Diabetes 37, 1157–1162.

Fehmann, H.C., Göke, R. and Göke, B. (1995). Cell and molecular biology of the incretin hormones glucagon-like peptide-1 and glucose-dependent insulin releasing polypeptide. Endocr. Rev. 16, 390–410.

Galbo, H., Richter, E.A., Christensen, N.J. and Holst, J.J. (1978). Sympathetic control of metabolic and hormonal responses to exercise in rats. Acta Physiol. Scand. 102, 441–449.

Gebre-Medhin, S., Mulder, H., Pekny, M., Zhang, Y.Z., Törnell, J., Westermark, P., Sundler, F., Ahrén B. and Betsholtz, C. (1997). Altered glucose homeostasis, body weight and nociception in IAPP (amylin) null mutant mice. Diabetes 46 (Suppl. 1), 29A.

Goldfine, I.D., Roth, J. and Birnbaumer, L. (1972). Glucagon receptors in the beta-cells. Binding of ^{125}I-glucagon and activation of adenylate cyclase. J. Biol. Chem. 247, 1211–1218.

Gregersen, S. and Ahrén, B. (1996). Studies on the mechanisms by which gastrin releasing peptide stimulates insulin secretion from mouse islets. Pancreas 12, 48–57.

Gremlich, S., Porret, A., Hani, E.H., Cherif, D., Vionnet, N., Froguel, P. and Thorens, B. (1995). Cloning, functional expression, and chromosomal localization of the human pancreatic islet glucose-dependent insulinotropic polypeptide receptor. Diabetes 44, 1202–1208.

Gutniak, M.K., Larsson, H., Heiber, S.J., Juneskans, O.T., Holst, J.J. and Ahrén, B. (1996). Potential therapeutic levels of glucagon-like peptide-1 achieved in humans by a buccal tablet. Diabet Care 19, 843–848.

Gutniak, M., Ørskov, C., Holst, J.J., Ahrén, B. and Efendic, S. (1992). Antidiabetogenic effect of glucagon-like peptide-1 (7-36) amide in normal subjects and patients with diabetes mellitus. N. Engl. J. Med. 326, 1316–1322.

Hertelendy, Z.I., Patel, D.G. and Knittel, J.J. (1996). Pancreastatin inhibits insulin secretion in RINm5F-cells through obstruction of G-protein mediated, calcium-directed exocytosis. Cell Calcium 19, 125–132.

Holst, J.J, Grönholt, R., Schaffalitzky de Muckadell, O.B. and Fahrenkrug, J. (1981). Nervous control of pancreatic endocrine secretion in pigs. II. The effect of pharmacological blockade on the response to vagal stimulation. Acta Physiol. Scand. 111, 9–14.

Holst, J.J., Ørskov, C. and Bersani, M. (1991). In: *The Endocrine Pancreas.* (Samols, E., Ed.), pp. 125–152. Raven Press, New York.

Houwing, H., Frankel, K.M., Strubbe, J.H., van Suylichem, P.T. and Steffens, A.B. (1995). Role of the sympathoadrenal system in exercise-induced inhibition of insulin secretion. Effects of islet transplantation. Diabetes 44, 565–571.

Kahn, S.E., D'Alessio, J.W., Schwartz, M.W., Fujimoto, W.Y., Ensinck, J.W., Taborsky, G.J. Jr. and Porte, D. Jr. (1990). Evidence of cosecretion of islet amyloid polypeptide by β-cells. Diabetes 39, 634–638.

Karam, J.H., Grodsky, G.M., Ching, K.N., Schmid, F., Burrill, K. and Forsham, P.H. (1974). "Staircase" glucose stimulation of insulin secretion in obesity. Measure of beta-cell sensitivity and capacity. Diabetes 23, 673–770.

Karlsson, S. and Ahrén, B. (1992). Cholecystokinin and the regulation of insulin secretion. Scand. J. Gastroenterol. 27, 161–165.

Karlsson, S. and Ahrén, B. (1993). Muscarinic receptor subtypes in carbachol-stimulated insulin and glucagon secretion in the mouse. J. Autonom. Pharmacol. 13, 439–446.

Karlsson, S. and Ahrén, B. (1996a). Gastrin-releasing peptide mobilizes calcium from intracellular stores in HIT-T15 cells. Peptides 17, 909–916.

Karlsson, S. and Ahrén, B. (1996b). A role for islet peptide YY in the regulation of insulin secretion Acta. Physiol. Scand. 157, 305–306.

Karlsson, S., Fridolf, T. and Ahrén, B. (1994). In: *Insulin Secretion and Pancreatic β-cell Research.* (Flatt, P.R. and Lenzen, S., Eds.), pp. 337–342. Smith-Gordon, London.

Konrad, R.J., Jolly, Y.C., Major, C. and Wolf, B.A. (1992). Carbachol stimulation of phospholipase A_2 and insulin secretion in pancreatic islets. Biochem. J. 287, 283–290.

Kreyman, B., Ghatei, M.A., Williams, G. and Bloom, S.R. (1987). Glucagon like peptide-1(7-36): A physiological incretin in man. Lancet 2, 1300–1304.

Lacey, R.J., Cable, H.C., James, R.F., London, N.J., Scarpello, J.H. and Morgan, N.G. (1993). Concentration-dependent effects of adrenaline on the profile of insulin secretion from isolated human islets. J. Endocrinol. 138, 555–563.

Larsson, L.I., Sundler, F. and Håkanson, R. (1976). Pancreatic polypeptide—a postulated new hormone: identification of its cellular storage site by light and electron immunocytochemistry. Diabetologia 12, 211–216.

Leighton, B. and Cooper, G.J. (1988). Pancreatic amylin and calcitonin gene-related peptide cause resistane to insulin in skeletal muscle in vitro. Nature 335, 632–635.

Lindskog, S., Skoglund, G. and Ahrén, B. (1992). Pancreastatin inhibits insulin secretion from isolated rat islets: studies on its mechanism of action. Diabet. Res. 19, 119–123.

Lorenzo, A., Razzaboni, B., Weir, G.C. and Yankner, B.A. (1994). Pancreatic islet cell toxicity of amylin asociated with type-2 diabetes mellitus. Nature 368, 756–760.

Ma, H.T., Kato, M. and Tatemoto, K. (1996). Effects of pancreastatin and somatostatin on secretagogue-induced rise in intracellular free calcium in single rat pancreatic islet cells. Regul. Pept. 61, 173–178.

Maeda, P. (1996). Gap junction involvement in secretion: the pancreas experience. Clin. Exp. Pharmacol. Physiol. 23, 1053–1057.

Mulder, H., Ahrén, B. and Sundler, F. (1995). Recent advances in localization and quantification of gene expression—A lesson from islet amyloid polypeptide. Cell Vision 2, 94–103.

Nieuwenhuizen, A.G., Karlsson, S., Fridolf, T. and Ahrén, B. (1994). Mechanisms underlying the insulinostatic effect of peptide YY in mouse pancreatic islets. Diabetologia 37, 871–878.

Östenson, C.G., Ahrén, B., Johansson, O., Karlsson, S., Hilliges, M. and Efendic, S. (1991). Diazepam binding inhibitor and the endocrine pancreas. Neuropharmacology 30, 1391–1398.

Pettersson, M. and Ahrén, B. (1990). Calcitonin gene-related peptide inhibits insulin secretion. Studies on ion fluxes and cyclic AMP in isolated rat islets. Diabet. Res. 15, 9–14.

Porte, D. Jr. (1967). A receptor mechanism for the inhibition of insulin release by epinephrine in man. J. Clin. Invest. 46, 86–94.

Porte, D. Jr. (1991). β-Cells in type II diabetes mellitus. Diabetes 40, 166–180.

Prentki, M. and Matschinsky, F. (1987). Ca^{2+}, cAMP and phospholipid-derived messengers in coupling mechanisms of insulin secretion. Physiol. Rev. 67, 1185–1248.

Rorsman, P., Bokvist, K., Ammala, C., Arkhammar, P., Berggren, P.O., Larsson, O. and Wahlander, K. (1991). Activation by adrenaline of a low-conductance G protein-dependent K^+ channel in mouse pancreatic β-cells. Nature 349, 77–79.

Samols, E and Stagner, J.I. (1991). In: The *Endocrine Pancreas.* (Samols, E., Ed.), pp. 93–124. Raven Press, New York.

Sanchez-Margalet, V., Lucas, M. and Goberna, R. (1996). Pancreastatin: Further evidence for its consideration as a regulatory peptide. J. Molecul. Endocrinol. 16, 1–8.

Sharp, G.W.G. (1996). Mechanisms of inhibition of insulin release. Am. J. Physiol. 271, C1781–1799.

Simonsson, E., Karlsson, S. and Ahrén, B. (1996). Involvement of phospholipase A_2 and arachidonic acid in cholecystokinin-8-induced insulin secretion in rat islets. Regul. Pept. 65,101–107.

Skoglund, G., Gross, R., Ahrén, B. and Loubatieres-Mariani, M.M. (1993). Different mechanisms are involved in neuropeptide Y-induced pancreatic vasoconstriction and inhibition of insulin secretion. Eur. J. Pharmacol. 236, 69–74.

Smith, D.M. and Bloom, S.R. (1995). Paracrine/autocrine control of the islet and the amylin family. Biochem. Soc. Trans. 23,336–340.

Straub, S.G. and Sharp, G.W.G. (1996). A wortmannin-sensitive signal transduction pathway is involved in the stimulation of insulin release by vasoactive intestinal polypeptide and pituitary adenylate cyclase-activating polypeptide. J. Biol. Chem. 271, 1660–1668.

Tatemoto, K., Carlquist, M. and Mutt, V. (1982). Neuropeptide Y—a novel brain peptide with structural similarities to peptide YY and pancreatic polypeptide. Nature 296,659–660.

Taylor, S.H. and Majid, P.A. (1971). Insulin and the heart. J. Mol. Cell. Cardiol. 2, 293–317.

Teff, K.L and Engelman, K. (1996). Oral sensory stimulation improves glucagon tolerance in humans: effects on insulin, C-peptide, and glucagon. Am. J. Physiol. 270, R1371–R1379.

Verchere, C.B., Kowalyk, S., Koerker, D.J., Baskin, D.G. and Taborsky Jr, G.J. (1996). Evidence that galanin is a parasympathetic, rather than a sympathetic, neurotransmitter in the baboon pancreas. Regul. Pept. 67, 93–101.

Wagoner, P.K., Chen, C., Worley, J.F., Dukes, I.D. and Oxford, G.S. (1993). Amylin modulates β-cell glucose sensing via effects on stimulus-secretion coupling. Proc. Natl. Acad. Sci. USA 90, 9145–9149.

Westermark, P., Johnson, K.H., O'Brien, T.D. and Betsholtz, C. (1992). Islet amyloid polypeptide—a novel controversy in diabetes research. Diabetologia 35, 297–303.

Wolf, B.A., Konrad, R.J. and Alter, C.A. (1994). In: *Insulin Secretion and Pancreatic β-cell Research.* (Flatt, R.R. and Lenzen, S., Eds.), pp. 297–304. Smith-Gordon, London.

Yada, T., Sakurada, M, Ihida, K., Nakata, M. Murata, F., Arimura, A. and Kikuchi, M. (1994). Pituitary adenylate cyclase activating polypeptide is an extraordinarily potent intra-pancreatic regulator of insulin secretion from islet beta-cells. J. Biol. Chem. 269, 1290–1293.

SIGNAL RECOGNITION
GLUCOSE AND PRIMARY STIMULI

Emile Van Schaftingen and Frans Schuit

Advances in Molecular and Cell Biology
Volume 29, pages 199-226.

ISBN: 0-7623-0547-9

I. INTRODUCTION

The β-cell secretes insulin in response to nutrients, principally glucose and amino acids, and to a number of hormones and pharmacological agents. Whereas the second class of compounds have rapid effects that are mediated by their binding to a receptor or to an ion channel, signal transduction from glucose and from most other nutrients requires the metabolism of these energy substrates and the generation of a signal—a rise in the adenosine triphosphate (ATP)/adenosine diphosphate (ADP) ratio—which in turn acts on K^+ channels to decrease the conductance of the plasma membrane to this cation. Ensuing depolarization of the plasma membrane leads to opening of Ca^{2+} channels and to an increase in the cytosolic concentration of this divalent ion, which triggers exocytosis.

This basic scenario will most likely have to be modified in the future as there is growing evidence for other factors intervening at various steps in this process (Detimary et al., 1994; Newgard and McGarry, 1995; Prentki, 1996). This is not really a surprise, because metabolic regulation is usually very complex and redundant. For instance, glucose utilization by the liver is affected directly by the blood glucose concentration (both glucokinase and glycogen phosphorylase are glucose receptors) but also indirectly by the changes in the concentrations of insulin and glucagon. The redundancy of regulation is probably a necessity: the sensitivity of an enzyme, defined as the relative change of its catalytic activity divided by the relative change in the concentration of a substrate or an effector, is at most equal to its Hill co-efficient. The value of the latter is usually not higher than 2 to 3 (for regulatory enzyme displaying positive cooperativity) but very often equal to 1 (for enzymes with Michaelis-Menten kinetics). It is, therefore, not surprising that fine-tuning of the blood glucose concentration between narrow limits requires the summing up of a number of different mechanisms whose individual sensitivity to glucose is not particularly elevated.

Many reviews have been written on the mechanism by which nutrients induce insulin secretion in the β-cells (Meglasson and Matschinsky, 1986; Matschinsky, 1990; Liang and Matschinsky, 1994; Newgard and McGarry, 1995; Prentki, 1996). The emphasis of this chapter is on the early steps of glucose metabolism in the β-cell and more particularly on the biochemical and molecular characteristics of the enzymes and other proteins involved in this process. The far-reaching goal is to explain what it is that gives the β-cell its unique capacity of converting the concentration of a nutrient into a suitable intracellular signal.

It is interesting in this respect to compare and contrast the β-cell with another cell type that shares certain metabolical features and is also very well studied: the liver parenchymal cell. The function of the liver is, however, very different: it is to maintain metabolic homeostasis by storing and degrading glycogen, by making glucose from a nonglucose precursor, by interconverting amino acids, by converting toxic compounds into water-soluble, excretable products, and so on. The metabolic capacity of the liver is, therefore, much more varied than that of the β-cell and much more intense, even when expressed per gram of tissue.

II. GLUCOSE ENTRY

During the three decades following the first description of facilitated diffusion of D-glucose across the β-cell plasma membrane by Hellman et al. (1971), the role of glucose transport in the control of glucose-induced insulin release has been intensively studied (Unger, 1991; Thorens, 1992; Newgard, 1996; Schuit, 1997). Both in whole mouse islets (Hellman et al., 1971) and in purified rat β-cells (Heimberg et al., 1993), the maximal velocity of glucose transport largely exceeds the flux of glucose metabolism, a situation similar to that observed in liver. It can therefore be proposed that the main physiological function of glucose transport in β-cells is a rapid equilibration between the intra- and extracellular glucose concentration in order to allow accurate "measurement" of blood glucose by "sensor" proteins located inside the cells. This mode of action is further assisted by the high blood flow of pancreatic islets (Svensson et al., 1988), preventing a "sink" in blood glucose that could result from the inevitable glucose usage by the cells lining the islet vessels.

While many molecular and cellular aspects of glucose transport as a possible step in the metabolic control of β-cells have been reviewed before (Unger, 1991; Thorens, 1992; Newgard, 1996; Schuit, 1997), some recent results remain intriguing. In rodent β-cells, the predominant transporter is the same as in the liver, glucose transporter type 2 (GLUT-2). This transporter can be kinetically distinguished from the other GLUT isoforms by a high Km for glucose and a high transport capacity (Thorens et al., 1988). In addition, GLUT-1 mRNA and protein is present in purified rat β-cells (Heimberg et al., 1995) but since no low Km transport is detected in these cells, its contribution to glucose sensing is not clear.

In agreement with the predominant—but possibly not exclusive—role of GLUT-2 in rodent β-cell glucose transport, Thorens and co-workers recently reported the phenotype of mice with a homozygous or heterozygous GLUT-2 null mutation (Guillam et al., 1997). While the heterozygous animals present no abnormalities, the homozygous GLUT-2 –/– animals die soon after birth with a phenotype comprising severe Type 1 diabetes, glycosuria and dehydration. Islets isolated from the neonatal –/– animals are poorly responsive to glucose but normally responsive to arginine and D-glyceraldehyde. Since this defect persists even after 24 h in tissue culture, its cause must be the absence of GLUT-2 and not the "diabetic environment" of the animals (Guillam et al., 1997). Interestingly, a slow and partial insulin secretion after acute stimulation with 20 mM glucose is still present in the –/– islets. This GLUT-2–independent secretion could be blocked by antimycin A, suggesting its dependence on glucose metabolism. It will be important to investigate in the near future if GLUT-1 is responsible for the remaining glucose-responsiveness in this model. It will also be interesting to study the effect of GLUT-1 overexpression in GLUT-2–negative β-cells.

In this perspective, it is intriguing that human islet cells (De Vos et al., 1995; Ferrer et al., 1995), but not human liver (Schuit, 1997), express GLUT-2 at much lower levels—approximately two orders of magnitude—than the corresponding cells or tissue in the rat. Moreover, human islet cells predominantly express GLUT-1 and GLUT-3 (De Vos et al., 1995), which may explain why glucose transport in human cells has a lower Km and Vmax than in the rat. Both theoretical modeling (Chen et al., 1994a; Sweet and Matschinsky, 1997) and expression studies in oocytes (Morita et al., 1994) support the idea that low Km transport is compatible with high Km glucose metabolism as long as glucose transport is not limiting for overall glycolytic flux. This condition is met in isolated human islet cells (De Vos et al., 1995).

What may then be the function of GLUT-2 in human islets? One possibility is that the low expression of GLUT-2 in human β-cells is required for a nontransport function. Some data in the literature are not well explained simply by the GLUT-2 transport function. For instance, Newgard and co-workers showed that insulin release from AtT-20ins cells becomes glucose-responsive after transfecting GLUT-2 cDNA (Hughes et al., 1992), but that this effect is not accompanied by an increase in glycolytic flux and can be dissociated from accelerated glucose transport (Hughes et al., 1993). In this respect, it may be worthwile to investigate if GLUT-2 can act as a signal-transducing protein, analogous to the Snf3p and Rgt2p glucose transporters, which act as true glucose receptors in *Saccharomyces cerevisiae* (Öczan et al., 1996). It should be noted, however, that any direct evidence for this concept is lacking in mammalian cells. A possible connection between GLUT-2 and signalling in β-cells is suggested by the fact that GLUT-2 can be phosphorylated in situ by cyclic AMP (cAMP)–dependent protein kinase (Thorens et al., 1996), but the physiological implication of this phenomenon remains to be assessed. Nontransport functions of GLUT-2 could involve its physical association with proteins implicated in the proximal part of glucose metabolism or in the effector function of insulin granule exocytosis. It may be speculated that the low expression of GLUT-2 in human pancreas (De Vos et al., 1995; Ferrer et al., 1995) is required for such an as yet undiscovered function of the protein. However, without further evidence for a nontransport function, it seems reasonable to summarize at this point that glucose transport in mammalian β-cells can be adequately carried out by various isoforms of the GLUT family as long as the velocity of glucose uptake is not rate-limiting for the overall metabolic flux in the cells.

III. GLUCOSE PHOSPHORYLATION

Present evidence indicates that glucokinase catalyzes the flux-determining step of glycolysis in β-cells and that it is responsible for the glucose sensitivity of these cells. It is, therefore, of interest to review the properties of this enzyme, which is present also in the liver, in α-cells and in other cell types.

A. Kinetic Properties of Glucokinase

Glucose phosphorylation is catalyzed in mammalian tissues by four hexokinases that all belong to the same protein family. Hexokinase IV, better known as glucokinase (Iynedjian, 1993; Van Schaftingen et al., 1994; Cardenas, 1995), differs from the three other isozymes not only by its smaller size (≈ 50 kDa vs. 100 kDa) but, more importantly, by its kinetic properties: (1) low affinity for glucose, with $S_{0.5}$ values typically in the 2 to 10 mM range, depending on the species; (2) sigmoidal saturation curve for its physiological substrate glucose, but also for mannose, with a Hill co-efficient of about 1.5; (3) lack of inhibition by physiological (< 1 mM) concentrations of glucose 6-phosphate; (4) sensitivity to inhibition by long chain acyl-CoAs; and (5) sensitivity to inhibition by a regulatory protein, which is controlled by fructose 6-phosphate and fructose-1-phosphate in mammalian liver. These properties have been conserved through evolution, indicating their potential importance for the function of this enzyme.

Liver and β-cell glucokinases are encoded by the same gene, which contains two different promoters and, hence two different first exons, the remaining nine exons being common (Magnuson and Shelton, 1989; Iynedjian, 1993). In both cases, the ATG codon resides in the first exon, which makes the two protein differ by their first 15 or so amino acids. However, the kinetic properties of the liver and islet form of enzyme are the same (Veiga-da-Cunha et al., 1996), which indicates that the difference in the amino-terminal region of this enzyme has no kinetic consequence. The use of two different promoters allows, however, a separate regulation of the expression of the two enzymes. The liver promoter is positively affected by insulin and negatively by glucagon, through a cAMP-dependent process (Nouspikel and Iynedjian, 1992). The expression of the islet enzyme is under the positive control of glucose by translational or post-translational mechanisms (Liang et al., 1992). It is worthy of note that the maximal activity of glucokinase is about fivefold more elevated in the liver than in β-cells. The reason for this difference is that the liver is an effector organ in which glucokinase has to process large quantities of substrate, whereas in islets, this enzyme serves to control the ATP/ADP ratio.

The first three kinetic properties of glucokinase mentioned above makes it a suitable glucose sensor. This is evident for the $S_{0.5}$ value, which is above the normal blood glucose concentration, and for the sigmoidal saturation curve, which increases the sensitivity to glucose. The lack of noncompetitive inhibition by glucose 6-phosphate is equally important, since such an inhibition would operate mainly at elevated glucose concentrations and would, therefore, reduce the sensitivity to glucose.

Inhibition by long chain acyl-CoAs is a well-conserved property of glucokinase and is particularly interesting since cytosolic acyl-CoAs are thought to play a role in the control of insulin secretion (Prentki, 1996). The effect of long chain acyl-CoAs on glucokinase is, however, not observed in liver (Dawson and Hales, 1969) or in INS-1 cell extracts, most likely due the presence of a fatty acyl-CoA binding protein (Moukil and Van Schaftingen, unpublished results). Furthermore, flux measurements in isolated hepatocytes also indicate that acyl-CoAs do not participate in the regulation of glucokinase activity in intact cells (Van Schaftingen et al., 1994). It is likely that the same conclusion applies to the islets.

B. The Regulatory Protein of Glucokinase

The mammalian liver regulatory protein acts as a competive inhibitor versus glucose and its effect (and its binding to glucokinase) is reinforced by fructose 6-phosphate and antagonized by fructose 1-phosphate (Van Schaftingen, 1989). Its presence thus explains that the affinity of glucokinase for glucose in intact hepatocytes incubated in the absence of fructose is lower than that of pure glucokinase, and that fructose stimulates glucose utilization at subsaturating concentrations of this aldose (Bontemps et al., 1978; Van Schaftingen and Vandercammen, 1989). Furthermore, the regulatory protein inhibits glucokinase when the concentration of fructose 6-phosphate is elevated; that is, during active gluconeogenesis or glycogenolysis (Van Schaftingen, 1995). It appears to be present in the livers of all species that contain glucokinase, but not in those (cat, ruminants) that do not contain this enzyme (Vandercammen and Van Schaftingen, 1993). The cDNAs encoding the regulatory protein from rat, human and xenopus liver have been cloned. They encode proteins that show substantial sequence homology with open-reading frames found in the genome of several bacteria and encoding proteins of unknown function (Van Schaftingen et al., 1997).

Evidence for the presence of regulatory protein in islets was reported by Malaisse et al. (1990). These authors found that 10 mM fructose 1-phosphate stimulated glucokinase activity in islet extracts, a finding recently confirmed by Becker et al. (1996). Furthermore they showed that purified liver glucokinase was (modestly) inhibited by islet extracts in the absence, but not in the presence of 10 mM fructose 1-phosphate. In addition, incubation of intact islets with D-glyceraldehyde (which condenses with dihydroxyacetone-phosphate to form fructose 1-phosphate) stimulated [5-^{3}H]glucose detritiation by about 20%. Fructose was devoid of significant effect, but this could be due to the fact that fructokinase activity is very low in islets (Malaisse et al., 1989) compared to the liver ($\approx$ 0.05 vs. 10 nmol/min/mg tissue). At variance with the liver regulatory protein, the inhibitor present in islets appears to be insensitive to fructose 6-phosphate.

Further biochemical characterization of the glucokinase inhibitor of glucokinase is difficult because of its low concentration in islets and the low amounts of available material. The mRNA encoding the liver protein appears, however, to be present in islets since it can be amplified by polymerase chain reaction (PCR) (Becker et al., 1996). The amount reported by these authors (≈ 25% of that present in liver) should be taken with caution since it is notoriously difficult to quantitate RNA abundance by simple PCR. Western blot analysis of rat islets with antibodies directed against the rat liver protein indeed gave negative results, suggesting that the low amounts of liver type mRNA are not significantly translated. The inhibitory protein found in islets must therefore be different from that present in liver. Attempts at cloning a homologous cDNA by screening islets libraries at low stringency with a probe derived from rat or xenopus liver have not met with success (unpublished results).

Taken together, these data suggest that the inhibitory protein present in islet differs from the liver regulatory protein by its poor sensitivity to fructose 1-phosphate and its lack of sensitivity to fructose 6-phosphate, and that it must be encoded by a different gene. The role of this regulatory protein remains to be established.

C. Glucokinase Translocation

Digitonin-permeabilization studies and immunohistochemical techniques have shown that glucokinase is present in the nucleus of hepatocytes incubated in the presence of a low concentration of glucose and that it translocates to the cytosol in response to elevated concentrations of glucose, to mannoheptulose (a glucokinase inhibitor), or to fructose, sorbitol and D-glyceraledehyde, all of which give rise to fructose 1-phosphate (Agius and Peak, 1993; Agius, 1994; Toyoda et al., 1994; Brown et al., 1997; Niculescu et al., 1997). These agents have in common to favor dissociation of the glucokinase/regulatory protein complex either by binding to glucokinase (glucose, mannoheptulose) or to its regulatory protein. Since the regulatory protein is a nuclear protein, whose intracellular localization is unaffected by elevated glucose concentration or by fructose (Agius et al., 1995; Brown et al., 1997), these data suggest that the regulatory protein serves as an anchor that maintains glucokinase in the nucleus under conditions favoring its association with this enzyme, but that it dissociates from this enzyme and allows it to migrate to the cytoplasm as soon as an effector promoting dissociation of the two proteins appears in the cell. Translocation of glucokinase to the cytosol correlates with the activity of the enzyme under a series of conditions, including adenovirus-mediated overexpression of glucokinase in cultured hepatocytes (Agius et al., 1996). The role of this translocation, which reflects the state of association of the two proteins, is not yet clear since there is ample evidence that binding of glucokinase to the regulatory protein is sufficient per se to inhibit this enzyme (Vandercammen and Van Schaftingen, 1990). One may hypothesize that glucokinase or its regulatory protein have a role to play in the regulation of gene expression.

There is also evidence for translocation of glucokinase occurring in β-cells. Using confocal microscopy, Noma et al. (1996) found by immunostaining that in β-cells of islets derived from euglycemic rats, glucokinase was present in a caplike perinuclear area that made up only a small fraction (about 9%) of total cytoplasm. In contrast, the staining of glucokinase was diffuse in most β-cells of rats that had been made acutely hyperglycemic by two consecutive glucose injections. More recently, glucokinase staining was shown to be essentially associated with the nucleus in MIN6 cells, a β-cell line (Murata et al., 1997). The technique used did not allow these authors to determine if glucokinase was intranuclear or if it was associated with the nuclear membrane.

By analogy with the liver, the effect of glucose to promote translocation in β-cells most likely requires a protein-anchor that binds to this enzyme in a glucose-sensitive manner; that is, a protein that behaves as a competitive inhibitor versus glucose. Due to its different subcellular localization, this hypothetical protein must be different from the liver regulatory protein. Translocation in islets argues, therefore, for the presence of a distinct regulatory protein in this tissue.

D. Role of Glucokinase as the Rate-limiting Enzyme of Glycolysis

In islet extracts, low Km hexokinases have a higher activity than glucokinase (Giroix et al., 1984), yet the low Km component of glucose usage in intact islets represents much less than the high Km component (Sener et al., 1976b). This observation is easily accounted for by the inhibition of low Km hexokinase by glucose 6-phosphate. When sorted cells are investigated, it appears that β-cells mainly contain glucokinase, the activity of hexokinase being at most equal to about 20% of the glucokinase Vmax and actually much less in a β-cell subpopulation with low glucose sensitivity (Heimberg et al., 1993).

Several arguments indicate that glucokinase catalyzes the rate-limiting step of glycolysis in pancreatic islets or in β-cells (Meglasson and Matschinsky, 1986). The rate of glucose entry as well as the Vmax of the other glycolytic enzymes that could potentially play a regulatory role (phosphofructokinase, pyruvate kinase) exceeds by at least one order of magnitude the activity of glucokinase. Furthermore, the apparent $S_{0.5}$ value and the degree of sigmoidicity of glucose usage in intact islets or in purified β-cells are close to those found for glucokinase in homogenates of these preparations. In addition, the preference of the α- over the β-anomer of glucose at subsaturating concentrations, the preference of glucose over mannose (another substrate of glucokinase), and the inhibitory effect of mannoheptulose on glucose metabolism are all easily accounted for by the properties of glucokinase. Since under these conditions, insulin release is affected in parallel with glucose metabolism, experiments with these agents strongly argue in favor of the concept that glucokinase is the glucose-sensing device in the control of insulin secretion (Matschinsky, 1996). This is also true of recent experiments showing that sorted β-cells displaying a higher sensitivity to glucose have a higher glucokinase activity and metabolize glucose more rapidly than cells with low glucose sensitivity (Heimberg et al., 1993).

Further proof for the critical role of glucose phosphorylation in glucose sensing derived from experiments in which yeast hexokinase is expressed in islets under the control of rat insulin II promoter, resulting in a twofold increase in islet hexokinase activity. These mice are mildly hypoglycemic (20% to 50% decrease in blood glucose concentration) and their islets show increased insulin secretion at low glucose concentrations (Epstein et al., 1992).

Experiments in which glucokinase has been overexpressed in β-cell lines confirm this view provided, however, that the degree of overexpression is moderate. Wang and Iynedjian (1997b) have engineered INS-1 cells to overexpress glucokinase in a graded manner in response to doxycycline. An up to eightfold increase in the expression of glucokinase provoked a commensurate increase in glycolytic flux at low physiological concentrations of glucose, but only a modest increase at elevated concentrations of this sugar, indicating, quite expectedly, that in the face of very high glucokinase fluxes, other steps of glycolysis become rate-limiting. This excessive activity can result in such an accumulation of glucose 6-phosphate (and most likely of other intermediates) that intracellular inorganic phosphate and ATP are depleted (Wang and Iynedjian, 1997a). This explains why moderate (e.g., threefold) degree of overexpression of glucokinase causes a leftward shift in the insulin secretory response curve whereas very high levels of overexpression make that glucose inhibits hormone release and actually intoxicates the cells. This toxicity most likely accounts for the observation by Becker et al. (1996) that a 20-fold level of overexpression of glucokinase in islets does not stimulate glucose utilization and barely increases glucose-dependent insulin secretion, whereas hexokinase overexpression has a marked effect on these parameters. The inhibition exerted by glucose 6-phosphate on hexokinase is expected to prevent this metabolite from accumulating to toxic levels.

E. MODY, Glucokinase Knockouts, and Transgenes

One major argument in favor of the critical role played by glucokinase in glucose sensing is the finding that MODY 2 (maturity onset diabetes of the young), a dominantly inherited form of diabetes, is due to mutations in the glucokinase gene (Froguel et al., 1993). This indeed means that a twofold decrease in liver and islet glucokinase activity is sufficient to cause a significant perturbation of glucose metabolism. Knockout mouse models have confirmed this: mice with only one functional glucokinase allele have elevated fasting blood glucose levels and reduced insulin secretion in response to a glucose challenge (Bali et al., 1995; Grupe et al., 1995) and reduced effect of hyperglycemia to suppress hepatic glucose production (Rossetti et al., 1997). The animals that are homozygous for the defect die perinatally with severe hyperglycemia (Grupe et al., 1995). Specific expression of glucokinase in β-cells (obtained by crossing glucokinase null mice with transgenic mice expressing this enzyme under the control of the proinsulin promoter) is nonetheless sufficient for survival, though the animals are still markedly hyperglycemic (Grupe et al., 1995).

Other experiments also allow us to appreciate the respective importance of hepatic and islet glucokinase for glucose homeostasis. A 70% decrease in the glucokinase level in pancreatic islet obtained with the ribozyme approach in transgenic mice impairs glucose-induced insulin release, but does not change the basal blood glucose and insulin levels (Efrat et al., 1994). Using the antisense approach to decrease the islet glucokinase expression, Ishihara et al. (1995) showed that the effect on glucose tolerance was strain-dependent, being observed in the C57BL/6 but not in the C3H background. This result indicates that other genetic factors are needed to cause diabetes, in agreement with the pleiotropic aspect of regulation mentioned in the first lines of this chapter. Of interest is the recent development of a mouse diabetes model in which homozygous IRS 1 (insulin receptor substrate 1) knockout is combined with β-cell–specific glucokinase knockout in the heterozygous state (Terauchi et al., 1997).

Hepatic overexpression of glucokinase whether under the control of the phosphoenolpyruvate carboxykinase (PEPCK) promoter (negatively controlled by insulin and positively by glucagon) or of the apolipoprotein AI promoter results in decreased blood glucose and insulin concentrations and improves glucose tolerance (Ferre et al., 1996b; Hariharan et al., 1997). In the case of the construct with the PEPCK promoter (Ferre et al., 1996a), the transgenic animals were "resistant" to diabetes: streptozotocin treatment, which suppressed insulin secretion, only modestly increased blood glucose and hydroxybutyrate levels.

The effect of additional copies of the entire glucokinase gene has also been investigated in transgenic mice (Niswender et al., 1997). One and two extra copies reduce the plasma glucose concentration by 25% and 37%, respectively. This reduction occurs without significant change in blood insulin concentration, suggesting that it results from changes in liver glucokinase levels. Other metabolical data supported this conclusion (Niswender et al., 1997). Increased insulin secretion was not observed most likely because hypoglycemia causes a downregulation of the islet glucokinase content.

Taken together, all data mentioned in this section indicate that both islet and liver glucokinases play an important role in glucose homeostasis.

IV. GLUCOSE 6-PHOSPHATE HANDLING

Under normal circumstances, the main fate of glucose 6-phosphate in β-cells is to be glycolyzed, other pathways, such as the pentose-phosphate pathway playing a minimal role (Meglasson and Matschinsky, 1986). Hydrolysis by glucose-6-phosphatase and conversion to glycogen are potentially important under pathological conditions and, therefore, deserve some comments.

A. Glucose-6-Phosphatase

Glucose-6-phosphatase, which is mainly expressed in gluconeogenic organs, is a complex enzymatic system associated with the endoplasmic reticulum. The enzyme appears to consist of a phosphohydrolase whose catalytic site faces the lumen of the endoplasmic reticulum and of translocases for the substrate (glucose 6-phosphate) and for the two products of the reaction (Arion et al., 1975). The specificity of the reaction appears to be dictated to a large extent by the glucose 6-phosphate translocase, explaining that detergents cause a loss in specificity. The cDNA encoding the hydrolase has been cloned (Shelly et al., 1993), as well as that of a putative glucose 6-phosphate translocase (Gerin et al., 1997). The enzyme has a Km for glucose 6-phosphate (about 2 mM), which is rather elevated compared to the intracellular glucose 6-phosphate concentration, and is therefore controlled by the concentration of its substrate. It is also inhibited by elevated concentrations of glucose (Ki $\approx$ 0.1 M).

The reason why nature has invented such a complex enzymatic system instead of a simple soluble glucose 6-phosphatase (a soluble 2-deoxyglucose-6-phosphatase exists in yeast [Sanz et al., 1994]) is not clear. Benedetti et al. (1986) showed that formation of inorganic phosphate by this enzyme favors Ca^{2+} pumping by the endoplasmic reticulum in the liver. In so doing, glucose-6-phosphatase would facilitate termination of the response to glycogenolytic agents acting through an elevation in the cytosolic concentration of Ca^{2+}.

Except for one report by Waddell and Burchell (1988), the results obtained by most authors indicate that the activity of glucose-6-phosphatase is low in islets from normal mice or rats (Ashcroft and Randle, 1968; Täljedal, 1969; Giroix et al., 1987; Perales et al., 1991; Khan et al., 1995) and indeed much lower than in liver (1/50 according to Perales et al., 1991). This conclusion is in agreement with the minimal activity of the glucose/glucose 6-phosphate cycle in islets from normal mice, only 3% to 4% of the phosphorylated glucose being dephosphorylated. It also agrees with the low rate of disappearance of 2-deoxyglucose 6-phosphate in rat islets following incubation with 2-deoxyglucose (Sweet et al., 1997). Thus, the impact of glucose-6-phosphatase on glucose sensing appears to be negligible in normal islets.

This conclusion may not necessarily apply to pathological situations. Glucose 6-phosphatase activity is several-fold higher in islets of ob/ob mice than of control mice and further increases following treatment with dexamethasone (Täljedal, 1969; Khan et al., 1995), in which case the activity becomes of the same order of magnitude as that in liver. When measured in the presence of 16.7 mM glucose, glucose cycling amounts to about 30% of the phosphorylated glucose in islets from ob/ob mice (Khan et al., 1990a). Glucose cycling was also increased by about 2.5-fold in islets from rats with diabetes induced with streptozotocin in the perinatal period (Khan et al., 1990b). Interestingly, glucose-6-phosphatase expression increases in liver in response to diabetes or to an increase in the concentration of glucose (Argaud et al., 1997); the same mechanism could apply to the islet enzyme.

By subtracting some of the glucose that is phosphorylated by glucokinase, or possibly, as suggested by Wolf et al. (1988), by increasing Ca^{2+} sequestration in the endoplasmic reticulum, the overexpression of glucose-6-phosphatase could participate in the alteration of insulin secretion found in diabetic animals. Adenovirus-mediated overexpression of glucose-6-phosphatase in INS-1 cells to levels about 10-fold more elevated than those in liver has been shown to decrease [5-^{3}H]glucose usage and insulin secretion by about 30% (Trinh et al., 1997). That the effect was relatively small compared to the degree of overexpression of the catalytic subunit could be due to the fact that one of the translocases becomes rate-limiting under these conditions.

B. Glycogen Metabolism and Pentose Phosphate Pathway

There is general agreement that glycogen metabolism normally contributes little to glucose 6-phosphate utilization in islets (Meglasson and Matschinsky, 1986). Nevertheless, prolonged incubation in the presence of elevated concentrations of glucose causes glycogen accumulation, which although it represents only a small fraction of glucose usage (0.3%, according to Malaisse et al. [1992], as compared to more than 10% in liver), could play a role in the so-called glucotoxicity phenomenon; that is, the lowered ability of glucose to stimulate insulin secretion in islets incubated with elevated glucose concentrations. Glucose withdrawal from such cells stimulates glycogenolysis, which, if the glycogen levels are elevated (e.g., ≈ 30 pmol/islet), can be of the same order of magnitude as the rate of glucose usage observed with 20 mM glucose. Under such conditions, glucose could, therefore, have no effect, and possibly even a negative effect on the glycolytic flux. It should be mentioned that prolonged incubation of β-cells with glucose causes a marked decrease in insulin content, which can also explain the decrease in insulin secretion (Ling et al., 1996)

The mechanism by which glycogen accumulates in islets in response to glucose is not known precisely. In the liver, glucose promotes glycogen formation by binding directly to glycogen phosphorylase *a* (phosphorylated form), making it a better substrate for phosphorylase phosphatase and therefore promoting its dephosphorylation and inactivation; when the level of phosphorylase *a*, an allosteric inhibitor of synthase phosphatase, is sufficiently low, synthase dephosphorylation can proceed (Hers, 1976). Glucose also acts by increasing the concentration of glucose 6-phosphate, a stimulator of synthase phosphatase (Cadefau et al., 1997). Since the regulation of glycogen metabolism appears to be qualitatively similar in islet and in liver (Zhang et al., 1994), the effect of glucose in β-cells could be similar to that described in the preceding lines.

C. Phosphofructokinase

Phosphofructokinase (PFK1) catalyzes a rate-controlling step of glycolysis in many cell types. Control can indeed move from the glucose phosphorylation step to PFK1 provided either (1) that glucose 6-phosphate has other (quantitatively) important fates (e.g., in liver: glycogen synthesis and hydrolysis by glucose 6-phosphatase) than glycolysis, or (2) that the glucose phosphorylating enzyme is retroinhibited by its reaction product, as is the case in the tissues containing low Km hexokinases. Thus PFK1 does not appear to be important in the control of glycolysis in β-cells, a conclusion supported by several observations (Meglasson and Matschinsky, 1986), including the fact that the Vmax of this kinase is more than 30-fold higher than that of glucokinase and that the release of tritium from [2-^{3}H] and [5-^{3}H] glucose (respectively catalyzed by phosphoglucose isomerase and triose phosphate isomerase) occur at similar rates in suspensions of islets.

There are, however, other reasons that make this enzyme interesting, one being that it has to keep pace with glucokinase following an increase in glucose concentration and another that it could, in this tissue as in other systems (Goldbeter and Caplan, 1976), be responsible for the generation of glycolytic oscillations. These, in turn, could be linked, via rhythmical changes in the ATP/ADP ratio, to oscillations in cytosolic Ca^{2+} and in insulin secretion (Longo et al., 1991; Nilsson et al., 1996; Tornheim, 1997).

Three different types of PFK1 subunits exist in mammalian tissues, called respectively L (liver), M (muscle), and C (brain, platelets). The corresponding tetrameric enzymes display similar kinetic properties: sigmoidal dependence versus fructose 6-phosphate; inhibition by elevated concentrations of ATP; stimulation by AMP, by fructose 2,6-bisphosphate, glucose 1,6-bisphosphate, and fructose 1,6-bisphosphate, the product of the reaction (Van Schaftingen, 1987). Hexose-bisphosphates appear to bind to the same site, which is more specific for fructose 2,6-bisphosphate with the liver enzyme than with the muscle enzyme; in the latter case fructose 1,6-bisphosphate is about 10-fold less potent than fructose 2,6-bisphosphate (Tornheim, 1985).

The feedback activation of the muscle enzyme by fructose 1,6-bisphosphate is responsible for kinetic instability and is thought to play an important role in the generation of glycolytic oscillations in muscle extracts. Accordingly, addition of saturating (0.5 μM) concentrations of fructose 2,6-bisphosphate suppresses glycolytic oscillations in this system (Tornheim, 1988). The M- and C-type PFK1 are found in isolated rat islets and in INS-1 cells in similar amounts (Yaney et al., 1995). The presence of the M-isozyme opens the possiblity that the mechanism of glycolytic oscillations could be the same in islets as in muscle extracts. This interpretation could provide an explanation for the paradoxical finding that overexpression of the L-isozyme in islets of transgenic mice, leading to a 30% to 70% increase in total islet PFK1 activity, causes hyperglycemia and decreased insulin secretion in response to glucose (Knobler et al., 1997). It is not known if glycolytic oscillations are still observed in this preparation.

Because of the competitive effects of the fructose-bisphosphates on PFK1-M, it is of interest to know their relative concentrations in β-cells. The concentration of fructose 2,6-bisphosphate in islets and in isolated β-cells depends on the glucose level (Sener et al., 1984; Burch et al., 1985), an effect that is presumably mediated by an increase in the concentration of fructose 6-phosphate, the substrate of PFK2 and a powerful inhibitor of fructose 2,6-bisphosphatase (Van Schaftingen, 1987). In β-cells or islets incubated in the presence of 5 mM glucose, concentrations of about 15 μM (in intracellular water) have been observed, which most likely allows PFK1 to readily phosphorylate fructose 6-phosphate (Sener et al., 1984). Such fructose 2,6-bisphosphate concentrations are of the same order of magnitude as those of fructose 1,6-bisphosphate, which would prevent this reaction product to participate in a positive feedback loop and in the generation of oscillations. Also of interest is the observation that following an increase in the concentration of glucose from 0 or 1.4 mM to 20 mM, the level of fructose 2,6-bisphosphate reaches a new steady state about twice as rapidly in β-cells as in hepatocytes. This indicates that the β-cells have a very active PFK2.

Glucagon, which in the liver causes a decrease in the concentration of fructose 2,6-bisphosphate, does not affect the level of this regulator in pancreatic islets (Sener et al., 1984). In the liver, fructose 2,6-bisphosphate is formed and degraded by the L-type of PFK2/FBPase2, a form of enzyme that is a substrate for cAMP-dependent protein kinase, which causes inactivation of the PFK2 activity and activation of FBPase2 (Rousseau and Hue, 1993). The heart form of bifunctional enzyme, which is not a substrate for cAMP-dependent protein kinase, was reported to be present in islets (Sakurai et al., 1996). When expressed in *Escherichia coli* and purified, this form of PFK2 proved to be more active than the testis, liver, and muscle isozymes. Thus, β-cells appear to be well equipped to allow PFK1 to adjust its rate to that of glucokinase.

V. LATE GLYCOLYTIC STEPS

The enzymes catalyzing the glycolytic steps beyond phosphofructokinase are very active, and, therefore, do not appear to play an important role in the control of glycolysis. Glyceraldehyde 3-phosphate dehydrogenase was reported to be 400-fold more active than glucokinase (Trus et al., 1981). This dehydrogenase catalyzes a thermodynamically unfavorable reaction, which can only proceed at elevated NAD^+/NADH ratios. Hence the importance of regenerating rapidly the NAD^+ that has been consumed by this reaction.

A lot of interest has recently arisen from the observation that pyruvate kinase is expressed in INS-1 cells and that its activity is controlled by glucose as in the liver (Marie et al., 1993). As recalled by MacDonald (1995b), the enzyme present in islet is the M_2 type (Chatterton et al., 1982; MacDonald and Chang, 1985b), a form of enzyme that is stimulated by fructose 1,6-bisphosphate, but which is not a substrate for cAMP-dependent protein kinase. The form of enzyme in mammalian liver is the L-type, which is also sensitive to fructose 1,6-bisphosphate, but which is inactivated by phosphorylation by cAMP-dependent protein kinase. This phosphorylation, therefore, participates in the inhibitory effect of glucagon and cAMP on glycolysis (Hers and Hue, 1983). Such an effect would not make sense in islets where agents acting through cAMP are known to potentiate the effect of glucose. Furthermore acute stimulation of β-cells with glucagon has no effect on the rate of glucose oxidation (Gorus et al., 1984).

Of the two enzymes that are needed to oppose the reaction catalyzed by pyruvate kinase, only one, pyruvate carboxylase, is present in islets (MacDonald and Chang, 1985a; MacDonald et al., 1992). As mentioned in section VII, this enzyme plays an important role in anaplerosis.

VI. ELECTRON SHUTTLES

In all cell types, the NADH that is generated by glyceraldehyde-3-phosphate dehydrogenase has to be reconverted to NAD^+ very efficiently in order to enable glycolysis to proceed. This goal can be reached either through the conversion of pyruvate to lactate, which can then be exported, or by transfer of the reducing equivalents to the respiratory chain. Since the internal membrane of the mitchondria is impermeable to NAD, this transfer is mediated by shuttles. The simplest one is the glycerol 3-phosphate shuttle, which consists of cytosolic and mitochondrial glycerol 3-phosphate dehydrogenases. The latter, which is present in the internal membrane of the mitochondrion, is flavin adenine dinucleotide (FAD)-linked and accepts electrons from its substrate at the outer face of the membrane, to transfer them to coenzyme Q. Subsequent reduction of O_2 yields 2 ATP per electron pair. A second shuttle comprises cytosolic and mitochondrial forms of aspartate transaminase and malate dehydrogenase, as well as appropriate translocases; the ATP yield is in this case of 3 mol per mol of NADH. A major advantage of the first shuttle over the second one is that it uses as electron carrier a metabolite (dihydroxyacetone-phosphate) that precedes the NADH-generating step of glycolysis. By contrast, the other system requires the accumulation of metabolites (pyruvate, malate, aspartate) that are beyond glyceraldehyde 3-phosphate dehydrogenase.

MacDonald (1981) first pointed out that islets are particularly rich in mitochondrial glycerol-3-phosphate dehydrogenase, which is about 50 times more active that in liver and skeletal muscle. It is now known that this is a feature of the β-cells. These cells are also characteristically poor in lactate dehydrogenase, whose activity is less than 1/1,000 that found in liver (Sekine et al., 1994). Mitochondrial glycerol 3-phosphate dehydrogenase displays a sigmoidal saturation curve for glycerol-3-phosphate and is stimulated by cytosolic Ca^{2+} at concentrations in the submicromolar range (Rutter et al., 1992), an effect mediated by calmodulinlike domains present in the enzyme (MacDonald and Brown, 1996). The effect of glucose to stimulate this shuttle can therefore be mediated both by increases in the concentration of glycerol 3-phosphate and of Ca^{2+} (Sener and Malaisse, 1992). The use of the glycerol 3-phosphate shuttle and the near absence of lactate dehydrogenase (which would otherwise delay increases in the concentration of pyruvate and divert reducing equivalents from the respiratory chain) contribute to ensure tight coupling between glycolysis and mitochondrial metabolism (Sekine et al., 1994).

Islets also possess the enzymatic arsenale required for the malate-aspartate shuttle (MacDonald, 1982). The inhibitory effect of aminooxyacetate on glucose oxidation and on glucose-stimulated insulin release (MacDonald, 1982; but see also Malaisse et al., 1982a) indicates the importance of this second shuttle, which is, however, difficult to appreciate in quantitative terms.

MacDonald (1995a) has provided evidence of the functioning of a third shuttle, the malate-pyruvate shuttle, which involves two mitochondrial enzymes, pyruvate carboxylase and malate dehydrogenase, as well as malic enzyme, which is present in the cytosol. This shuttle allows the formation of NADPH in the cytosol at the expense of mitochondrial NADH and ATP. Its functioning would account for the elevated level of pyruvate carboxylase in islets and for the fact that glucose increases the cytosolic NADPH/NADP ratio, as indicated by the change in the malate/pyruvate ratio (Hedeskov et al., 1987). The exact role of the reducing equivalents so generated is not known.

VII. MITOCHONDRIAL GLUCOSE METABOLISM

The conceptual links between mitochondrial glucose metabolism and the control of insulin release are manifold. The first relationship was established by Hellerström (1967) who observed increased oxygen consumption by glucose-stimulated islets. Second, Ascroft et al. (1973) demonstrated that both glucose metabolism and insulin release correlate with changes in total islet ATP content, reflecting the activity of oxidative phosphorylation. Third, some of the effects of glucose are mimicked by substrates that are converted into citric acid cycle intermediates. Examples of such substrates are leucine or its α-keto acid 2-ketoisocaproate (Hutton et al., 1980) and succinate, at least when added to the cells as mono- or dimethyl esters (MacDonald and Fahien, 1988). Leucine is transaminated to 2-ketoisocaproate (Malaisse et al., 1982b) and further oxidized by the branched-chain keto-acid dehydrogenase (BCKD) before entering the Krebs cycle. The expression of BCKD is regulated by glucose (MacDonald et al., 1991). Fourth, the process of glucose-induced insulin release is acutely and exquisitely sensitive to the addition of pharmacological inhibitors of the respiratory chain or mitochondrial ATP synthase (MacDonald and Fahien, 1990). Fifth, the chronic exposure of rat islets to interleukin-1β causes a parallel inhibition of insulin release and depression of the activity of aconitase, a heme-containing enzyme of the citric acid cycle (Welsh et al., 1991). Finally, glucose-induced insulin release in the mouse insulinoma cell line MIN6 is critically dependent on the preservation of an intact mitochondrial genome (Soejima et al., 1996), which encodes several proteins of the respiratory chain. This last experiment provides an in vitro model to study the insulin secretory defects observed in diabetic patients with mtDNA mutations (Maassen and Kadowaki, 1996). Together, these observations have fueled the concept that mitochondrial catabolic pathways are crucial, not only because they provide the energy required for the secretory process, but also because they are at the basis of one or several metabolic coupling factors. There are, however, some observations that are not easy to explain with this hypothesis. For instance, it is unclear why pyruvate is a poor secretagogue when administered in the absence of glucose, since pyruvate is rapidly taken up in islets and appears to be quite metabolized to CO_2 (Sener et al., 1976a).

The most studied metabolic coupling factor is [ATP] which closes K_{ATP}-channels (Ashcroft et al., 1984), resulting in membrane depolarization, influx of Ca^{2+}, and stimulation of exocytosis (Wollheim et al., 1996) when glucose metabolism accelerates. The characteristics of this coupling mechanism have been described earlier in the introduction. However, some discrepancies between the effects of glucose on cellular [ATP] on the one hand and effects of ATP on the K_{ATP}-channels on the other, were noted. These discrepancies can be summarized as follows: (1) in situ measurements of islet [ATP] (Ghosh et al., 1991) disclosed modest changes in [ATP] after stimulation with glucose; (2) the estimated basal [ATP] in islets, in the millimolar range, is much higher than the concentrations required for the closure of the channels (in the high micromolar range); (3) the concentration dependency of glucose in the stimulation of insulin release and in the rise in islet [ATP] are different. Further work has clarified at least part of these issues. First, the K_{ATP}-channels are not only regulated by ATP but also opened by ADP via its interaction with the SUR1 subunit of the channel (Nichols et al., 1996). As a consequence, the cellular [ATP]/[ADP] ratio is a more important signal for β-cell activation than [ATP] per se. Second, the islet [ADP] decreases when medium glucose is raised from 3 to 10 mM; that is, in the range of glucose concentrations wherein secretion is stimulated (Detimary et al., 1995). Third, the total islet content of adenine nucleotides is much influenced by a large glucose-unresponsive pool in the secretory granules; subtraction of this inert pool clearly shows that glucose stimulates insulin release and the cellular [ATP]/[ADP] ratio with comparable concentration and time dependence (Detimary et al., 1995). It can be proposed that the local changes in β-cell [ATP]/[ADP] close to the K_{ATP}-channels are even more pronounced during glucose stimulation than what is seen in whole cells.

It seems now well established, however, that the "general model" cannot explain all aspects of glucose-induced insulin release. Henquin and colleagues observed that ATP may have other targets than K_{ATP}-channels in the control of insulin release, since islets exposed to diazoxide and high $[K^+]$ secrete insulin after glucose stimulation (Gembal et al., 1993). Evidence for ATP- and Ca^{2+}-independent signal transduction after glucose stimulation of β-cells is accumulating as well. In a recent study (Maechler et al., 1997), insulin release from α-toxin permeabilized INS-1 cells was shown to be stimulated by nutrients even when cytosolic [ATP] and $[Ca^{2+}]$ were kept constant by a clamping technique. The measurement of mitochondrial changes in $[Ca^{2+}]$ was possible in this system by engineering the cells to express aequorin in the mitochondrial matrix (Kennedy et al., 1996). The salient observations in this study (Maechler et al., 1997) were as follows: (1) Incubation with succinate, a Krebs cycle intermediate which directly delivers electrons to complex II of the respiratory chain, led to a parallel rise in mitochondrial $[Ca^{2+}]$ and in stimulation of insulin release. (2) Glycerol 3-phosphate, which delivers electrons to coenzyme Q but no carbon to the Krebs cycle, and the Krebs cycle intermediate citrate were not effective in stimulating insulin secretion when added separately but well when they were combined. Therefore, stimulation of exocytosis requires other factor(s) than cytosolic [ATP] and $[Ca^{2+}]$; the generation of such mitochondrial factor(s) correlates with the rise in mitochondrial $[Ca^{2+}]$ and depends on the simultaneous supply of electrons to the respiratory chain and of carbon into the citric acid cycle.

On the basis of metabolic experiments, two candidate coupling factors have been proposed to act in addition to ATP: malonyl-CoA (Prentki et al., 1992; Prentki, 1996) and NADPH (MacDonald, 1995a). Prentki and co-workers examined in islets and β-cell lines the production of malonyl-CoA by acetyl-CoA carboxylase, which catalyzes the committed step of fatty acid synthesis. In β-cells, acetyl-CoA carboxylase is regulated by glucose both at the transcriptional level (Brun et al., 1993) and by allosteric effects such as the glucose-induced accumulation of cytoplasmic citrate (Brun et al., 1996; Schuit et al., 1997). The resulting glucose-induced accumulation of malonyl-CoA has important effects on the metabolism of β-cells. First, being a substrate for fatty acid synthase, it contributes to enhanced de novo synthesis of acyl-CoAs (Corkey et al., 1989). Second, being a strong allosteric inhibitor of CPT-I (Newgard and McGarry, 1995), it prevents the β-oxidation of already existing acyl-CoAs in the cell. Both effects could synergize in the rapid accumulation of long-chain acyl-CoAs and diacylglycerol, which may be directly implicated in the control of exocytosis (Prentki and Matschinsky, 1987). The efflux of mitochondrial citrate requires an intact pyruvate-citrate shuttle, which depends on cytosolic citrate lyase, malic enzyme, and mitochondrial pyruvate carboxylase. Interestingly, the inhibition of citrate lyase by hydroxycitrate has a negative effect on the glucose-stimulated insulin release from the perfused rat pancreas (Chen et al., 1994b). Furthermore, the importance of anaplerotic pyruvate carboxylation has been illustrated by the high expression of pyruvate carboxylase in β-cells, which contrasts with a much lower expression (by one order of magnitude or more) in isolated α-cells (Schuit et al., 1997). Pyruvate carboxylase may also contribute to the electron flux through the pyruvate/malate shuttle (see the preceding section).

How can the generation of metabolic coupling factors by glucose be regulated at the mitochondrial level? As discussed before, the glucokinase "glucose sensor" operates under normal conditions as the overall flux-generating enzyme with a high degree of metabolic control in β-cells (Wang and Iynedjian, 1997b). At the mitochondrial level, it is remarkable that glucose oxidation accelerates by more than one order of magnitude when its concentration is increased from 1 to 10 mM glucose (Schuit et al., 1997) whereas the cellular energy charge rises, as evidenced by a sharp increase in the [ATP]/[ADP] ratio (Detimary et al., 1995). In many cell types, the flux of glucose catabolism slows down, both at the level of the citric acid cycle and at the level of oxidative phosphorylation, when the energy charge of the cell increases. How, then, can β-cells display increased [ATP]/[ADP] ratio and glucose oxidation at the same time? A possible (partial) explanation may be the control of mitochondrial dehydrogenases by a rise in mitochondrial [Ca^{2+}] (McCormack et al., 1990a). This regulation can account for the activation of the pyruvate dehydrogenase complex (PDH) by a Ca^{2+}-activated PDH phosphatase, which dephosphorylates the catalytic subunits of the complex. In rat islets, this process appears to be quantitatively important and occurs between 5 and 15 mM extracellular glucose (McCormack et al., 1990b). Interestingly, this regulation explains two remarkable aspects of metabolic organization in β-cells. First, in rat β-cells the ratio of glucose oxidation over total glucose utilization *increases* at higher substrate levels (De Vos et al., 1991) whereas the same ratio decreases in most other cell types (Sener and Malaisse, 1987). Second, anaplerosis of glucose carbon into the Krebs cycle was recently observed to increase sharply between 5 and 10 mM glucose (Schuit et al., 1997), which may be explained by activation of PDH by glucose, resulting in increased concentration of acetyl-CoA, an allosteric stimulator of pyruvate carboxylase.

VIII. CONCLUSION

While many cell types tend to maintain their ATP/ADP ratio as independent as possible from the external supply of substrates, the β-cell has the peculiarity of using the ATP/ADP ratio as a signal of glucose abundance. To achieve this goal, the β-cell has to express isozymic species at unusual low or high levels, so that, once phosphorylated in the β-cell, glucose molecules are committed to be burnt in the "mitochondrial furnace." This special enzymatic equipment involves: (1) a nonrate-limiting glucose transport system; (2) a glucose-phosphorylating enzyme with low affinity for its substrate, and expressed at rather low levels so that it can be rate-limiting for glycolysis and ATP generation; (3) highly active enzymes permitting the conversion of glucose 6-phosphate to pyruvate, with low activities of enzymes opening to side-pathways (glycogen synthesis, pentose-phosphate pathway, lactate formation); (4) efficient shuttles, permitting the transfer of reducing equivalents from the cytosol to the mitochondria and preventing inhibition of glycolysis in the near-absence of lactate dehydrogenase; and (5) highly active anaplerotic pathways, enabling the cell to "warm up" the mitochondrial furnace, as soon as glucose is available.

Many questions still remain about the existence of other signals generated by glucose (pyridine nucleotides, malonyl-CoA, long-chain acyl-CoAs) as well as about the impact of the ATP/ADP ratio on target(s) other than the potassium channel. The effect of this ratio on mitochondrial respiration also needs to be assessed and compared to what occurs in other cell types. Hopefully, increasing knowledge of the proteins involved in β-cell metabolism and of the genetic defects in various forms of diabetes, together with the availablitity of efficient tools to overexpress the proteins of interest, will help to solve these questions in the near future.

ACKNOWLEDGMENTS

Work in the authors' laboratories is supported by the Belgian Funds for Medical Research; the Flemish Community; the "Actions de Recherche Concertées"; the Belgian Federal Service for Scientific, Technical and Cultural Affairs; and the Juvenile Diabetes Foundation International.

REFERENCES

Agius, L. (1994). Control of glucokinase translocation in rat hepatocytes by sorbitol and the cytosolic redox state. Biochem. J. 298, 237–243.

Agius, L. and Peak, M. (1993). Intracellular binding of glucokinase in hepatocytes and translocation by glucose, fructose and insulin. Biochem. J. 296, 785–796.

Agius, L., Peak, M., Newgard, C.B., Gomez-Foix, A.M. and Guinovart, J.J. (1996). Evidence for a role of glucose-induced translocation of glucokinase in the control of hepatic glycogen synthesis. J. Biol. Chem. 271, 30479–30486.

Agius, L., Peak, M. and Van Schaftingen, E. (1995). The regulatory protein of glucokinase binds to the hepatocyte matrix, but, unlike glucokinase, does not translocate during substrate stimulation. Biochem. J. 309, 711–713.

Argaud, D., Kirby, T.L., Newgard, C.B. and Lange, A.J. (1997). Stimulation of glucose-6-phosphatase gene expression by glucose and fructose-2,6-bisphosphate. J. Biol. Chem. 272, 12854–12861.

Arion, W.J., Wallin, B.K., Lange, A.J. and Ballas, L.M. (1975). On the involvement of a glucose 6-phosphate transport system in the function of microsomal glucose 6-phosphatase. Mol. Cell. Biochem. 6, 75–83.

Ashcroft, F.M., Harrison, D.E. and Ashcroft, S.J.H. (1984). Glucose induces closure of single potassium channels in isolated rat pancreatic-β cells. Nature 312, 446–448.

Ashcroft, S.J.H. and Randle, P.J. (1968). Glucose-6-phosphatase activity of mouse pancreatic islets. Nature 219, 857–858.

Ashcroft, S.J.H., Weerasinghe, L.C.C. and Randle, P.J. (1973). Interrelationship of islet metabolism, adenosine triphosphate content and insulin release. Biochem. J. 132, 223–231.

Bali, D., Svetlanov, A., Lee, H.W., Fusco-DeMane, D., Leiser, M., Li, B., Barzilai, N., Surana, M., Hou, H., Fleischer, N., DePinho, R., Rossetti, L. and Efrat, S. (1995). Animal model for maturity-onset diabetes of the young generated by disruption of the mouse glucokinase gene. J. Biol. Chem. 270, 21464–21467.

Becker, T.C., Noel, R.J., Johnson, J.H., Lynch, R.M., Hirose, H., Tokuyama, Y., Bell, G.I. and Newgard, C.B. (1996). Differential effects of overexpressed glucokinase and hexokinase I in isolated islets. Evidence for functional segregation of the high and low Km enzymes. J. Biol. Chem. 271, 390–394.

Benedetti, A.R., Fulceri, M., Ferro, M. and Comporti, M. (1986). On a possible role of glucose-6-phosphatase in the regulation of liver cell cytosolic Ca^{++} concentration. Trends Biochem. Sci, 11, 284–285.

Bontemps, F., Hue, L. and Hers, H.G. (1978). Phosphorylation of glucose in isolated rat hepatocytes. Sigmoidal kinetics explained by the activity of glucokinase alone. Biochem. J. 174, 603–611.

Brown, K.S., Kalinowski, S.S., Megill, J.R., Durham, S.K. and Mookhtiar, K.A. (1997). Glucokinase regulatory protein may interact with glucokinase in the hepatocyte nucleus. Diabetes, 46, 179–186.

Brun, T., Roche, E., Assimacopoulos-Jeannet, F., Corkey, B.E., Kim, K.H. and Prentki, M. (1996). Evidence for an anaplerotic/malonyl-CoA pathway in pancreatic β-cell nutrient signalling. Diabetes, 45, 190–198.

Brun, T., Roche, E., Kim, K. H. and Prentki, M. (1993). Glucose regulates acetyl-CoA carboxylase gene expression in a pancreatic β-cell line (INS-1). J. Biol. Chem. 268, 18905–18911.

Burch, P.T., Berner, D.K., Najafi, H., Meglasson, M.D. and Matschinsky, F.M. (1985). Regulatory role of fructose-2,6-bisphosphate in pancreatic islet glucose metabolism remains unsettled. Diabetes 34, 1014–1018.

Cadefau, J., Bollen, M. and Stalmans, W. (1997). Glucose-induced glycogenesis in the liver involves glucose-6-phosphate-dependent dephosphorylation of glycogen synthase. Biochem. J., 322, 745–750.

Cardenas, M.L. (1995) *"Glucokinase": Its Regulation and Role in Liver Metabolism.* Springer Verlag, New York/R.G. Landes, Austin.

Chatterton, T.A., Reynolds, C.H., Lazarus, N.R. and Pogson, C.I. (1982). Immunological and kinetic properties of pyruvate kinase in rat pancreatic islets. Biochem. J. 204, 605–608.

Chen, C., Hosokawa, H., Bumbalo, L. M. and Leahy, J. L. (1994a). Regulatory effects of glucose on the catalytic activity and cellular content of glucokinase in the pancreatic β-cells. J. Clin. Invest. 94, 1616–1620.

Chen, S., Ogawa, A., Ohneda, M., Unger, R. H., Foster, D. W. and McGarry, J. D. (1994b). More direct evidence for a malonyl-CoA-carnitine palmitoyltransferase I interaction as a key event in pancreatic β-cell signaling. Diabetes 43, 878–883.

Corkey, B. E., Glennon, M. C., Chen, K. S., Deeney, J. T., Matschinsky, F. M. and Prentki, M. (1989). A role for malonyl-CoA in glucose-stimulated insulin secretion from clonal pancreatic β-cells. J. Biol. Chem. 264, 21608–21612.

Dawson, C.M. and Hales, C.N. (1969). The inhibition of rat liver glucokinase by palmitoyl-CoA. Biochim. Biophys. Acta 176, 657–659.

Detimary, P., Gilon, P., Nenquin, M. and Henquin, J.C. (1994). Two sites of glucose control of insulin release with distinct dependence on the energy state in pancreatic β-cells. Biochem. J. 297, 455–461.

Detimary, P., Jonas, J.C. and Henquin, J.C. (1995). Possible links between glucose-induced changes in the energy state of pancreatic β cells and insulin release. Unmasking by decreasing a stable pool of adenine nucleotides in mouse islets. J. Clin. Invest. 96, 1738–1745.

De Vos, A., Heimberg, H., Quartier, E., Huypens, P., Bouwens, L., Pipeleers, D. and Schuit, F. (1995). Human and rat β-cells differ in glucose transporter but not in glucokinase gene expression. J. Clin. Invest. 96, 2489–2495.

De Vos, A., Schuit, F.C. and Malaisse, W.J. (1991). Preferential stimulation by glucose of its oxidation relative to glycolysis in purified insulin-producing cells. Biochem. Int. 24, 117–121.

Efrat, S., Leiser, M., Wu, Y.J., Fusco-DeMane, D., Emran, O.A., Surana, M., Jetton, T.L., Magnuson, M.A., Weir, G. and Fleischer, N. (1994). Ribozyme-mediated attenuation of pancreatic β-cell glucokinase expression in transgenic mice results in impaired glucose-induced insulin secretion. Proc. Natl. Acad. Sci. USA 91, 2051–2055.

Epstein, P.N., Boschero, A.C., Atwater, I., Cai, X. and Overbeek, P.A. (1992). Expression of yeast hexokinase in pancreatic β-cells of transgenic mice reduces blood glucose, enhances insulin secretion and decreases diabetes. Proc. Natl. Acad. Sci. USA 89, 12038–12042.

Ferre, T., Pujol, A., Riu, E., Bosch, F. and Valera, A. (1996a). Correction of diabetic alterations by glucokinase. Proc. Natl. Acad. Sci. USA 93, 7225–7230.

Ferre, T., Riu, E., Bosch, F. and Valera, A. (1996b). Evidence from transgenic mice that glucokinase is rate limiting for glucose utilization in the liver. FASEB J. 10, 1213–1218.

Ferrer, J., Benito, C. and Gomis, R. (1995). Pancreatic islet GLUT2 glucose transporter mRNA and protein expression in humans with and without NIDDM. Diabetes 44, 1369–1374.

Froguel, P., Zouali, H., Vionnet, N., Velho, G., Vaxillaire, M., Sun, F., Lesage, S., Stoffel, M., Takeda, J., Passa, P., Permutt, M.A., Bell, G.I. and Cohen, D. (1993). Familial hyperglycemia due to mutations in glucokinase: Definition of a subtype of diabetes mellitus. N. Eng. J. Med. 328, 697–702.

Gembal, M., Detimary, P., Gilon, P., Gao, Z. Y. and Henquin, J.C. (1993). Mechanisms by which glucose can control insulin release independently from its action on adenosine triphosphate-sensitive K^+ channels in mouse B cells. J. Clin. Invest. 91, 871–880.

Gerin, I., Veiga-da-Cunha, M., Achouri, Y., Collet, J.F. and Van Schaftingen, E. (1997). Sequence of a putative glucose 6-phosphate translocase, mutated in glycogen storage disease type Ib. FEBS Lett. 419, 235–238.

Ghosh, A., Ronner, P., Cheong, E., Khalid, P. and Matschinsky, F.M. (1991). The role of ATP and free ADP in metabolic coupling during fuel-stimulated insulin release from islet β-cells in the isolated perfused rat pancreas. J. Biol. Chem. 266, 22887–22892.

Giroix, M.H., Sener, A. and Malaisse, W.J. (1987). Hexose metabolism in pancreatic islets. Absence of glucose 6-phosphatase in rat islets. Mol. Cell. Endocrinol. 49, 219–225.

Giroix, M.H., Sener, A., Pipeleers, D.G. and Malaisse, W.J. (1984). Hexose metabolism in pancreatic islets. Inhibition of hexokinase. Biochem. J. 223, 447–453.

Goldbeter, A. and Caplan, S.R. (1976). Oscillatory enzymes. Ann. Rev. Biophys. Bioeng. 5, 449–476.

Gorus, F.K., Malaisse, W.J., and Pipeleers, D.G. (1984). Differences in glucose handling by pancreatic A- and B-cells. J. Biol. Chem. 259, 1196–1200.

Grupe, A., Hultgren, B., Ryan, A., Ma, Y.H., Bauer, M. and Stewart, T.A. (1995). Transgenic knockouts reveal a critical requirement for pancreatic β-cell glucokinase in maintaining glucose homeostasis. Cell 83, 69–78.

Guillam, M.T., Hümmler, E., Schaerer, E., Wu, J.Y., Birnbaum, M.J., Beermann, F., Schmidt, A., Dériaz, N. and Thorens, B. (1997). Early diabetes and abnormal postnatal pancreatic islet development in mice lacking Glut-2. Nat. Genet. 17, 327–330.

Hariharan, N., Farrelly, D., Hagan, D., Hillyer, D., Arbeeny, C., Sabrah, T., Treloar, A., Brown, K., Kalinowski, S. and Mookhtiar, K. (1997). Expression of human hepatic glucokinase in transgenic mice liver results in decreased glucose levels and reduced body weight. Diabetes 46, 11–16.

Hedeskov, C.J., Capito, K. and Thams, P. (1987). Cytosolic ratios of free [NADPH]/[$NADP^+$] and [NADH]/[NAD^+] in mouse pancreatic islets, and nutrient-induced insulin secretion. Biochem. J. 241, 161–167.

Heimberg, H., De Vos, A., Pipeleers, D., Thorens, B. and Schuit, F. (1995). Differences in glucose transporter gene expression between rat pancreatic α- and β-cells are correlated to differences in glucose transport but not in glucose utilization. J. Biol. Chem. 270, 8971–8975.

Heimberg, H., De Vos, A., Vandercammen, A., Van Schaftingen, E., Pipeleers, D. and Schuit, F. (1993). Heterogeneity in glucose sensitivity among pancreatic β-cells is correlated to differences in glucose phosphorylation rather than glucose transport. EMBO J. 12, 2873–2879.

Hellerström, C. (1967). Effects of carbohydrates on the oxygen consumption of isolated pancreatic islets of mice. Endocrinology 81, 105–112.

Hellman, B., Sehlin, J. and Täljedal, I.B. (1971). Evidence for mediated transport of glucose in mammalian pancreatic β-cells. Biochim. Biophys. Acta. 241, 147–154.

Hers, H.G. (1976). The control of glycogen metabolism in the liver. Ann. Rev. Biochem. 45, 167–189.

Hers, H.G. and Hue, L. (1983). Gluconeogenesis and related aspects of glycolysis. Ann. Rev. Biochem. 52, 617–653.

Hughes, S.D., Johnson, J.H., Quaade, C. and Newgard, C.B. (1992). Engineering of glucose-stimulated insulin secretion and biosynthesis in non-islet cells. Proc. Natl. Acad. Sci. USA 89, 688–692.

Hughes, S.D., Quaade, C., Johnson, J.H., Ferber, S. and Newgard, C.B. (1993). Transfection of AtT-20ins cells with GLUT-2 but not GLUT-1 confers glucose-stimulated insulin secretion. Relationship to glucose metabolism. J. Biol. Chem. 268, 15205–15212.

Hutton, J.C., Sener, A., Herchuelz, A., Atwater, I., Kawazu, S., Boschero, C., Somers, G., Devis, G. and Malaisse, W.J. (1980). Similarities in the stimulus-secretion coupling mechanisms of glucose- and 2-keto acid-induced insulin release. Endocrinology 106, 203–219.

Ishihara, H., Tashiro, F., Ikuta, K., Asano, T., Katagiri, H., Inukai, K., Kikuchi, M., Yazaki, Y., Oka, Y. and Miyazaki, J.-I. (1995). Inhibition of pancreatic β-cell glucokinase by antisense RNA expression in transgenic mice: mouse-strain dependent alteration of glucose tolerance. FEBS Lett. 371, 329–332.

Iynedjian, P.B. (1993). Mammalian glucokinase and its gene. Biochem. J. 293, 1–13.

Kennedy, E.D., Rizzuto, R., Theler, J.M., Pralong, W.F., Bastianutto, C. and Pozzan, T. (1996). Glucose-stimulated insulin secretion correlates with changes in mitochondrial and cytosolic Ca^{2+} in aequorin-expressing INS-1 cells. J. Clin. Invest. 98, 2524–2538.

Khan, A., Chandramouli, V., Östenson, C.G., Berggren, P.O., Löw, H., Landau, B.R. and Efendic, S. (1990a). Glucose cycling is markedly enhanced in pancreatic islets of obese hyperglycemic mice. Endocrinology 126, 2413–2416.

Khan, A., Chandramouli, V., Östenson, C.G., Löw, H., Landau, B.R. and Efendic, S. (1990b). Glucose cycling in islets from healthy and diabetic rats. Diabetes 39, 456–459.

Khan, A., Hong-Lie, C. and Landau, B.R. (1995). Glucose-6-phosphatase activity in islets from ob/ob and lean mice and the effect of dexamethasone. Endocrinology 136, 1934–1938.

Knobler, H., Weiss, Y., Peled, M. and Groner, Y. (1997). Impaired glucose-induced insulin response in transgenic mice overexpressing the L-phosphofructokinase gene. Diabetes 46, 1414–1418.

Liang, Y. and Matschinsky, F.M. (1994). Mechanisms of action of nonglucose insulin secretagogues. Ann. Rev. Nutr. 14, 59–81.

Liang, Y., Najafi, H., Smith, R.M., Zimmerman, E.C., Magnuson, M.A., Tal, M. and Matschinsky, F.M. (1992). Concordant glucose induction of glucokinase, glucose usage, and glucose-stimulated insulin release in pancreatic islets maintained in organ culture. Diabetes 41, 792–806.

Ling, Z., Kiekens, R., Mahler, T., Schuit, F.C., Pipeleers-Marichal, M., Sener, A., Klöppel, G., Malaisse, W.J. and Pipeleers, D.G. (1996). Effects of chronically elevated glucose levels on the functional properties of rat pancreatic β-cells. Diabetes, 45, 1774–1782

Longo, E.A., Tornheim, K., Deeney, J.T., Varnum, B.A., Tillotson, D., Prentki, M. and Corkey, B.E. (1991). Oscillations in cytosolic free Ca^{2+}, oxygen consumption and insulin secretion in glucose-stimulated rat pancreatic islets. J. Biol. Chem. 266, 9314–9319.

Maassen, J. A. and Kadowaki, T. (1996). Maternally inherited diabetes and deafness: a new diabetes subtype. Diabetologia 39, 375–382.

MacDonald, M.J. (1981). High content of mitochondrial glycerol-3-phosphate dehydrogenase in pancreatic islets and its inhibition by diazoxide. J. Biol. Chem. 256, 8287–8290.

MacDonald, M.J. (1982). Evidence for the malate aspartate shuttle in pancreatic islets. Arch. Biochem. Biophys. 213, 643–649.

MacDonald, M.J. (1995a). Feasibility of a mitochondrial pyruvate malate shuttle in pancreatic islets. Further implications of cytosolic NADPH in insulin secretion. J. Biol. Chem. 270, 20051–20058.

MacDonald, M.J. (1995b). Is liver type pyruvate kinase present in the normal pancreatic islet? Diabetologia 38, 125.

MacDonald, M.J. and Brown, L.J. (1996). Calcium activation of mitochondrial glycerol phosphate dehydrogenase. Arch. Biochem. Biophys. 326, 79–84

MacDonald, M.J. and Chang, C.M. (1985a). Do pancreatic islets contain significant amounts of phosphoenolpyruvate carboxykinase or ferroactivator activity? Diabetes 34, 246–250.

MacDonald, M.J. and Chang, C.H. (1985b). Pancreatic islets contain the M2 isoenzyme of pyruvate kinase. Its phosphorylation has no effect on enzyme activity. Mol. Cell. Biochem. 68, 115–120.

MacDonald, M.J. and Fahien, L.A. (1988). Glyceraldehyde phosphate and methyl esters of succinic acid. Two "new" potent insulin secretagogues. Diabetes 37, 997–999.

MacDonald, M.J. and Fahien, L.A. (1990). Insulin release in pancreatic islets by a glycolytic and a Krebs cycle intermediate: Contrasting patterns of glyceraldehyde phosphate and succinate. Arch. Biochem. Biophys. 279, 104–108.

MacDonald, M.J., McKenzie, D.I., Kaysen, J.H., Walker, T.M., Moran, S.M., Fahien, L.A. and Towle, H.C. (1991). Glucose regulates leucine-induced insulin release and the expression of the branched chain ketoacid dehydrogenase E1α subunit gene in pancreatic islets. J. Biol. Chem. 266, 1335–1340.

MacDonald, M.J., McKenzie, D.I., Walker, T.M., Kaysen, J.H. (1992). Lack of glyconeogenesis in pancreatic islets: Expression of gluconeogenic enzyme genes in islets. Horm. Metab. Res. 24, 158–160.

Maechler, P., Kennedy, E.D., Pozzan, T. and Wollheim, C.B. (1997). Mitochondrial activation directly triggers the exocytosis of insulin in permeabilized pancreatic β-cells. EMBO J. 16, 3833–3841.

Magnuson, M.A. and Shelton, K.D. (1989). An alternate promoter in the glucokinase gene is active in the pancreatic β-cells. J. Biol. Chem. 264, 15936–15942.

Malaisse, W.J., Malaisse-Lagae, F. and Sener, A. (1982a). The stimulus secretion coupling of glucose-induced insulin release: Effect of aminooxyacetate upon nutrient-stimulated insulin secretion. Endocrinology, 111, 392–397.

Malaisse, W.J., Malaisse-Lagae, F., Davies, D.R. and Van Schaftingen, E. (1989). Presence of fructokinase in pancreatic islets. FEBS Lett. 255, 175–178.

Malaisse, W.J., Malaisse-Lagae, F., Davies, D.R., Vandercammen, A. and Van Schaftingen (1990). Regulation of glucokinase by a fructose-1-phosphate-sensitive protein in pancreatic islets. Eur. J. Biochem. 190, 539–545.

Malaisse, W.J., Marynissen, G. and Sener, A. (1992). Possible role of glycogen accumulation in B-cell toxicity. Metabolism 41, 814–819.

Malaisse, W.J., Sener, A., Malaisse-Lagae, F., Welsh, M., Matthews, D.E., Bier, D.M. and Hellerström, C. (1982b). The stimulus-secretion coupling of amino acid-induced insulin release. Metabolic response of pancreatic islets to L-glutamine and L-leucine. J. Biol. Chem. 257, 8731–8737.

Marie, S., Diaz-Guerra, M.J., Miquerol, L., Kahn, A. and Iynedjian, P.B. (1993). The pyruvate kinase gene as a model for studies of glucose-dependent regulation in the endocrine pancreatic islets. J. Biol. Chem. 268, 23881–23890.

Matschinsky, F.M. (1990). Glucokinase as glucose sensor and metabolic signal generator in pancreatic β-cells and hepatocytes. Diabetes 39, 647–652.

Matschinsky, F.M. (1996). A lesson in metabolic regulation inspired by the glucokinase glucose sensor paradigm. Diabetes 45, 223–241.

McCormack, J.G., Halestrap, A.P. and Denton, R.M. (1990a). Role of calcium ions in regulation of mammalian intramitochondrial metabolism. Physiol. Rev. 70, 391–425.

McCormack, J.G., Longo, E.A. and Corkey, B.E. (1990b). Glucose-induced activation of pyruvate dehydrogenase in isolated rat pancreatic islets. Biochem. J. 267, 527–530.

Meglasson, M.D. and Matschinsky, F.M. (1986). Pancreatic islet glucose metabolism and regulation of insulin secretion. Diabetes Metab. Rev. 2, 163–214.

Morita, H., Yano, Y., Niswender, K.D., May, J.M., Whitesell, R.R. , Wu, L., Printz, R.L., Granner, D.K., Magnuson, M.A. and Powers, A.C. (1994). Coexpression of glucose transporters and glucokinase in Xenopus oocytes indicates that both glucose transport and phosphorylation determine glucose utilization. J. Clin. Invest. 94, 1373–1382.

Murata, T., Katagiri, H., Ishihara, H., Shibasaki, Y., Asano, T., Toyoda, Y., Pekiner, B., Pekiner, C., Miwa, I. and Oka, Y. (1997). Co-localization of glucokinase with actin filaments. FEBS Lett. 406, 109–113.

Newgard, C.B. (1996). Regulatory role of glucose transport and phosphorylation in pancreatic islet β-cells. Diabetes Rev. 4, 191–206.

Newgard, C.B. and McGarry, J.D. (1995). Metabolic coupling factors in pancreatic β-cell signal transduction. Ann. Rev. Biochem. 64, 689–719.

Nichols, C.G., Shyng, S.L., Nestorowicz, A., Glaser, B., Clement, J.P., Gonzales, G., Aguilar-Bryan, L., Permutt, M.A. and Bryan, J. (1996). Adenosine diphosphate as an intracellular regulator of insulin secretion. Science 272, 1785–1787.

Niculescu, L., Veiga-da-Cunha, M. and Van Schaftingen, E. (1997). Investigation on the mechanism by which fructose, hexitols and other compounds regulate the translocation of glucokinase in rat hepatocytes. Biochem. J. 321, 239–246.

Nilsson, T., Schultz, V., Berggren, P.O., Corkey, B.E. and Tornheim, K. (1996). Temporal patterns of changes in ATP/ADP ratio, glucose 6-phosphate and cytoplasmic free Ca2+ in glucose-stimulated pancreatic β-cells. Biochem. J. 314, 91–94.

Niswender, K.D., Shiota, M., Postic, C., Cherrington, A.D. and Magnuson, M.A. (1997). Effects of increased glucokinase gene copy number on glucose homeostasis and hepatic glucose metabolism. J. Biol. Chem. 272, 22570–22575.

Noma, Y., Bonner-Weir, S., Latimer, J.B., Davalli, A.M., and Weir, G.C. (1996). Translocation of glucokinase in pancreatic β-cells during acute and chronic hyperglycemia. Endocrinology 137, 1485–1491.

Nouspikel, T. and Iynedjian, P.B. (1992). Insulin signalling and regulation of glucokinase gene expression in cultured hepatocytes. Eur. J. Biochem. 210, 365–373

Öczan, S., Dover, J., Rosenwald, A.G., Wölfl, S. and Johnston, M. (1996). Two glucose transporters in *Saccharomyces cerevisiae* are glucose sensors that generate a signal for induction of gene expression. Proc. Natl. Acad. Sci. USA 93, 1248–1232.

Perales, M.A., Sener, A. and Malaisse, W.J. (1991). Hexose metabolism in pancreatic islets: the glucose-6-phosphatase riddle. Mol. Cell. Biochem. 101, 67–71.

Prentki, M. (1996). New insights into pancreatic β-cell metabolic signaling in insulin secretion. Eur. J. Endocrinol. 134, 272–286.

Prentki, M. and Matschinsky, F.M. (1987). Ca^{2+}, cAMP, and phospholipid-derived messengers in coupling mechanisms of insulin secretion. Physiol. Rev. 67, 1185–1248.

Prentki, M., Vischer, S., Glennon, M.C., Regazzi, R., Deeney, J.T. and Corkey, B.E. (1992). Malonyl-CoA and long chain acyl-CoA esters as metabolic coupling factors in nutrient-induced insulin secretion. J. Biol. Chem. 267, 5802–5810.

Rossetti, L., Chen, W., Hu, M.Z., Hawkins, M., Barzilai, N. and Efrat, S. (1997). Abnormal regulation of HGP by hyperglycemia in mice with a dirupted glucokinase allele. Am. J. Physiol. Endocrinol. Metab. 36, E743–E750.

Rousseau, G.G. and Hue, L. (1993). Mammalian 6-phosphofructo-2-kinase/fructose-2,6-bisphosphatase: a bifunctional enzyme that controls glycolysis. Progr. Nucleic Acid Res. Mol. Biol. 45, 99–127.

Rutter, G.A., Pralong, W.F. and Wollheim, C.B. (1992). Regulation of mitochondrial glycerol-phosphate dehydrogenase by Ca^{2+} within electropermeabilized insulin secreting cells (INS-1). Biochim. Biophys. Acta 1175, 107–113.

Sakurai, T., Johnson, J.H., and Uyeda, K. (1996). Islet fructose 6-phosphate, 2-kinase:fructose 2,6-bisphosphatase: isozymic form, expression, and characterization. Biochem. Biophys. Res. Commun. 218, 159–163.

Sanz, P., Randez-Gil, F. and Prieto, J.A. (1994). Molecular characterization of a gene that confers 2-deoxyglucose resistance in yeast. Yeast 10, 1195–1202.

Schuit, F. (1997). Is GLUT2 required for glucose sensing? Diabetologia 40, 104–111.

Schuit, F., De Vos, A., Farfari, S., Moens, K., Pipeleers, D., Brun, T. and Prentki, M. (1997). Metabolic fate of glucose in purified islet cells. Glucose-regulated anaplerosis in β cells. J. Biol. Chem. 272, 18572–18579.

Sekine, N., Cirulli, V., Regazzi, R., Brown, L.J., Gine, E., Tamarit-Rodriguez, J., Girotti, M., Marie, S., MacDonald, M.J., Wollheim, C.B. and Rutter, G.A. (1994). Low lactate dehydrogenase and high mitochondrial glycerol phosphate dehydrogenase in pancreatic β-cells. Potential role in nutrient sensing. J. Biol. Chem. 269, 4895–4902.

Sener, A., Kawazu, S., Hutton, J.C., Boschero, A.C., Devis, G., Somers, G., Herchuelz, A. and Malaisse, W.J. (1976a). The stimulus-secretion coupling of glucose-induced insulin release. Effect of exogenous pyruvate on islet function. Biochem. J. 176, 217–232.

Sener, A., Levy, J. and Malaisse, W.J. (1976b). The stimulus-secretion coupling of glucose-induced insulin release. Does glycolysis control calcium transport in the B-cell? Biochem. J. 156, 521–525.

Sener, A. and Malaisse, W.J. (1987). Stimulation by D-glucose of mitochondrial oxidative events in islet cells. Biochem. J. 246, 89–95.

Sener, A. and Malaisse, W.J. (1992). Hexose metabolism in pancreatic islets. Ca^{2+}-dependent activation of the glycerol phosphate shuttle by nutrient secretagogues. J. Biol. Chem. 267, 13251–13256.

Sener, A., Van Schaftingen, E., Van de Winkel, M., Pipeleers, D.G., Malaisse-Lagae, F., Malaisse, W.J., and Hers, H.G. (1984). Effects of glucose and glucagon on the fructose 2,6-bisphosphate content of pancreatic islets and purified pancreatic B-cells. Biochem. J. 221, 759–764.

Shelly, L.L., Lei, K.J., Pan, C.J., Sakata, S.F., Ruppert, S., Schutz, G. and Chou, J.Y. (1993). Isolation of the gene for murine glucose-6-phosphatase, the enzyme deficient in glycogen storage disease type 1A. J. Biol. Chem. 268, 21482–21485.

Soejima, A., Inoue, K., Takai, D., Kaneko, M., Ishihara, H., Oka, Y. and Hayashi, J. I. (1996). Mitochondrial DNA is required for regulation of glucose-stimulated insulin secretion in a mouse pancreatic β-cell line, MIN6. J. Biol. Chem. 271, 26194–26199.

Svensson, A.M., Jansson, L. and Hellerström, C. (1988). The volume area of the capillaries in the endocrine and exocrine pancreas of the rat. Histochemistry 90, 43–46.

Sweet, I.R. and Matschinsky, F.M. (1997). Are there kinetic advantages of GLUT2 in pancreatic glucose sensing? Diabetologia 40, 112–119.

Sweet, I.R., Najafi, H., Li, G., Grodberg, J. and Matschinsky, F.M. (1997). Measurement and modeling of glucose-6-phosphatase in pancreatic islets. Am. J. Physiol. 272, E696–E711.

Täljedal, I.B., (1969). Presence, induction and possible role of glucose 6-phosphatase in mammalian pancreatic islets. Biochem. J. 114, 387–394.

Terauchi, Y., Iwamoto, K., Tamemoto, H., Komeda, K., Ishii, C., Kanazawa, Y., Asanuma, N., Aizawa, T., Akanuma, Y., Yasuda, K., Kodama, T., Tobe, K., Yazaki, Y. and Kadowaki, T. (1997). Development of non-insulin dependent diabetes mellitus in the double knockout mice with disruption of insulin receptor substrate 1 and β-cell glucokinase genes: Genetic reconstitution of diabetes as a polygenic disease. J. Clin. Invest. 99, 861–866.

Thorens, B. (1992). Molecular and cellular physiology of GLUT-2, a High-Km facilitated diffusion glucose transporter. Int. Rev. Cytol. 137, 209–283.

Thorens, B., Dériaz, N., Bosco, D., De Vos A., Pipeleers, D., Schuit, F., Meda, P., and Porret, A. (1996). Protein kinase A-dependent phosphorylation of GLUT2 in pancreatic b cells. J. Biol. Chem. 271, 8075–8081.

Thorens, B., Sarkar, H.K., Kaback, H.R. and Lodish, H.F. (1988). Cloning and functional expression in bacteria of a novel glucose transporter present in liver, intestine, kidney, and β-pancreatic islet cells. Cell 55, 281–290.

Tornheim, K. (1985). Activation of muscle phosphofructokinase by fructose 2,6-bisphosphate and fructose 1,6-bisphosphate is differently affected by other regulatory metabolites. J. Biol. Chem. 260, 7985–7989.

Tornheim, K. (1988). Fructose 2,6-bisphosphate and glycolytic oscillations in skeletal muscle extracts. J. Biol. Chem. 263, 2619–2624.

Tornheim, K. (1997). Are metabolic oscillations responsible for normal oscillatory insulin secretion? Diabetes 46, 1375–1380.

Toyoda, Y., Miwa, I., Kamiya, M., Ogiso, S., Nonogaki, T., Aoki, S. and Okuda, J. (1994). Evidence for glucokinase translocation by glucose in rat hepatocytes. Biochem. Biophys. Res. Commun. 204, 252–256.

Trinh, K., Minassian, C., Lange, A.J., O'Doherty, R.M. and Newgard, C.B. (1997). Adenovirus-mediated expression of the catalytic subunit of glucose-6-phosphatase in INS-1 cells. Effects on glucose cycling, glucose usage and insulin secretion. J. Biol. Chem. 272, 24837–24842.

Trus, M.D., Zawalich, W.S., Burch, P.T., Berner, D.K., Weill, V.A. and Matschinsky, F.M. (1981). Regulation of glucose metabolism in pancreatic islets. Diabetes 30, 911–922.

Unger, R.H. (1991). Diabetic hyperglycemia: Link to impaired glucose transport in pancreatic b cells. Science 251, 1200–1205.

Vandercammen, A. and Van Schaftingen, E. (1990). The mechanism by which rat liver glucokinase is inhibited by the regulatory protein. Eur. J. Biochem. 191, 483–489.

Vandercammen, A. and Van Schaftingen, E. (1993). Species and tissue distribution of the regulatory protein of glucokinase. Biochem. J. 294, 551–556.

Van Schaftingen, E. (1987). Fructose 2,6-bisphosphate. Adv. Enzymol. Rel. Areas Mol. Biol. 59, 315–395.

Van Schaftingen, E. (1989). A protein from rat liver confers to glucokinase the property of being antagonistically regulated by fructose 6-phosphate and fructose 1-phosphate. Eur. J. Biochem. 179, 179–184.

Van Schaftingen, E. (1995). Glucosamine-sensitive and -insensitive detritiation of [2-^{3}H] glucose in isolated rat hepatocytes: A study of the contributions of glucokinase and glucose-6-phosphatase. Biochem. J. 308, 23–29.

Van Schaftingen, E., Detheux, M. and Veiga-da-Cunha, M. (1994). Short-term control of glucokinase activity: role of a regulatory protein. FASEB J. 8, 414–419.

Van Schaftingen, E. and Vandercammen, A. (1989). Stimulation of glucose phosphorylation by fructose in isolated rat hepatocytes. Eur. J. Biochem. 179, 173–177.

Van Schaftingen, E., Veiga-da-Cunha and Niculescu, L. (1997). The regulatory protein of glucokinase. Biochem. Soc. Trans. 25, 136–140.

Veiga-da-Cunha, M., Xu, L.Z., Lee, Y.H., Marotta, D., Pilkis, S.J. and Van Schaftingen, E. (1996). Effect of mutations on the sensitivity of human beta-cell glucokinase to liver regulatory protein. Diabetologia 39, 1173–1179.

Waddell, I.D. and Burchell, A. (1988). The microsomal glucose-6-phosphatase enzyme of pancreatic islets. Biochem. J. 255, 471–476.

Wang, H. and Iynedjian, P.B. (1997a). Acute glucose intolerance in insulinoma cells with unbalanced overexpression of glucokinase. J. Biol. Chem. 272, 25731–25736.

Wang, H. and Iynedjian, P.B. (1997b). Modulation of glucose responsiveness of insulinoma β-cells by graded overexpression of glucokinase. Proc. Natl. Acad. Sci. USA 94, 4372–4377.

Welsh, N., Eizirik, D. L., Bendtzen, K. and Sandler, S. (1991). Interleukin-1β-induced nitric oxide production in isolated rat pancreatic islets requires gene transcription and may lead to inhibition of the Krebs cycle enzyme aconitase. Endocrinology 129, 3167–3173.

Wolf, B.A., Colca, J.R., Turk, J., Florholmen, J., McDaniel, M.L. (1988). Regulation of Ca^{2+} homeostasis by islet endoplasmic reticulum and its role in insulin secretion. Am. J. Physiol. 254, E121–E136.

Wollheim, C.B., Lang, J. and Regazzi, R. (1996). The exocytotic process of insulin secretion and its regulation by Ca2+ and G-proteins. Diabetes Rev. 4, 276–297.

Yaney, G.C., Schultz, V., Cunningham, B.A., Dunaway, G.A., Corkey, B.E. and Tornheim, K. (1995). Phosphofructokinase isozymes in pancreatic islets and clonal beta-cells (INS-1). Diabetes 44, 1285–1289.

Zhang, T.M., Maggetto, C. and Malaisse, W.J. (1994). Hexose metabolism in pancreatic islets: Glycogen synthase and glycogen phosphorylase activities. Biochem. Med. Metab. Biol. 51, 129–139.

ION CHANNEL SIGNAL TRANSDUCTION IN PANCREATIC β-CELLS

Arun S. Rajan and Lydia Aguilar-Bryan

Advances in Molecular and Cell Biology
Volume 29, pages 227-246.

ISBN: 0-7623-0547-9

I. INTRODUCTION

Plasma membrane ion channels are protein tunnels on the surface of cells that permit cellular entry and exit of electrically charged ions. Depending on the signal that makes them open and close ("gating"), ion channels can be broadly classified into ligand-gated and voltage-gated channels. Although these molecules are better known for their role in generating the spark of the heartbeat or in the lightninglike conduction of nerve action potentials, it is now evident that ion channels play a critical role in igniting the physiological process of insulin secretion in pancreatic β-cells. Recognition of a signal at the cell surface and its translation into an intracellular gesture defines the process of "signal transduction." The preceding chapters have described the external stimuli identified by the β-cell and succeeding chapters will detail the responsive cytoplasmic pathways that control insulin release. Here, we depict how signal transduction by specific ion channels *mediates* the stimulus-secretion coupling process in insulin-secreting β-cells.

II. ELECTRICITY IN THE β-CELL

The primary physiological function of the pancreatic β-cell is to sense changes in circulating glucose levels and respond with an appropriate insulin secretory response. The significance of this apparently simple function is of profound importance in health because its failure disrupts normal blood glucose homeostasis and results in important insulin secretory abnormalities such as diabetes mellitus and familial hyperinsulinism.

The β-cell is rather unique because its electrical activity changes in response to stimuli that are primarily metabolic in nature (Dean and Matthews, 1974). Thus, examination of the electrical events in β-cell stimulus-secretion coupling are best understood by following the changes in response to its primary metabolic stimulus, glucose (see Figure 1, p. 240). In vitro experiments demonstrate that when β-cells are impaled with fine glass microelectrodes in the absence of glucose, a stable resting membrane potential of approximately –70 mV is recorded. The electrogenic Na^+/K^+-ATPase pump and the high K^+ permeability of open K^+ channels in the resting state maintain this negative membrane potential.

As blood glucose levels increase (e.g., after a meal), a reduction in K^+ permeability caused by closure of K^+ channels results in gradual cell depolarization to a more positive (less negative) membrane potential (–60 to –50 mV). When glucose levels exceed a threshold for stimulating insulin secretion (> 120 mg/dl or 7 mM), an action potential is triggered in the β-cell. The upstroke of this action potential is characterized by the cell rapidly depolarizing from a threshold potential (~ –50 mV) to a plateau potential (–40 to –30 mV) mediated by increased conductance of Ca^{2+} into the cells through voltage-gated Ca^{2+} channels. After a plateau period that lasts for several seconds, the cells repolarize rapidly to the resting potential through a combination of mechanisms that include enhanced K^+ efflux through K^+ channels and reduced Ca^{2+} influx through Ca^{2+} channels. A *stable* resting state does not really exist and, like the nodal cells of the cardiac pacemaker, the cell again undergoes a gradual depolarization. This cycle of depolarization and repolarization is repeated and constitute the "slow waves" or "bursts" that typify glucose-induced electrical activity in β-cells. An additional characteristic of β-cell action potentials is the presence of recurrent "spikes" (depolarizations to –20 to –10 mV) that are superimposed upon the plateau potentials and represent voltage-dependent Ca^{2+} influx through open Ca^{2+} channels.

Normally, changes in blood glucose levels are not sudden. Rather, the shifts are more gradual or subtle in nature. How does the electrical activity of β-cells reflect these changes? Synchronization of electrical activity with the fine variations in blood glucose occurs as follows. As ambient glucose increases past a threshold level (>7 mM) to supranormal levels (>11 mM), there is a progressive reduction in the inter-burst period (i.e., interval between slow waves) and an increase in the duration of the plateau potential resulting in a high action potential firing rate. At excessive glucose levels (>20 mM), incessant firing and fusion of the Ca^{2+} spikes occurs with no slow waves present (resembling the electrophysiological state of tetanus in skeletal muscle), leaving the β-cells perpetually depolarized. Thus, changes in membrane electrical activity link blood glucose excursions to oscillations in free cytoplasmic Ca^{2+} levels that ultimately control the cellular exocytosis of insulin (Wollheim and Sharp, 1981). The major regulators of β-cell electrical activity evidently are the K^+ and Ca^{2+} channels, and this chapter focuses on notable representatives of these classes of channels.

III. POTASSIUM CHANNELS: ATP-SENSITIVE K^+ (K_{ATP}) CHANNELS

A. Structure and Function

We begin with a discussion of a ligand-gated K^+ channel, the adenosine triphosphate (ATP)-sensitive K^+ (K_{ATP}) channel, which has been well characterized in β-cells (Rorsman and Trube, 1985). As its name implies, this is an ion channel that fluxes K^+, and its conductance is regulated by changes in levels of nucleotides such as ATP and adenosine diphosphate (ADP). It is worth clarifying the nomenclature of this entity at this juncture, since it appears to be a source of confusion even for scientists working in this field. The binding of ATP may result in *activation* of a channel resulting in increased flux of K^+ out of the cells—this has been labeled as an ATP-*regulated* K^+ channel (e.g., $ROMK_1$; Ho et al., 1993). This channel is *not* synonymous with the β-cell ATP-sensitive (or ATP-dependent) K^+ (K_{ATP}) channel, which is characterized by ATP or sulfonylurea-dependent binding and closure of the channel with inhibition of K^+ efflux from the cells. In addition,

K_{ATP} channels are activated by MgADP and the drug diazoxide. Another identified channel that is confused with the K_{ATP} channel is the cardiac inward rectifier (CIR; Ashford et al., 1995), which is a member of an inward rectifier K^+ channel family called GIRK. These other (non-K_{ATP}) channels are not the objects of our discussion.

Another important concept that needs to be understood is that the K_{ATP} channel is a heteromultimer formed by *two* different molecules: in the pancreatic β-cell this is a union of a protein called the high affinity sulfonylurea receptor (SUR1) *and* a second molecule named the inward rectifier (K_{IR} 6.2 or BIR; Inagaki et al., 1995a), the latter constituting the ion transporter or pore. Unfortunately, the inward rectifier and the K_{ATP} channel are terms often used interchangeably and it must be recognized that the inward rectifier is only *one* constituent of the fully functioning K_{ATP} channel. To emphasize this concept, one could consider the analogy of a car (K_{ATP}) consisting of an engine (K_{IR} 6.2) and a chassis (SUR1) with the two components being inexorably linked in the normally functioning vehicle (K_{ATP}). Therefore, a description of the structure and function K_{ATP} channels should address the β-cell high affinity sulfonylurea receptor (SUR1) and inward rectifier (K_{IR} 6.2) molecules, which together constitute the complete K_{ATP} channel (Inagaki et al., 1995a; Aguilar-Bryan et al., 1998 in press).

B. High Affinity Sulfonylurea Receptors

During World War II, a group of sulfonamide-derived compounds used to treat typhoid fever were found to cause hypoglycemia. This class of blood glucose lowering agents—the sulfonylureas—now constitute the mainstay of oral diabetes management. It is now well established that the pancreas plays a central role in the hypoglycemic action of sulfonylureas.

Studies in a number of laboratories have proven the existence of a specific, high affinity (K_D 0.5 to 10 nM) binding protein for sulfonylureas on the surface of β-cells. In an ongoing project that has lasted for more than a decade in the Aguilar-Bryan and Bryan laboratories, we have characterized, isolated, and cloned the β-cell sulfonylurea receptor (Aguilar-Bryan et al., 1995). Briefly, our experimental strategy was to tag the sulfonylurea receptor using a ^{125}I-radiolabelled derivative of glibenclamide (a potent second-generation sulfonylurea). The "tagged receptor" in β-cell membranes was then followed sequentially through a series of chromatographic steps to isolate and purify the β-cell high affinity sulfonylurea receptor (SUR1). After a partial amino-acid sequence was derived from enzymatic digestion of the isolated protein, classical molecular biology techniques were used to clone the gene encoding the β-cell sulfonylurea receptor. The "average" receptor is 1,582 amino acids long with an approximate molecular mass of 177 kDa (from sequence analysis).

The human gene specifying SUR1 is found in pancreatic islet cells (β-, α-, and δ-cells), brain, and other neuroendocrine cells. Homology cloning has uncovered a second sulfonylurea receptor gene with at least two functional splice variants (SUR2A, SUR2B; Inagaki et al., 1996; Isomoto et al., 1996) encoding low affinity receptors (K_D in the μM range) believed to constitute ATP-sensitive K^+ channels in other tissues such as cardiac, skeletal, vascular and nonvascular smooth muscle, when coupled to inward rectifiers, K_{IR} 6.2 or K_{IR} 6.1 (see the discussion that follows). In this chapter, we confine our discussion to the β-cell high affinity sulfonylurea receptor (SUR1).

It was surprising to discover that SUR1 is not an ion channellike protein but rather a member of an ubiquitous family of proteins, the ATP-binding cassette (ABC) superfamily (Higgins, 1992), which transport or pump a wide variety of molecules across cell membranes. Multiple transmembrane-spanning domains and two nucleotide-binding folds characterize this family of proteins. The greatest similarities between the sulfonylurea receptors and other ABC proteins (e.g., multidrug resistance protein [MRP], cystic fibrosis transmembrane conductance regulator [CFTR]) are found within the nucleotide binding folds. The exact topology of the class of proteins closely related to SUR1 continues to evolve, with current thinking indicating 11 transmembrane-spanning domains in the first region and 6 in the second group, associated with the nucleotide-binding folds (Tusnady et al., 1997).

Earlier studies of ionic flux and K_{ATP} channel activity showing channel closure by sulfonylureas (Sturgess et al., 1985; Schmid-Antomarchi et al., 1987) suggested that the sulfonylurea receptor *was* the K_{ATP} channel or closely associated with it. The presence of a high affinity sulfonylurea-binding site and nucleotide binding regions in SUR1 allows it to be a "sensor" of surrounding changes in nucleotide levels. These features and the potential of performing transporter functions (by virtue of it being a member of the ABC transporter protein family) also supported the possibility of SUR1 being K_{ATP} channels. However, while the recombinant SUR1 receptors (expressed in COS cells) showed high affinity sulfonylurea binding, they displayed no K^+ channel activity (Aguilar-Bryan et al., 1995). This argued that another missing ingredient was required for fully functional K_{ATP} channels.

C. Inward Rectifiers

The inward rectifiers are a broad class of potassium channels that differ from voltage-gated K^+ channels by not being activated by membrane potential and exhibiting the property of inward rectification (Doupnik et al., 1995). This is a phenomenon of preferential conduction of ions in an inward (rather than outward) direction when compared at comparable positive and negative membrane potentials with symmetrical K^+ distribution. β-cells (like most cells) maintain a membrane potential that is positive to the potassium equilibrium potential (E_K) so that K^+ currents are mainly outward, and subject to physiological modulation by rectification. Much like the categorization of ABC transporters and sulfonylurea receptors, the inward rectifier family of potassium channels has been classified into six groups (K_{IR} 1.x–6.x) on the basis of electrophysiological and molecular biological characterization (Doupnik et al., 1995). The K_{IR} 6.x group represents the second subunit of the K_{ATP} channel subfamily of inward rectifiers, and we focus our discussion on one of its members, K_{IR} 6.2, which couples to SUR1 to yield K_{ATP} channels.

This subunit K_{IR} 6.2 (also known as β-cell inward rectifier, BIR) was isolated by homology to another inward rectifier K_{IR} 6.1, and is distributed abundantly in β-cells and brain (Inagaki et al., 1995b). It is characterized by moderate inward rectification (in contrast to strong inward rectification (K_{IR} 2.x) where channel conductance *rapidly* approaches zero at potentials slightly positive to E_K). The inward rectifier K^+ channel has two transmembrane spanning domains (M1 and M2) that flank a highly conserved P or H5 domain (also present in voltage-dependent K^+ channels), which is thought to form the K^+ selective pore (Kubo et al., 1993; Lopez et al., 1994). A mutation in one residue of M2, Asn to Asp (N160D mutant) induces steep voltage dependence to rectification and converts K_{IR} 6.2 into a strong inward rectifier (Aguilar-Bryan et al., 1998 in press). Co-expression experiments of mixtures of inward rectifiers demonstrate a tetrameric pore structure in functional channels (Glowatzki et al., 1995).

Although several reports claimed the identification of K_{ATP} channels in the inward rectifier K^+ channel family, detailed analysis of these putative candidates led to the conclusion that the conductances and pharmacology described were not that of the classic β-cell K_{ATP} channels (Rorsman and Trube, 1985; Ho et al., 1993). Expression of K_{IR} 6.2 alone did not produce a potassium conductance, suggesting that it required another subunit for activation. Previous studies in photolabeling the high affinity sulfonylurea receptor had identified a 140 kDa species in β-cell membranes. Using a novel azido-iodo derivative of glibenclamide, more recent studies demonstrated high affinity co-photolabeling of SUR1 (140-kDa) and an unknown 38- to 40-kDa protein (Schwanstecher et al., 1994a; Schwanstecher et al., 1994b). While the 140-kDa receptor is confined to internal membrane compartments, the glycosylated 150- to 170-kDa species is apparently found on the plasma membrane (Ozanne et al., 1995). The stage was set to determine if K_{IR} 6.2 represented the small unknown photolabeled species and whether a conjoined SUR1–K_{IR} 6.2 complex constituted the β-cell K_{ATP} channel.

The co-expression of SUR1 and K_{IR} 6.2 produced immediately gratifying results. Both subunits are photolabeled by azido-iodoglibenclamide. The 140-kDa species is SUR1; the small 38- to 40-kDa species has been identified as K_{IR} 6.2; and the highly glycosylated species reflects the combination of both species, constituting K_{ATP} channels in the plasma membrane (Schwanstecher et al., 1994a; Schwanstecher et al., 1994b; Ozanne et al., 1995). Importantly, co-expression of SUR1 and K_{IR} 6.2 (in COS cells) generated weakly rectifying K^+ channels that are inhibited by nanomolar concentrations of glibenclamide and half-maximally inhibited by 8- to 10-μM ATP (Inagaki et al., 1995a). The single channel conductance is about 76 picoSiemens at –60 mV in symmetrical KCl (140 mM). These features of the recombinant channels mimic the classical β-cell K_{ATP} channels and have established that a union of SUR1 and K_{IR} 6.2 units constitutes functional, native β-cell K_{ATP} channels (Inagaki et al., 1995a). Furthermore, there is evidence that an octameric complex with a 1:1 stoichiometry of SUR1 and K_{IR} 6.2 units $(SUR1/K_{IR}\ 6.2)_4$ and a molecular mass of 950 kDa represents the β-cell K_{ATP} channel (Clement et al., 1997).

D. Modulation

K_{ATP} channels are present in several tissues and are of different classes, so we reemphasize here that our discussion relates to the β-cell variant. K_{ATP} channels in β-cells are characteristically closed by intracellular ATP (at levels in μM range) coupled to the metabolic state of the cells. Channel blockade at high ATP concentrations is mediated by interaction with binding sites likely located in the sulfonylurea receptor protein, as opposed to a physical blockade of the pore (Kakei et al., 1985). MgADP opens K_{ATP} channels blocked by ATP. In addition, intracellular Mg^{2+} in the presence of physiological levels of ATP (Findlay, 1987) prevents "rundown" or loss of electrical activity during recording from cell membrane patches. The sensitivity of β K_{ATP} channels to ATP estimated by channel closure is in the low micromolar (K_i 15 to 50 μM) range, suggesting that these channels will be closed at ATP concentrations (>2 mM) estimated for β-cells. The spare channel hypothesis (Cook et al., 1988) was proposed, arguing that β-cells have a large excess of channels of which only a few (~10%) need to be open to maintain the resting membrane potential. Thus, high concentrations of ATP would be required to induce closure of all the channels in response to metabolic activity. Despite the attractiveness of this hypothesis, differences in β-cell ATP levels during cell metabolism have been difficult to discern, and a simple role for ATP in K_{ATP} channel regulation hard to defend (Ghosh et al., 1991). It has been noted that higher concentrations of ATP are required to evoke channel closure in the presence of ADP, suggesting that ADP (Nichols and Lederer, 1991; Terzic et al., 1994) modulates channel regulation by ATP. Thus, ATP/ADP ratio rather than absolute ATP levels may be more important in regulating channel activity. Although high concentrations of ADP may compete directly with ATP for the ATP-binding site, it is likely that the modulatory effects of ADP occur by binding to a different site. There is evidence that mutations in SUR1 result in the loss of response to ADP (in the presence of ATP), clearly indicating the importance of ADP as a regulator (Aguilar-Bryan and Bryan, 1996). Thus, the presence of multiple nucleotide binding sites (e.g., ATP, ADP) with differing specificity in the SUR1 protein offer a means for regulation of K_{ATP} channel activity by cell metabolism. Furthermore, since the K_{ATP} channel can now be reconstituted, it will be possible to more precisely examine its mode of regulation (i.e., by ATP, ADP, ATP/ADP, etc.).

ATP plays a dual role because a low level of ATP is also required for activation (presumably by phosphorylation upon ATP hydrolysis) during channel opening (Ribalet et al., 1989). Channel activity is enhanced by cyclic AMP (cAMP)–dependent protein kinase-mediated phosphorylation. The role of protein kinase C in K_{ATP} channel modulation in β-cells remains inconclusive. In studies with insulin-secreting cell lines, both inhibition and activation of channel activity by phorbol esters (activators of protein kinase C) have been reported whereas no decrease in channel activity has been observed in normal β-cells (Wollheim et al., 1988; Henquin et al., 1989; Ribalet et al., 1989). Protein phosphorylation has also been shown to modulate ligand interactions at the channel with both sulfonylurea-induced closure and diazoxide-induced opening of K_{ATP} channels requiring phosphorylation of the channel protein (Dunne, 1989; Schwanstecher et al., 1991). In addition, the inhibition of glibenclamide binding to pancreatic islets in the presence of Mg^{2+} by trinucleotides (ATP, guanosine triphosphate [GTP]), in contrast to mononucleotides (AMP-PNP, guanosine monophosphate [GMP]-PNP) also suggests a role for protein phosphorylation in these inhibitory effects (Schwanstecher et al., 1991). The presence of multiple protein phosphorylation sites in the sulfonylurea-binding protein provides for phosphorylation of the receptor as a modulatory event in K_{ATP} channel function. With the recent demonstration of a reconstitutable K_{ATP} channel in vitro, the role and mechanisms of K_{ATP} channel phosphorylation may be clarified in the future.

While it is well established that sulfonylureas block K_{ATP} channels, our knowledge of exactly how these drugs achieve their effects continues to evolve. The ability of different sulfonylureas to bind to the β-cell SUR1 protein is well correlated with the drug's potency to inhibit channel activity and stimulate insulin secretion (Boyd et al., 1990). Thus, glibenclamide (a second generation sulfonylurea) binds tightly and more irreversibly than tolbutamide (a first generation drug), resulting in more potent effects of glibenclamide on K_{ATP} channels and in patients. The functional effects of sulfonylureas on K_{ATP} channels may be determined by receptor density (downregulation), lipid solubility of the drugs, protein binding, pH, nucleotides, Mg^{2+}, and phosphorylation. These variables probably influence sulfonylurea binding and conformational changes, inducing alterations in channel activity. An abbreviated list of other drugs including general anesthetics (barbiturates, pentobarbitone, and thiopentane), quinine, sparteine, amantadine, and ligustrazine (a Chinese herbal remedy inducing vasodilatation) and imidazoline-type α-adrenoreceptor antagonists have less specific inhibitory effects on β-cell K_{ATP} channels, and have not been as well characterized (Gopalakrishnan et al., 1993). Likewise, several drugs are known to increase β K_{ATP} channel activity. The best-documented effects are those of diazoxide, which specifically activates K_{ATP} channels (Dunne, 1989). The resulting inhibition of insulin release is responsible for its potent hyperglycemic effect. The action of diazoxide is dependent on the presence of ATP and reversed by glibenclamide (Dunne, 1989). Other compounds demonstrated to activate K_{ATP} channel activators in β-cells include cromakalim, pinacidil, and minoxidil, but their effects are less potent than diazoxide (Gopalakrishnan et al., 1993). Similarly, endogenous peptides such as vasoactive intestinal peptide (VIP) and calcitonin gene-related peptide (CGRP) have been suggested as channel activators (Nelson et al., 1990; Lippton et al., 1991). While diazoxide is believed to bind to sulfonylurea receptor, the actions of the array of other agents on channel activation and closure are secondary to binding to their own receptor sites. There is presently no evidence for direct binding of peptides to sulfonylurea receptors.

G-protein modulation of K_{ATP} channels may be important in mediating the effects of certain hormones and neurotransmitters (Dunne and Petersen, 1986). Guanine nucleotides (GTP and guanosine diphosphate [GDP]) evoke reversible dose-dependent increase in channel activity in the presence of Mg^{2+}. It was also found that the nonhydrolyzable analog GTPγS and AlF^{4-} (which mimics GTPγS action) produce K_{ATP} channel activation in β-cells. In β-cells, galanin, somatostatin, and α_2-adrenoreceptor agonists inhibit insulin secretion through a G-protein mediated pathway (de Weille et al., 1988; Fosset et al., 1988; Dunne, 1989). This is evident from the pertussis toxin-sensitivity of ligand-mediated channel activation and resultant effects. Although there is some data showing somatostatin and galanin activation of K_{ATP} channels in insulin-secreting cell lines, these studies are still incomplete because comparable results in normal β-cells have not been obtained. There is also evidence for effects at multiple sites for all these agents. Recently, the incretin, glucagonlike peptide 1 (GLP-1), has been shown to have a potent effect on insulin secretion. Although the actions of this agent on β-cells appear to include multiple pathways, there is convincing evidence for involvement of K_{ATP} channels in GLP-1 signal transduction (Leech et al., 1996). Furthermore, G-protein mediation of the *inhibition* of K_{ATP} channels by GLP-1 is emerging as a plausible modulatory mechanism.

IV. VOLTAGE-DEPENDENT CA^{2+} CHANNELS

A. Structure and Function

Intracellular free cytosolic Ca^{2+} is the primary second messenger regulating β-cell insulin secretion. Depolarization of the cell membrane results in opening of voltage-sensitive Ca^{2+} channels and Ca^{2+} influx into β-cells, triggering insulin release (Rajan et al., 1990). As is the case with K^+ channels, several classes of voltage-dependent Ca^{2+} channels exist (McCleskey et al., 1986). In pancreatic β-cells the types identified include the L-type and T-type channels (Satin and Cook, 1988). Since the L-type channels predominate in β-cells, our discussion is primarily focussed on this group of channels.

The L-type Ca^{2+} channel is so named because of the long-lasting nature of the current (McCleskey et al., 1986). This channel has been observed by investigators in every type of β-cell preparation (Rorsman and Trube, 1986; Satin and Cook, 1988; Hiriart and Matteson, 1988; Keahey et al., 1989). The distinguishing feature of this class of channels is its sensitivity to dihydropyridines, which act as Ca^{2+} channel antagonists and agonists (Triggle and Janis, 1987). Analogous to the utility of sulfonylureas in the study of K_{ATP} channels, the dihydropyridines have proven to be useful ligands that bind to specific receptor sites in the Ca^{2+} channels, thereby enabling insight into L-type Ca^{2+} channel structure (Triggle and Janis, 1987).

Using radiolabeled dihydropyridine probes (e.g., [^{3}H]-PN 200), several laboratories have tagged the dihydropyridine-binding protein and purified it from a number of tissues. The abundance of this protein in skeletal muscle T-tubules (10- to 100-fold greater than other cells) made that tissue the preferred source for dihydropyridine-binding protein purification (Catterall et al., 1988). Similar to the isolation of the sulfonylurea receptor, the labeled dihydropyridine-binding protein was followed through multiple chromatographic steps (Curtis and Catterall, 1984; Borsotto et al., 1985; Flockerzi et al., 1986). The subunit composition of the isolated protein was then determined by electrophoretic separation and DNA coding sequences identified by cloning techniques. Based on these data, the L-type Ca^{2+} channel is believed to be a multimer of five distinct transmembrane subunits: α_1, α_2, β, γ, and δ (Catterall et al., 1988). This general structure appears to be conserved across all members of the L-type Ca^{2+} channels identified (Catterall et al., 1988). The molecular weight of the oligomeric complex in skeletal muscle Ca^{2+} channels is approximately 400,000 M_r and stoichiometric analysis suggests a 1:1:1:1:1 interaction of the subunits in the functional Ca^{2+} channel (Campbell et al., 1988). Elegant expression studies of the α_1 subunit (in mammalian cells) have demonstrated that it can bind dihydropyridines, sense voltage changes *and* flow Ca^{2+} current (in contrast to the comparable role of sulfonylurea receptors in K_{ATP} channels), suggesting that it is the major functional component of Ca^{2+} channels (Tanabe et al., 1988; Perez-Reyes et al., 1989). The a_1 subunit (similar to other Na^+ and K^+ channels) contains four homologous repeating units, each containing six transmembranal α-helical segments, S1 to S6 (Catterall et al., 1988). The other Ca^{2+} channel subunits play an important role in Ca^{2+} channel regulation (e.g., phosphorylation, Catterall et al., 1988).

Oligonucleotide primers based on a conserved amino-acid sequence in the S6 region of repeat III of the skeletal muscle α_1 subunit were employed in the screening of a β-cell cDNA library. These studies resulted in the identification of a novel α_1 subunit that is expressed in β-cells and other tissues such as brain (Yaney et al., 1992). Although the predicted structure of the β-cell L-type Ca^{2+} channel is similar to that of skeletal muscle Ca^{2+} channels (i.e., four units with S1–S6 regions), there are subtle variations, particularly in the C-terminal region (Yaney et al., 1992). These changes may potentially confer different regulatory characteristics to β-cell Ca^{2+} channels.

B. Modulation

The primary modulator of voltage-dependent Ca^{2+} channels is membrane potential. In β-cells, an inward Ca^{2+} current through L-type Ca^{2+} channels is observed when cells held ("voltage-clamped") at a potential of –70 to –50 mV are depolarized to membrane potentials greater than –40 mV (Rajan et al., 1990). Peak current is generally observed between –20 and 0 mV. The size of the current varies with the charge carrier, with the largest current (60 to 400 pA) being observed with Ba^{2+} rather than Ca^{2+}. The inward current through L-type Ca^{2+} channels is long-lasting (i.e., slow inactivation), observed in the –50 to +50 mV range over a 200- to 300-ms recording time frame, although a transient, rapidly inactivating component has been observed at lower potentials (–30 to 0 mV). In contrast, the T-type ("transient") Ca^{2+} channel is activated at –50 to –40 mV, reaches its peak at –10 to 0 mV, and disappears above 20 mV, lasting only for 10 to 100 ms. The S4 region of the α_1 subunit contains a sequence of charged amino acids that is highly conserved across Na^+, K^+, and Ca^{2+} channels and highlights the functional importance of this region as the putative voltage sensor for this protein (Stuhmer et al., 1989).

A hallmark of L-type Ca^{2+} channels is their sensitivity to dihydropyridines (Triggle and Janis, 1987). Thus, antagonists such as nimodipine, nifedipine, PN 200, and nitrendipine bind to specific receptor sites on the L-type Ca^{2+} channel to inhibit Ca^{2+} current, whereas dihydropyridine agonists such as Bay K 8644 increase L-type Ca^{2+} current. Other (nondihydropyridine) L-type Ca^{2+} channel antagonists include verapamil and diltiazem, which bind to L-type Ca^{2+} channel to inhibit Ca^{2+} current. These Ca^{2+} channel blockers (verapamil, diltiazem, and nifedipine) are used clinically in the management of hypertension. Interestingly, the binding of these three classes of drugs to distinct specific receptor sites is allosterically regulated. Hence, binding of one drug (e.g., verapamil) to its site decreases dihydropyridine binding to its site whereas diltiazem binding enhances dihydropyridine to L-type Ca^{2+} channels.

An important feature of L-type Ca^{2+} channel regulation is the effect of intracellular Ca^{2+} itself on Ca^{2+} channel current. The rise in cytosolic Ca^{2+} after Ca^{2+} influx acts as a signal to inactivate the Ca^{2+} channel to restrict further Ca^{2+} influx (Chay, 1987). Ca^{2+} currents resist inactivation by intracellular Ca^{2+} and hence last longer than Ca^{2+} currents. Modulation of Ca^{2+} current is induced by binding of Ca^{2+} to a distinct site on the Ca^{2+} channel protein. Ca^{2+} binding also allosterically regulates Ca^{2+} dihdropyridine binding. Thus, negative feedback by elevated intracellular Ca^{2+} could be important in regulation of membrane electrical activity and insulin secretion, and in the effects of other drugs affecting Ca^{2+} channel activity.

The voltage-dependent Ca^{2+} channels are regulated by second messenger pathways including phosphorylation. There is considerable evidence to support that cAMP enhances Ca^{2+} influx and insulin secretion in β-cells (Malaisse and Malaisse-Lagae, 1984). Since all the actions of cAMP in cells are mediated by activation of cAMP-dependent protein kinase, phosphorylation of the Ca^{2+} channels (as noted in cardiac Ca^{2+} channels; Osterrieder et al., 1982) is a likely mechanism of cAMP action. Indeed, direct effects of glucose on Ca^{2+} channels are presumably mediated by effects of cAMP on Ca^{2+} channels (Ammala et al., 1993), although a potential role of cAMP-dependent phosphorylation of ATP-sensitive K^+ channels upon glucose (or GLP-1) stimulation is also a plausible effect. There is also evidence of calcium-calmodulin protein kinase and protein kinase C-dependent phosphorylation of β-cell Ca^{2+} channels to regulate insulin release (Rajan et al., 1990). In this case, both enhancement and inhibition of Ca^{2+} conductance by protein kinase C has been observed. Five phosphorylation sites have been identified within the putative cytoplasmic domains of the α_1 subunit of the β-cell L-type Ca^{2+} channels. In addition, the β subunit of Ca^{2+} channels also possesses cAMP phosphorylation sites. Thus, phosphorylation of Ca^{2+} channels represents a modulatory mechanism for controlling Ca^{2+} channel activity in β-cells.

Like K_{ATP} channels, L-type Ca^{2+} channels are also regulated by G-proteins (Rajan et al., 1990). There is evidence that a variety of hormones and neurotransmitters modulate L-type Ca^{2+} channel function through the interaction of G-proteins. Thus catecholamines (e.g., α_2-agonists), somatostatin and galanin reduce Ca^{2+} influx into β-cells via L-type Ca^{2+} channels, an action that is inhibited by pertussis toxin pretreatment. Likewise, GLP-1 has been shown to enhance Ca^{2+} influx into β-cells through L-type Ca^{2+} channels, presumably by G-protein activation (Leech et al., 1996).

V. OTHER ION CHANNELS

The Ca^{2+}-activated K^+ (maxi-K) channels comprise a family of high conductance K^+ channels (Kaczorowski et al., 1996). These channels are activated in β-cells (as in other cell types) by depolarization and binding of Ca^{2+} to promote transition to the open or conducting state. The presumed role of this channel is in cell repolarization (Dunne and Petersen, 1991). Ion conduction through maxi-K channels is blocked by tetraethyl ammonium (TEA) and venom peptides such as charybdotoxin. In contrast to the high conductance (~250 pS) of the maxi-K channel, a lower conductance (sub pS), Ca^{2+}-activated K^+ channel activated by GTPγS is present in β-cells (Ammala et al., 1991). This channel is insensitive to apamin, TEA, or sulfonylureas. A receptor-operated K^+ channel (K_I) identified in mouse β-cells is activated by adrenalin in a G-protein–dependent manner and clonidine to induce membrane hyperpolarization may be identical to the apamin-insensitive, low-conductance Ca^{2+}-activated K^+ channel (Rorsman et al., 1991). A TEA-sensitive, delayed rectifier K^+ channel (~10 pS) is activated by progressive membrane depolarization and may also contribute to membrane repolarization (Rorsman and Trube, 1986). Nonselective cation channels (K_{NS}) have been identified that are

Ca^{2+}-sensitive and blocked by adenosine nucleotides (Ashcroft and Rorsman, 1989).

T-type ("transient") voltage-gated Ca^{2+} channels similar to those observed in other tissues have been found in β-cells (Satin and Cook, 1988). These channels are rapidly activated at more negative potentials than L-type Ca^{2+} channels and are insensitive to dihydropyridine blockers. However, a definitive role for T-type Ca^{2+} channels is yet to be established (Rajan et al., 1990). Na^+ channels have been identified and may regulate electrical activity in β-cells (Hiriart and Matteson, 1988; Pressel and Misler, 1990). Tetrodotoxin, an inhibitor of voltage-gated Na^+ channels, has been reported to inhibit glucose induced insulin secretion or be ineffective, suggesting species- or cell model-based differences in regulation (Dunne et al., 1990). In general, however, Na^+ and Cl^- channels are the least characterized channels in β-cells.

VI. ROLE OF ION CHANNELS IN THE PATHOPHYSIOLOGY OF INSULIN SECRETION

A. Regulation of β-Cell Electrical Activity

We began this chapter with a description of the electrical activity in β-cells. The regulation of glucose-induced electrical activity and insulin secretion involves the integrated control of several types of ion channel. What role do the ion channels we have described serve in this process? The following discussion is enhanced by reference to Figures 1 and 2.

The resting membrane potential (observed at subthreshold glucose concentrations) is characterized by silent cells with a high resting K^+ permeability. The only single channel currents recorded in this state are those of K_{ATP} channels. However, less than 10% of these channels are open at existing ATP levels in the cells. At stimulatory glucose concentrations, activity from these open channels ceases in response to changing ATP/ADP ratios upon glucose metabolism. The membrane potential rises from its resting value toward the threshold potential for action potential firing. Thus, metabolic events are coupled to the generation of electrical activity in the β-cell membrane.

The inward current associated with the upstroke of the action potential results from Ca^{2+} entering the L-type voltage-gated Ca^{2+} channels, which are maximally active at approximately −20 to 0 mV. Termination of the Ca^{2+}-spike potential ("downstroke") is achieved by Ca^{2+}-dependent inactivation of the Ca^{2+} channels and activation of the maxi-K channels. The latter are recruited because of the existing favorable state of cell depolarization and Ca^{2+} influx. It is also believed that the delayed rectifier K^+ channel may contribute to repolarization with the maxi-K channel, with the effects of both these channels inhibited by TEA blockade. The Ca^{2+} channel activity underlies the slow wave

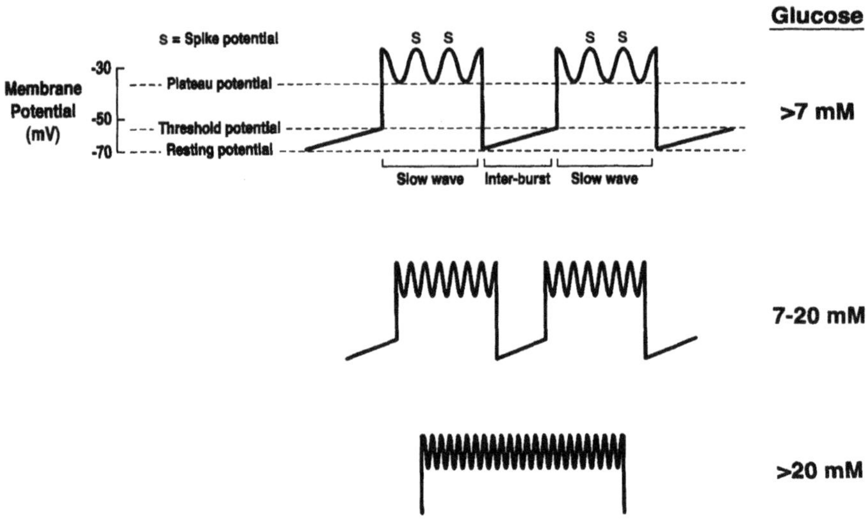

Figure 1. Electrical activity in the beta cell. The primary metabolic stimulus, glucose, initiates signaling in the beta cell by generating characteristic action potentials (see text for details) that ultimately trigger insulin release.

formation, which may additionally involve the lower conductance Ca^{2+}-activated K^+ channels, in response to oscillations in Ca^{2+} and membrane potential. The slow, voltage-dependent inactivation of L-type voltage-gated Ca^{2+} channels may also explain the glucose-induced, pacemakerlike pattern of electrical activity.

In pancreatic islets, glucose stimulation is associated with Ca^{2+}oscillations that are synchronized with electrical activity. Different processes such as oscillations in ATP/ADP ratio in close proximity to the K_{ATP} channels or oscillations in Ca^{2+} influx through coordinated actions of the L-type Ca^{2+} channels and Ca^{2+}-activated K^+ channels may account for the observed Ca^{2+} oscillations and putative underpinning of pulsatile insulin release. While the roles of a handful of β-cell ion channels are beginning to be understood, the functions of several identified channels in β-cell physiology remain to be determined. Finally, the complex modulation of the surface ion channels by extra- and intracellular ligands, voltage, intracellular regulation (e.g., by phosphorylation or G-proteins), and hormones and neurotransmitters must be considered in the regulation of β-cell electrical activity and stimulus-secretion coupling of insulin release.

In summary, glucose stimulation of β-cells involves metabolic changes (ATP/ADP) that inhibit K^+ efflux and trigger action potentials. The resulting voltage-gated Ca^{2+} influx changes intracellular $Ca^{2+}$and promotes insulin secretion.

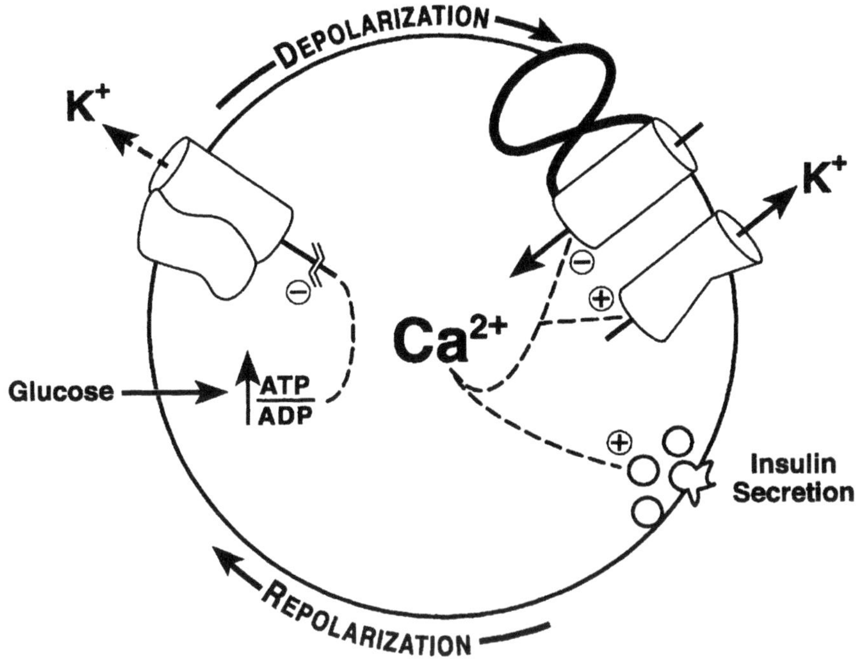

Figure 2. Ion channel signal transduction in the beta cell. Glucose metabolism results in cell depolarization by closure of ATP-sensitive K^+ channels. Voltage-dependent Ca^{2+} influx triggers insulin secretion and promotes K^+ efflux to repolarize the cell to the resting state (see text for details).

B. Species Variance

For the most part, there is little variation in the properties of ion channels in β-cells in rodents or humans. The conductance, pharmacology, and regulation of K_{ATP} channels studied in rodents (rats, mice) and humans are nearly identical. This is confirmed by highly homologous sequences of SUR1 and K_{IR} 6.2 in rodents and humans. Likewise, the characteristics of L-type Ca^{2+}channels remain consistent across species. However, the T-type Ca^{2+}channels are not observed in mice, and though present in rats, are unlikely to play a significant role because of the rapid activation kinetics (Rajan et al., 1990).

The nonselective cation (K_{NS}) channels are more frequently found in human than in rodent cells, may be more sensitive to nucleotides than previously thought, and may be able to contribute to the resting membrane potential or background K^+ or Na^+ current. The contribution of Na^+ channels to β-cell electrical events is believed to vary significantly across species. Thus, while Na^+ currents are inactivated at the resting membrane potential and play no role in mouse β-cell electrical activity, there is evidence that Na^+ channel activity may contribute to action potential depolarization in human β-cells (Pressel and Misler, 1990). These differences may account for the conflicting observed effects of the Na^+ channel blocker tetrodotoxin on glucose-stimulated insulin secretion.

The lack of significant distinctions across species between the major ion channels involved in β-cell electrical activity lends confidence to the results obtained in rodent models and their application in the human domain.

C. Disease Implications

Ionic flux studies in rodent diabetic models demonstrated abnormalities of K^+ flux associated with defective insulin secretion (Frankel and Sehlin, 1987). However, direct studies of K_{ATP} channels in diabetic rats showed impaired glucose modulation of the K_{ATP} channels but no impairment of the channels themselves (Tsuura et al., 1992). The most compelling evidence to date linking ion channel defects to abnormalities in insulin secretion (which underlies the critical role of K_{ATP} channels in insulin secretion) is provided by the molecular characterization of the disease, persistent hyperinsulinemic hypoglycemia of infancy (PHHI) (Aguilar-Bryan and Bryan, 1996; Permutt et al., 1996).

PHHI is a genetic disease that is the most common cause of hypoglycemia in newborns and is estimated to occur about once in every 50,000 births in random populations and once in 2,500 births in inbred populations. Pioneering studies on β-cells from these patients indicated that uncoupling of insulin secretion from glucose metabolism caused PHHI. Thus the β-cells constitutively ("persistently") release insulin with no sensitivity to glucose. Exploratory studies were initiated based on the hypothesis that K_{ATP} channel mutations may give rise to the PHHI phenotype. Linkage analysis had mapped PHHI to chromosome 11p14–15.1 and exciting new results demonstrated that the SUR1 and K_{IR} 6.2 genes mapped to 11p15.1. Genetic screening looking for mutations in PHHI patients has identified about 40 mutations in the sulfonylurea receptor (SUR1) and K_{IR} 6.2 that cover a wide range from point mutations to severe truncations. One very intriguing mutation points toward altered nucleotide sensing (lack of activation by ADP). This would explain the tonic inhibition of K_{ATP} channels by ATP, resulting in persistent firing, Ca^{2+} influx and insulin release. The direct link between a defect in the K_{ATP} channel and abnormal insulin secretion has provided impetus to the search for similar linkage in type II diabetes, which to date has not been fruitful. Along the same lines, the secondary failure of insulin secretory response to sulfonylureas is a common occurrence in diabetic patients and it remains to be determined whether altered expression, structure, or function of β-cell K_{ATP} channels may contribute to this state. Abnormal regulation of Ca^{2+} fluxes and intracellular Ca^{2+} have been observed in diabetic rodents, but the loci of the defects contributing to this state have not been identified (Flatt, 1992). While it is obvious that perturbations in

Ca^{2+} flux will play a critical role in insulin secretion, site-specific lesions in voltage-dependent Ca^{2+} channels associated with diabetes remain to be established.

REFERENCES

Aguilar-Bryan, L. and Bryan, J. (1996). ATP-sensitive potassium channels, sulfonylurea receptors, and persistent hyperinsulinemic hypoglycemia of infancy. Diab. Rev. 4, 336–346.

Aguilar-Bryan, L., Clement, J.P. IV, Gonzalez, G., Kunjilwar, K., Babenko, A. and Bryan, J. (1998). Towards understanding the assembly and structure of K_{ATP} channels. Physiol. Rev. (in press).

Aguilar-Bryan, L., Nichols, C.G., Wechsler, S.W., Clement, , J.P. IV, Boyd, A. E. III, Gonzalez, G., Herrera-Sosa, H., Nguy, K., Bryan, J. and Nelson, D.A. (1995). Cloning of the beta cell high-affinity sulfonylurea receptor: A regulator of insulin secretion. Science 268, 423–426.

Ammala, C., Ashcroft, F.M.. and Rorsman, P. (1993). Calcium independent potentiation of insulin release by cyclic AMP in single beta cells. Nature 363, 356–358.

Ammala, C., Larsson, O., Berggren, P.-O., Bokvist, K., Juntti-Berggren, L., Kindmark, H. and Rorsman, P. (1991). Inositol triphosphate-dependent periodic activation of a Ca^{2+}-activated K^+ conductance in glucose stimulated pancreatic beta-cells. Nature 353, 849–852.

Ashcroft, F.M. and Rorsman, P. (1989). Electrophysiology of the pancreatic beta cell. Prog. Biophys. Mol. Biol. 54, 87–143.

Ashford, M.L.J., Bond, C.T., Blair, T.A. and Adelman, J.P. (1995). Cloning and functional expression of a rat heart K_{ATP} channel—Retraction. Nature 378, 792.

Borsotto, M., Barhanin, J., Fosset, M. and Lazdunski, M. (1985). The 1,4-dihydropyridine-receptor associated with the skeletal muscle voltage-dependent Ca^{2+} channel: Purificiation and subunit composition. J. Biol. Chem. 260, 14255–14263.

Boyd, A.E. III, Aguilar-Bryan, L. and Nelson, D.A. (1990). Molecular mechanisms of action of glyburide on the beta cell. Am. J. Med. 89 (Suppl. 2A), 3S–10S.

Campbell, K.P., Leung, A.T. and Sharp, A.H. (1988). The biochemistry and molecular biology of the dihydropyridine-sensitive calcium channel. Trends Neurosci. 11, 425–430.

Catterall, W.A., Seagar, M.J. and Takahashi, M. (1988). Molecular properties of dihydropyridine-sensitive calcium channels in skeletal muscle. J. Biol. Chem. 263, 3535–3538.

Chay, T.R. (1987). The effects of inactivation of calcium channels by intracellular Ca^{2+} ions in bursting pancreatic β cells. Cell Biophys. 11, 77–90.

Clement, J.P. IV, Kunjilwar, K., Gonzalez, G., Schwantecher, M., Panten, U., Aguilar-Bryan, L. and Bryan, J. (1997). Association and stoichiometry of K_{ATP} channel subunits. Neuron. 18, 827–838.

Cook, D.L., Satin, L.S., Ashford, M.L.J. and Hales, C.N. (1988). ATP-sensitive K channels in pancreatic beta-cells. Spare channel hypothesis. Diabetes 37, 495–498.

Curtis, B.M. and Catterall, W.A. (1984). Purification of the calcium antagonist receptor of the voltage-sensitive calcium channel from skeletal muscle transverse tubules. Biochemistry. 23, 2113–2114.

de Weille, J.R., Schmid-Antomarche, H., Fosset, M and Lazdunski, M. (1988). ATP-sensitive K^+ channels that are blocked by hypoglycemia-inducing sulfonylureas in insulin secreting cells are activated by galanin, a hyperglycemia-inducing hormone. Proc. Natl. Acad. Sci. USA. 85, 1312–1316.

Dean, P.M. and Matthews, E.K. (1974). Glucose-induced electrical activity in pancreatic islet cells. J. Physiol. 210, 255–264.

Doupnik, C.A., Davidson, N. and Lester, H.A. (1995). The inward rectifier potassium channel family. Curr. Opinion Neurobiol. 5, 268–277.

Dunne, M.J. (1989). Protein phosphorylation is required for diazoxide to open ATP-sensitive potassium channels in insulin (RINm5F) secreting cells. FEBS Lett. 250, 262–266.

Dunne M.J. and Petersen, O.H. (1986). GTP and GDP activation of K^+ channels that are inhibited by ATP. Pflugers Arch. 407, 564–565.

Dunne, M.J and Petersen, O.H. (1991). Potassium selective ion channels in insulin-secreting cells: physiology, pharmacology, and their role in stimulus-secretion coupling. Biochim. Biophys. Acta. 1071, 67–82.

Dunne, M.J., Yule, D.I., Gallacher, D.V. and Petersen, O.H. (1990). Stimulant-evoked depolarization and increase in $[Ca^{2+}]_i$ in insulin-secreting cells is dependent on external Na^+. J. Membr. Biol. 113, 131–138.

Findlay, I. (1987). The effects of magnesium upon adenosine triphosphate-sensitive potassium channels in a rat insulin-secreting cell line. J. Physiol. 391, 611–629.

Flatt, P.R. (1992). *Nutrient Regulation of Insulin Secretion.* (Flatt, P.R. Ed), pp. 321–386. Portland Press.

Flockerzi, V., Oeken, H.J. and Hoffman, F. (1986). Purification of a functional receptor for calcium-channel blockers from rabbit skeletal-muscle microsomes. Eur. J. Biochem. 161, 217–214.

Fosset, M., Schmid-Antomarchi, H., de Weille, J.R. and Lazdunski, M. (1988). Somatostain activates glibenclamide-sensitive and ATP-regulated K^+ channels in insulinoma cells via a G protein. FEBS Lett. 242, 94–96.

Frankel, B.J. and Sehlin, J. (1987). Abnormalities in glucose stimulated insulin release, ^{45}Ca uptake, and ^{86}Rb efflux in diabetic Chinese hamster islets. Diabetes 36, 648–653.

Ghosh, A., Ronner, P., Cheong, E., Khalid, P. and Matschinsky, F.M. (1991). The role of ATP and free ADP in metabolic coupling during fuel-stimulated insulin release from islet beta cells in the isolated perfused rat pancreas. J. Biol. Chem. 266, 22887–22892.

Glowatzki, E., Fakler, G., Brandle, U., Rexhausen, U., Zenner, H.P., Ruppersberg, J.P. and Fakler, B. (1995). Subunit assembly-dependent assembly of inward-rectifier K^+ channels. Proc. Royal Soc. Lond. 261, 251–261.

Gopalakrishnan, M., Janis, R.A. and Triggle, D.J. (1993). ATP-sensitive K^+ Channels: Pharmacological properties, regulation, and therapeutic potential. Drug Dev. Res. 28, 95–127.

Henquin, J.C., Schmeer, W., Nenquin, M. and Plant, T.D. (1989). Does protein kinase C link glucose metabolism to beta-cell membrane depolarization. Diabetologia 32, 496A.

Higgins, C.F. (1992). ABC transporters: from microorganisms to man. Annu. Rev. Cell Biol. 8, 67–113.

Hiriart, M. and Matteson, D.R. (1988). Na channels and two types of Ca channels in rat pancreatic β cells identified with the reverse hemolytic plaque assay. J. Gen. Physiol. 91, 617–639.

Ho, K., Nichols, C.G., Lederer, W.J., Lytton, J., Vassilev, P.M., Kanazirske, M.V. and Hebert, S.C. (1993). Cloning and expression of an inwardly rectifying ATP-regulated potassium channel. Nature 362, 31–38.

Inagaki, N., Gonoi, T., Clement J.P. IV, Namba, J., Inazawa, J., Gonzalez, G., Aguilar-Bryan, L., Seino, S. and Bryan, J. (1995a). Reconstitution of IK_{ATP}: An inward rectifier subunit plus the sulfonylurea receptor. Science 270, 1166–1170.

Inagaki, N., Gonoi, T., Clement, J.P. IV, Namba, N., Inazawa, J., Gonzalez, G., Aguilar-Bryan, L., Seino, S. and Bryan, J. (1996). A family of sulfonylurea receptors determines the pharmacological properties of ATP-sensitive K^+ channels. Neuron 16, 1011–1017.

Inagaki, N., Tsuura, Y., Namba, N., Masuda, K., Gonoi, T., Horie, M., Seino, Y., Mizuta, M. and Seino, S. (1995b). Cloning and functional characterization of a novel ATP-sensitive potassium channel ubiquitously expressed in rat tissues, including pancreatic islets, pituitary, skeletal muscle, and heart. J. Biol. Chem. 270, 5691–5704.

Isomoto, S., Kondo, C., Yamada, M., Matsumoto, S., Higashiguchi, O., Horio, Y., Matsuzawa, Y. and Kurachi, Y. (1996). A novel sulfonylurea receptor forms with BIR (Kir6.2) a smooth muscle type ATP-sensitive K^+ channel. J. Biol. Chem. 271, 24321–24324.

Kaczorowski G.J., Knaus, H.G., Leonard, R.J., McManus, O.B. and Garcia, M.L. (1996). High-conductance calcium-activated potassium channels; structure, pharmacology and function. J. Bioenerg. Biomembr. 28, 255–267.

Kakei, M., Noma, A. and Shibasaki, T. (1985). Properties of adenosine-triphosphate-regulated potassium channels in guinea pig ventricular cells. J. Physiol. 254, 905–913.

Keahey, H.H., Rajan, A.S., Boyd, A.E. III and Kunze, D.L. (1989). Characterization of voltage-dependent Ca^{2+} channels in β-cell line. Diabetes 38, 188–193.

Kubo, Y., Baldwin, T.J., Jan, Y.N. and Jan, LY. (1993). Primary structure and functional expression of a mouse inward rectifier potassium channel. Nature 362, 127–133.

Leech, C.A., Holz, G.G. and Habener, J.F. (1996). Signal transduction of PACAP and GLP-1 in pancreatic beta cells. Ann. NY Acad. Sci. 805, 81–92.

Lippton, H.L., Cohen, G.A., McMurtry, I.F. and Hyman, A.L. (1991). Pulmonary vasodilation to endothelin isopeptides in vivo is mediated by potassium channel activation. J. App. Physiol. 70, 947–952.

Lopez, G.A., Jan, Y.N. and Jan, LY. (1994). Evidence that the S6 segment of the Shaker voltage-gated K^+ channel comprises part of the pore. Neuron 367, 179–182.

Malaisse, W.J. and Malaisse-Lagae, F. (1984). The role of cyclic AMP in insulin release. Experientia 40, 1068–1075.

McCleskey, E.W., Fox, A.P., Feldman, D. and Tsien, R.W. (1986). Different types of calcium channels. J. Exp. Biol. 124, 177–190.

Nelson, M.T., Huang, Y., Brayden, J.E., Hescheler, J. and Standen, N.B. (1990). Arterial dilations in response to calcitonin gene-related peptide involve activation of K^+ channels. Nature 344, 770–773.

Nichols, C.G. and Lederer, W.J. (1991). Adenosine triphosphate-sensitive potassium channels in the cardiovascular system. Am. J. Physiol. 261, H1675–1686.

Osterrieder, W., Brum, A., Hescheler, J. and Trautwein, W. (1982). Injection of subunits of cyclic AMP-dependent protein kinase into cardiac myocytes Ca^{2+} current. Nature 298, 576–578.

Ozanne, S.E., Guest, P.C., Hutton, J.C. and Hales, C.N. (1995). Intracellular localization and molecular heterogeneity of the sulfonylurea receptor in insulin-secreting cells. Diabetologia. 38, 277–282.

Perez-Reyes, E., Kim, H.S., Lacerda, A.E., Horne, W., Wei, X., Rampe, D., Campbell, K.P., Brown, A.M. and Birnbaumer, L. (1989). Induction of calcium currents by the expression of the α_1-subunit of the dihydropyridine receptor from skeletal muscle. Nature 340, 233–236.

Permutt, M.A., Nestorowicz, A. and Glaser, B. (1996). Familial hyperinsulinism: An inherited disorder of spontaneous hypoglycemia in neonates and infants. Diab. Rev. 4, 347–355.

Pressel, D.M. and Misler, S. (1990). Sodium channels contribute to action potential generation in canine and human pancreatic islet B cells. J. Membr. Biol. 116, 273–280.

Rajan, A.S., Aguilar-Bryan, L., Nelson, D.A., Yaney, G.C., Hsu, W.H., Kunze, D.L. and Boyd, A.E. III (1990). Ion channels and insulin secretion. Diabetes Care 13, 340–363.

Ribalet, B., Ciani, S. and Eddlestone, G.T. (1989). ATP mediates both activation and inhibition of $K_{(ATP)}$ channel activity via cAMP-dependent protein kinase in insulin secreting cell lines. J. Gen. Physiol. 94, 693–717.

Rorsman, P., Bokvist, K., Ammala, C., Arkhammar, P., Berggren, P.-O., Larsson, O. and Wahlander, K. (1991). Activation by adrenaline of a low conductance G protein-dependent K^+ channel in mouse pancreatic β-cells. Nature 349, 77–79.

Rorsman, P. and Trube, G. (1985). Glucose-dependent K^+ channels in pancreatic b-cells are regulated by intracellular ATP. Pflugers Arch. 405, 305–309.

Rorsman, P. and Trube, G. (1986). Calcium and delayed potassium currents in mouse pancreatic beta-cells under voltage-clamp conditions. J. Physiol. 347, 531–550.

Satin, L.S. and Cook, D.L. (1988). Evidence for two calcium currents in insulin-secreting cells. Pflugers Arch. 411, 401–409.

Schmid-Antomarchi, H., DeWeille, J., Fosset, M. and Lazdunski, M. (1987). The receptor for the antidiabetic sulfonylureas controls the activity of the ATP-modulated K^+ channel. J. Biol. Chem. 262, 15840–15844.

Schwanstecher, M., Loser, S., Rietze, I. and Panten, U. (1991). Phosphate and thiophosphate group donating adenine and guanine nucleotides inhibit glibenclamide binding to membranes from pancreatic islets. Naunyn-Schmiedebergs Arch. Pharmacol. 343, 83–89.

Schwanstecher, M., Loser, S., Chudziak, F., and Panten, U. (1994a). Identification of a 38-kDa high affinity sulfonylurea-binding peptide in insulin-secreting cells and cerebral cortex. J. Biol. Chem. 269, 17768–17771.

Schwanstecher, M., Loser, S., Chudziak, F. and Panten, U. (1994b). Photoaffinity labeling of the cerebral sulfonylurea receptor using a novel radioiodinated azidoglibenclamide analogue. J. Neurochem. 63, 698–708.

Stuhmer, W., Conti, F., Suzuki, H., Wang, X., Noda, M., Yahagi, N., Kubo, H. and Numa, S. (1989). Structural parts involved in activation and inactivation of the sodium channel. Nature 339, 597–603.

Sturgess, N.C., Ashford, M.L.J., Cook, D.L. and Hales, C.N. (1985). The sulfonylurea receptor may be an ATP-sensitive potassium channel. Lancet 11, 474–475.

Tanabe, T., Beam, K.G., Powell, J.A. and Numa, S. (1988). Restoration of excitation-contraction coupling a slow calcium current in dysgenic mice by dihydropyridine receptor complementary DNA. Nature 336, 134–139.

Terzic, A., Findlay, I., Hosoya, Y. and Kurachi, Y. (1994). Dualistic behavior of ATP-sensitive K^+ channels toward intracellular nucleoside diphosphates. Neuron 12, 1049–1058.

Triggle, D.J. and Janis, R.A. (1987). Calcium channel ligands. Annu. Rev. Pharmacol Toxicol. 27, 345–369.

Tsuura, Y., Ishida, H., Okamoto, Y., Tsuji, K., Kurose, T., Horie, M., Imura, H., Okada, Y. and Seino, Y. (1992). Impaired glucose sensitivity of ATP-sensitive K^+ channels in pancreatic β-cells in streptozotocin-induced NIDDM rats. Diabetes 41, 861–865.

Tusnady, G.E., Bakos, E., Varadi, A. and Sarkadi, B. (1997). Membrane topology distinguishes a subfamily of the ATP-binding cassette (ABC) transporters. FEBS Lett. 402, 1–3.

Wollheim, C.B., Dunne, M.J., Peter-Riesch, B., Bruzonne, R., Pozzan, T. and Patersen, O.H. (1988). Activators of protein kinase C depolarize insulin-secreting cells by closing K^+ channels. EMBO J. 7, 2443–2449.

Wollheim, C.B. and Sharp, G.W.G. (1981). Regulation of insulin release by calcium. Physiol. Rev. 61, 914–973.

Yaney, G.C., Wheeler, M.B., Wei, X., Perez-Reyes, E., Birnbaumer, L., Boyd, A.E. III and Moss, L.G. (1992). Cloning of a novel α1-subunit of the voltage-dependent calcium channel from the β-cell. Mol. Endo. 6, 2143–2152.

SIGNAL TRANSDUCTION

REGULATION OF INSULIN SECRETION BY CHANGES IN CA^{2+} CONCENTRATION AND ACTION IN PANCREATIC β-CELLS

Jean-Claude Henquin, Jean-Christophe Jonas, Yoshihiko Sato, Philippe Detimary, and Patrick Gilon

Advances in Molecular and Cell Biology
Volume 29, pages 247-275.

ISBN: 0-7623-0547-9

I. INTRODUCTION

Defects in the regulation of insulin secretion from pancreatic β-cells largely contribute to the perturbations of glucose homeostasis in noninsulin-dependent diabetes (Type 2 diabetes). Under physiological conditions, the rate of insulin release is controlled by variations in the concentration of circulating nutrients and by various hormonal and neural signals. Hormones and neurotransmitters modulate β-cell function by binding to membrane receptors and activating transduction pathways essentially similar to those existing in other cell types. In contrast, glucose, the major physiological stimulus, does not bind to a receptor but must be metabolized to induce insulin secretion. The acceleration of β-cell metabolism results in the production of a number of potential second messengers that might trigger or modulate exocytosis of storage granules containing insulin. Another peculiarity of β-cells is not so much that they are electrically excitable, but that their membrane potential is controlled by metabolism. The electrical activity induced by glucose involves marked changes in the flux of many ions across the plasma membrane. One of these ions, Ca^{2+}, plays a pre-eminent role to which this chapter is devoted.

The first indication that Ca^{2+} ions are essential players in stimulus-secretion coupling in β-cells dates back to the observation that the stimulation of insulin secretion by glucose and other agents was abrogated by omission of extracellular Ca^{2+} (Grodsky and Bennett, 1966; Milner and Hales, 1967). It was then demonstrated that glucose augments ^{45}Ca influx (Hellman et al., 1971), efflux (Malaisse et al., 1973) and net uptake (Malaisse-Lagae and Malaisse, 1971) in isolated islets. These and many subsequent studies led to the concept that an elevation in the cytoplasmic Ca^{2+} concentration ($[Ca^{2+}]_i$) in β-cells serves as second messenger in glucose-induced insulin secretion (Hedeskov, 1980; Wollheim and Sharp, 1981).

II. Ca^{2+} AS SECOND MESSENGER

The second messenger property of Ca^{2+} rests on the possibility of changing $[Ca^{2+}]_i$ very rapidly and markedly during stimulation. Like other cell types, unstimulated β-cells maintain $[Ca^{2+}]_i$ between 50 and 100 nM by several mechanisms: pumping of Ca^{2+} into the endoplasmic reticulum by a sarco(endo)plasmic reticulum Ca^{2+}-ATPase (SERCA) (Prentki et al., 1984; Váradi et al., 1996b), and extrusion of Ca^{2+} across the plasma membrane by a calmodulin-sensitive Ca^{2+} ATPase (Pershadsing et al., 1980; Váradi et al., 1996a) and an Na^+/Ca^{2+} countertransport that couples the uphill, outward movement of Ca^{2+} to the entry of Na^+ down its electrochemical gradient (Herchuelz and Plasman, 1991). There exist thus large (~10,000-fold) gradients of Ca^{2+} concentration between the cytoplasm and either the lumen of the endoplasmic reticulum or the extracellular medium. Opening of Ca^{2+} channels in the plasma membrane or in the endoplasmic reticulum membranes allows rapid movements of Ca^{2+} and causes increases in cytoplasmic $[Ca^{2+}]_i$.

What is the evidence that cytoplasmic Ca^{2+} can regulate insulin secretion? A direct demonstration that Ca^{2+} triggers exocytosis of insulin granules was obtained from experiments using islet cells of which the plasma membrane was permeabilized by different means. In these cells, the membrane potential is dissipated and the cytosolic concentration of small diffusible molecules (including ATP, Ca^{2+}) is clamped by the composition of the incubation medium. An increase in free $[Ca^{2+}]$ induced insulin release independently from any other change (Jones et al., 1985; Tamagawa et al., 1985). More recently, exocytosis was estimated by capacitance measurements in voltage-clamped single β-cells; increases in capacitance reflect increases in plasma membrane surface caused by fusion of secretory granules. This technique confirmed that Ca^{2+} directly triggers exocytosis (Gillis and Misler, 1993; Ammala et al., 1993; Bokvist et al., 1995). The mechanisms underlying the action of Ca^{2+} are partially identified (Wollheim et al., 1996; Regazzi, 1999).

Although studies using permeabilized cells (Vallar et al., 1987; Jonas et al., 1994) and capacitance measurements (Proks et al., 1996) have shown that poorly hydrolyzable GTPγS can increase exocytosis in the presence of low $[Ca^{2+}]_i$, the concept that cytoplasmic Ca^{2+} serves as second messenger in stimulus-secretion coupling in pancreatic β-cells seems to be correct. However, it is also too restrictive. This will already be apparent from the following description of three pathways of glucose-induced insulin secretion.

III. MAJOR PATHWAYS OF GLUCOSE STIMULATION

A. K^+-ATP Channel- and Ca^{2+}-Dependent Pathway

The best characterized pathway for the stimulation of insulin release by glucose involves depolarization of β-cells with acceleration of Ca^{2+} influx and subsequent rise in $[Ca^{2+}]_i$ (Figure 1). The depolarization is underlain by the closure of ATP-sensitive K^+ channels (K^+-ATP channels). These channels are tetramers of a complex of two proteins: SUR1, a high-affinity sulfonylurea receptor and Kir 6.2, an inward rectifier K^+ channel (Inagaki et al., 1995; Sakura et al., 1995; Clement et

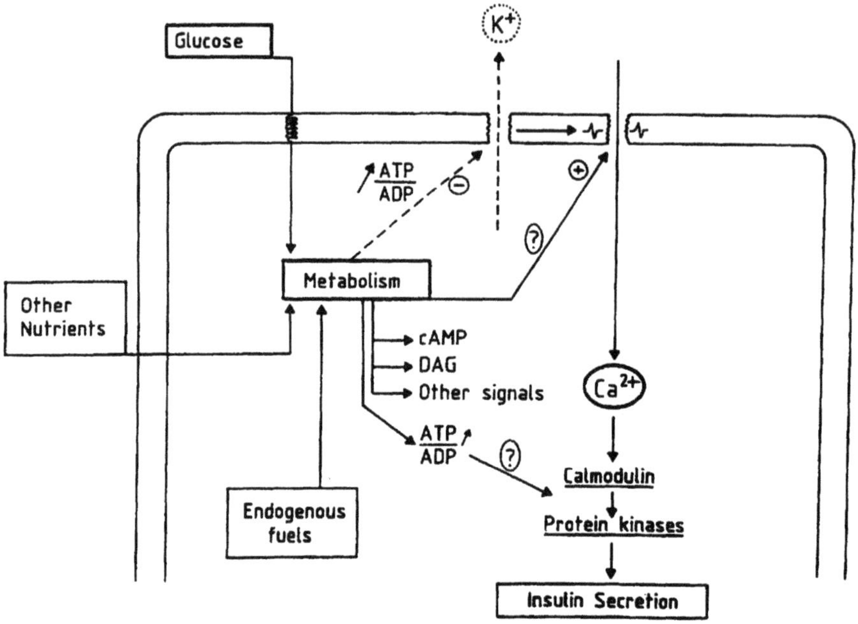

Figure 1. Schematic representation of the major pathways by which glucose regulates insulin secretion. (From Henquin et al., 1994.) ADP, adenosine diphosphate; ATP, adenosine triphosphate; cAMP, cyclic adenosine monophosphate; DAG, diacyl glycerol.

al., 1997). SUR1 endows the pore-forming Kir 6.2 with sensitivity to sulfonylureas and diazoxide, which respectively close and open the channel. As discussed more extensively in the preceding chapter by A.S. Rajan and L. Aguilar-Bryan, SUR1 also mediates the stimulatory (opening) action of Mg^{2+}-adenosine diphosphate (ADP) (Nichols et al., 1996), whereas the inhibitory (closing) site(s) of adenosine triphosphate (ATP) could be located on Kir 6.2 itself (Tucker et al., 1997). Although the channel owes its name to its sensitivity to ATP, it has not yet been proved that the nucleotide is its (sole) regulator in intact cells. Many indirect arguments, including the evidence that the ATP/ADP (and guanosine triphosphate [GTP]/guanosine diphosphate [GDP]) ratio changes over a wide range of glucose concentrations (Detimary et al., 1996), support this concept. Because K^+-ATP channels are the major determinants of the resting membrane potential in β-cells, their closure is followed by a depolarization brought about by a background current of a yet unidentified nature (Henquin and Meissner, 1984; Satin and Smolen, 1994). When the depolarization is sufficient, voltage-dependent Ca^{2+} channels open, allowing influx of Ca^{2+} into the cell, down the electrochemical gradient. This opening of Ca^{2+} channels is intermittent, oscillating with the slow waves of membrane potential (Henquin and Meissner, 1984; Cook et al., 1991; Atwater et al., 1994; Satin and Smolen, 1994). The result is a rise in $[Ca^{2+}]_i$, which parallels the electrical activity of β-cells and constitutes the triggering signal for exocytosis of insulin granules (Valdeolmillos et al., 1989; Gilon and Henquin, 1992) (see Figure 1).

B. K^+-ATP Channel-Independent, Ca^{2+}-Dependent Pathway

When K^+-ATP channels are opened by diazoxide, the β-cell membrane potential is held at the resting level, close to the equilibrium potential for K^+ (E_K), even in the presence of a stimulatory concentration of glucose. Ca^{2+} influx through voltage-dependent Ca^{2+} channels does not occur, $[Ca^{2+}]_i$ remains low, and insulin secretion is not stimulated (Gilon and Henquin, 1992). However, depolarization of the membrane by increasing the concentration of extracellular K^+, a maneuver that simply shifts E_K, restores Ca^{2+} influx and insulin release. Under these conditions (high K^+ + diazoxide), glucose still exerts a concentration-dependent stimulation of insulin secretion (Gembal et al.; 1992; Sato et al., 1992). This effect is independent of K^+-ATP channels and of any change in the membrane potential, but is dependent on $[Ca^{2+}]_i$, which has to be elevated, but has not to change (Gembal et al., 1992). Glucose does not increase $[Ca^{2+}]_i$ but amplifies the action of the ion on the releasing process (Gembal et al., 1993; Jonas et al., 1998) (see Figure 1). The existence of this mechanism is now widely accepted (Aizawa et al., 1994; Kelley et al., 1994; Meredith et al., 1995; Chan and MacPhail, 1996), but the underlying mechanisms are still unclear and under investigation. The metabolism of glucose is required (Gembal et al., 1992, 1993) and changes in adenine and guanine nucleotides may be implicated (Gembal et al., 1993; Meredith et al., 1995; Detimary et al., 1996; Eliasson et al., 1997). In contrast, the protein kinase A, protein kinase C, nitric oxide-protein kinase G, long chain acyl coenzyme A, and phospholipase A_2-arachidonic acid pathways seem to play no or only a minor role (Gembal et al., 1993; Sato and Henquin, 1998).

It is not so well recognized that a second K^+-ATP channel-independent, Ca^{2+}-dependent pathway may also exist, which, in contrast to the first one, involves changes in $[Ca^{2+}]_i$. When K^+-ATP channels are blocked by maximally effective concentrations of sulfonylureas, they are unlikely to remain a site for glucose action. Yet, the sugar increases insulin secretion under these conditions (Henquin, 1980; Panten et al., 1988; Best et al., 1992; Jijikali and Malaisse, 1997). Although the amplification of the action of Ca^{2+} by glucose contributes to this increase, a slightly further rise in $[Ca^{2+}]_i$ is also involved. Glucose increases the frequency of Ca^{2+} spikes in β-cells persistently depolarized by 100 μM tolbutamide (Bozem and Henquin, 1988) and accelerates the Ca^{2+}-dependent ^{45}Ca efflux from islets perifused with 2 μM glibenclamide (Jijikali and Malaisse, 1997). Although a full understanding of the phenomenon requires further studies, a direct modulation of voltage-dependent Ca^{2+} channels by glucose metabolism (Smith et al., 1989) is a plausible explanation (see Figure 1).

C. Ca^{2+}-Independent Pathway

It has recently been reported that glucose can increase insulin secretion in a concentration-dependent manner even under conditions of stringent Ca^{2+}-depletion (preincubation and incubation of the islets in a Ca^{2+}-free medium supplemented with the calcium-chelating agent EGTA). The effect of the sugar, however, was detected only during strong activation of both protein kinase A (by cAMP formed in response to forskolin) and protein kinase C (by a phorbol ester) (Komatsu et al., 1995). Under less severe Ca^{2+}-deprivation, stimulation of the kinases by pituitary adenylate cyclase-activating peptide and carbachol was sufficient to disclose the effect of glucose (Komatsu et al., 1996). This stimulatory action of glucose requires metabolism of the sugar, can be blocked by activation of α_2-adrenoceptors and is not accompanied by any detectable change in $[Ca^{2+}]_i$ (Komatsu et al., 1995, 1996). In contrast to these results, we recently reported that, even under optimal experimental conditions, the amount of insulin that can be released in the absence of Ca^{2+} does not exceed a small percentage of appropriate controls in the presence of Ca^{2+}. Moreover, this Ca^{2+}-independent release showed no or only poor glucose-dependence (Sato et al., 1998b).

D. Respective Importance of the Different Pathways

There is little doubt that generation of the triggering signal (the rise in $[Ca^{2+}]_i$) by closure of K^+-ATP channels, depolarization, and stimulation of Ca^{2+} influx is the primary mechanism of β-cell stimulation by glucose. Its importance is attested, for instance, by the ability of sulfonylureas and diazoxide respectively to mimic and antagonize the effects of glucose in vivo.

The second mechanism of action of glucose, the amplification of the action of Ca^{2+}, has been identified under conditions where $[Ca^{2+}]_i$ was clamped. However, there is no obvious reason why this amplification should not also take place under physiological conditions. Thus, we recently observed it when $[Ca^{2+}]_i$ was forced to oscillate (by repetitive depolarizations with KCl) as it does spontaneously during stimulation by glucose alone (Jonas et al., 1998). It is also known that this amplification shows a higher sensitivity to glucose and lesser dependence on the ATP/ADP ratio than the rise in $[Ca^{2+}]_i$ (Gembal et al., 1993; Detimary et al., 1994). We suggest that the modulation of the Ca^{2+} action by glucose serves to optimize the efficacy of the increase in $[Ca^{2+}]_i$. According to an older terminology, these two pathways correspond to the initiating and potentiating signals that glucose was hypothesized to produce in β-cells (Cerasi, 1975). It has long been known that glucose not only stimulates insulin release, but also influences the β-cell response to other secretagogues like arginine (Efendic et al., 1971; Levin et al., 1972). This potentiating (amplifying) action of glucose has been attributed to both an increase in Ca^{2+} influx and an increase in the efficacy of the triggering signal (Hermans et al., 1987). The impairment of this potentiation by glucose in Type

2 diabetes (Hollander et al., 1982; Ward et al., 1984) indicates that the second pathway of glucose action could have a major pathophysiological importance.

At present, it is not possible to evaluate the contribution of a glucose action on voltage-dependent Ca^{2+} channels because the information on the subject is too scanty.

In contrast, the Ca^{2+}-independent pathway does not seem to contribute significantly to the physiological regulation of insulin secretion by glucose (Sato et al., 1998b). If the underlying mechanism proves to be operative only in the presence of extremely low $[Ca^{2+}]_i$, much lower than that occurring in β-cells in vivo, the phenomenon is a pure curiosity. If it is also operative in the presence of physiological $[Ca^{2+}]_i$, one may wonder whether it is not one and the same as the second mechanism discussed previously. We do not believe this to be the case because the K^+-ATP channel-independent, Ca^{2+}-dependent pathway does not require or involve activation of protein kinases A and C. Finally, one should bear in mind that strong and simultaneous activation of both kinases is necessary to weakly duplicate certain effects of Ca^{2+} itself, whereas mild activation of each kinase individually suffices to augment markedly the efficacy of Ca^{2+} (Sato et al., 1998b).

IV. MECHANISMS OF THE $[Ca^{2+}]_i$ RISE PRODUCED BY GLUCOSE IN β-CELLS

A. Inhibition of Ca^{2+} Efflux?

A decrease in the rate of Ca^{2+} extrusion from the cytoplasm to the extracellular medium or the endoplasmic reticulum lumen could theoretically contribute to the increase in $[Ca^{2+}]_i$ produced by glucose. This was one plausible interpretation of measurements of ^{45}Ca efflux from prelabeled islets. Thus, when Ca^{2+} influx is not possible (e.g., in a Ca^{2+}-free medium), glucose decreases the efflux of the tracer (Malaisse et al., 1973). However, direct measurements subsequently showed that $[Ca^{2+}]_i$ decreases under these conditions, the fall in ^{45}Ca efflux rate being the consequence of this decrease.

B. Mobilization of Ca^{2+} from Intracellular Stores

This is the major mechanism by which many stimuli induce a rapid rise in $[Ca^{2+}]_i$ in electrically nonexcitable cells. Although the extracellular (Henquin, 1978) or intracellular (Kikuchi et al., 1979) origin of Ca^{2+} triggering the first phase of glucose-induced insulin release has remained controversial for a while, mobilization of intracellular Ca^{2+} by glucose has never been demonstrated by direct measurements of $[Ca^{2+}]_i$ in normal β-cells.

Most investigators agree that glucose lowers $[Ca^{2+}]_i$ when it does not promote Ca^{2+} influx. Under control conditions, this decrease occurs transiently, before the entry of Ca^{2+} increases $[Ca^{2+}]_i$. When Ca^{2+} influx is prevented, the decrease in

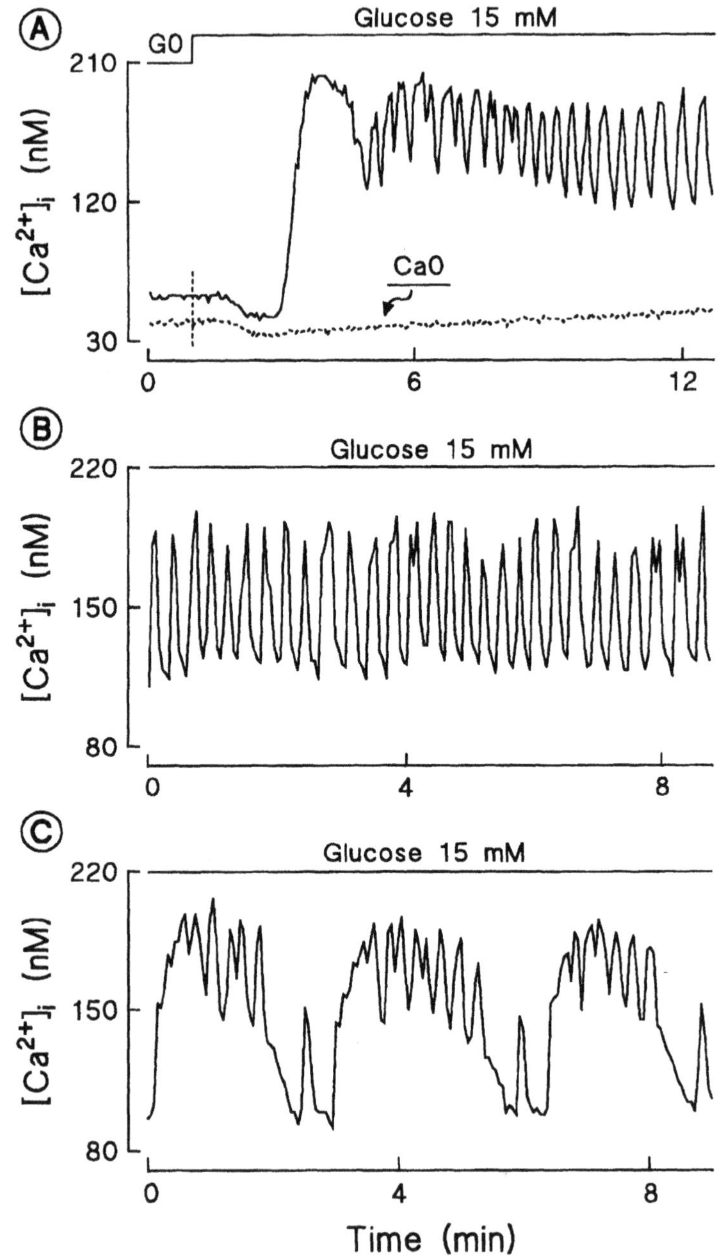

Figure 2. Effects of glucose on $[Ca^{2+}]_i$ in intact mouse islets. (**A**) After an initial period of perifusion without glucose (G0) the islet was stimulated with 15 mM glucose. The medium contained 2.5 mM or 0 mM $CaCl_2$. (**B**) Regular, fast oscillations of $[Ca^{2+}]_i$ in an islet continuously stimulated with 15 mM glucose. (**C**) Mixed (fast and slow) oscillations in an islet continuously stimulated with 15 mM glucose.

$[Ca^{2+}]_i$ is sustained (Figure 2A). The phenomenon is ascribed to a stimulation of Ca^{2+} sequestration in the endoplasmic reticulum because of its blockade by thapsigargin, an inhibitor of the SERCA pump (Roe et al., 1994; Hamakawa and Yada, 1995). How glucose stimulates the pump is unknown except for the fact that the sugar must be metabolized, and produces this effect at concentrations below those which stimulate Ca^{2+} influx (Gylfe, 1988). Moreover, a similar effect is produced by nonglucidic fuels exclusively metabolized in mitochondria (Gylfe, 1988). Although the ATP dependency of the SERCA pump has a Km that is about two orders of magnitude lower than the concentration of ATP in the cytoplasm, experimental evidence suggests that an increase in the ATP/ADP ratio may be implicated (Corkey et al., 1988). The rapidity of the changes in this ratio upon stimulation with glucose (Detimary et al., 1998b) is compatible with this hypothesis.

In 1993, however, two provocative studies fostered the debate by describing novel effects of glucose. Takasawa et al. (1993) suggested that cyclic ADP-ribose formed in glucose-stimulated β-cells could serve as second messenger to mobilize Ca^{2+} from an inositol trisphosphate-insensitive pool and, thereby, to stimulate insulin release. More recent studies have failed to find any evidence for a role of cyclic ADP-ribose in Ca^{2+} homeostasis and stimulus-secretion coupling in β-cells (Rutter et al., 1994; Willmott et al., 1995; Webb et al., 1996). Another study concluded that β-cell membrane depolarization by glucose or KCl is sufficient to mobilize intracellular Ca^{2+} (in absence of Ca^{2+} entry), possibly through an increase in inositol 1,4,5-trisphosphate production (Roe et al., 1993). This conclusion has received some support (Gromada et al., 1996; Liu et al., 1996) but remains at variance with results obtained in other laboratories, including ours. As an example, Figure 3B shows that the Ca^{2+}-channel blocker D600 completely abrogated glucose- and KCl-induced $[Ca^{2+}]_i$ rises in clusters of islet cells (Miura et al., 1997).

C. Influx of Ca^{2+} from the Extracellular Medium

Three pathways might, theoretically at least, be involved in the entry of Ca^{2+}: voltage-dependent Ca^{2+} channels, voltage-independent Ca^{2+} channels, and the Na^+/Ca^{2+} countertransport working in the reverse mode.

The influx of Ca^{2+} from the extracellular medium essentially takes place through voltage-sensitive Ca^{2+} channels that open upon depolarization of the membrane. These channels mainly belong to the "long-lasting" or "L-type" (Rorsman and Trube, 1986; Plant, 1988). They are blocked by dihydropyridines and phenylalkylamines, which explains that these drugs prevent the glucose-induced rise in $[Ca^{2+}]_i$ and stimulation of insulin release. The isoforms of the α and β subunits composing these channels have recently been identified (Ihara et al., 1995; see also chapter by Rajan and Aguilar-Bryan earlier in this volume). In certain species, voltage-dependent Ca^{2+} channels of the "transient" or "T-type" are also present, but their functional role is unclear.

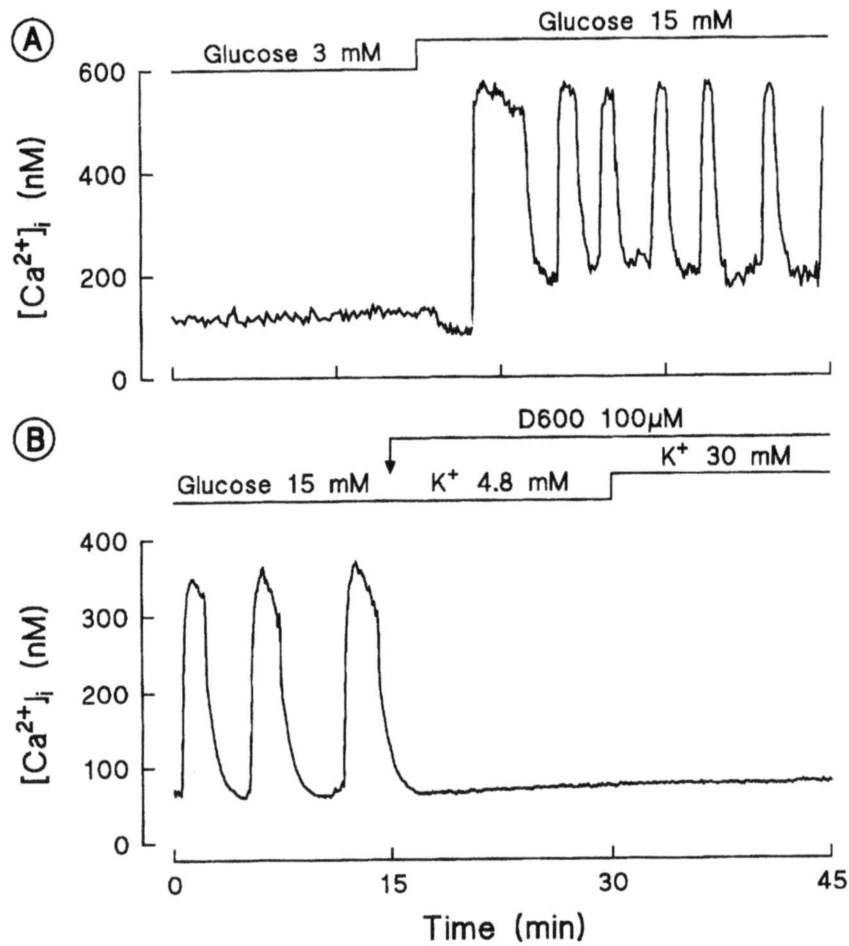

Figure 3. Effects of glucose on $[Ca^{2+}]_i$ in clusters of five to eight mouse β-cells. (**A**) The concentration of glucose was increased from 3 to 15 mM as indicated. (**B**) The cluster was continuously stimulated with 15 mM glucose. The Ca^{2+}-channel blocker D600 was then added, and the concentration of KCl in the medium eventually raised to 30 mM. (Adapted from Miura et al., 1997.)

In electrically nonexcitable cells, the emptying of intracellular Ca^{2+} stores triggers Ca^{2+} influx through dihydropyridine-insensitive, voltage-independent Ca^{2+} channels. The process is termed "capacitative Ca^{2+} entry" (Putney and Bird, 1993), and the current that the channels carry, "calcium release-activated Ca^{2+} current" or I_{CRAC} (Hoth and Penner, 1993). Ca^{2+} store depletion may also activate nonselective cationic channels (Friel, 1996) and produce a "calcium release-activated nonselective cation current" or I_{CRAN}. Several studies have recently shown that a similar Ca^{2+} entry pathway exists in β-cells (Leech et al., 1994; Silva et al., 1994; Worley et al., 1994a, 1994b; Bertram et al., 1995; Mears et al., 1997; Miura et al., 1997), and the channels possibly carrying these currents have recently been cloned in the insulin-secreting MIN 6 cells (Sakura and Ashcroft, 1997).

The capacitative Ca^{2+} entry itself appears to contribute very little to the $[Ca^{2+}]_i$ rise in β-cells, but the inward current (I_{CRAC} and I_{CRAN}) could potentiate glucose-induced depolarization and augment Ca^{2+} influx through voltage-dependent Ca^{2+} channels (Miura et al., 1997). This sequence of events is fully compatible with the known mode of action of agents which, like muscarinic-agonists, mobilize Ca^{2+} stores by producing inositol-1,4,5-triphosphate (IP_3). Whether it also plays a role in the response to glucose is less clear and still at an early stage of investigation. Both theoretical arguments and experimental results suggest that I_{CRAN} might contribute to the first phase of glucose-induced electrical activity (Dukes et al., 1997; Mears et al., 1997).

In β-cells as in other cell types, the Na^+/Ca^{2+} countertransport normally extrudes Ca^{2+} from the cytoplasm to the extracellular medium but it may also work in a reverse mode and promote Ca^{2+} influx when the membrane depolarizes (Herchuelz and Plasman, 1991). The importance of the exchange in the lowering of $[Ca^{2+}]_i$ is undisputed, but its contribution to Ca^{2+} influx, in particular in response to glucose, is less evident (Van Eylen et al., 1998). In our hands, even when islets were placed under conditions selected to facilitate this pathway, no evidence could be obtained that $[Ca^{2+}]_i$ in β-cells can be raised by an influx of Ca^{2+} through reverse Na^+/Ca^{2+} exchange (Garcia-Barrado et al., 1996).

V. SUBCELLULAR LOCALIZATION OF $[Ca^{2+}]_i$ CHANGES

Most measurements of cytoplasmic Ca^{2+} average the values over the whole surface of the cell and do not identify possible gradients of $[Ca^{2+}]_i$ between regions. This information has been provided by studies using digital image analysis or targetting of the Ca^{2+} indicator.

The rise in $[Ca^{2+}]_i$ brought about by glucose stimulation is more marked close to the plasma membrane than in the center of the cell (Gylfe et al., 1991; Theler et al., 1992), where the increase is somewhat delayed. A recent study even indicates that this submembrane preferential rise is polarized (Martin et al., 1997). A similar observation has been made in single voltage-clamped β-cells, in which labeling of insulin granules with quinacrine revealed that the largest rise in $[Ca^{2+}]_i$ occurs in the region that is more rich in granules (Bokvist et al., 1995). Studies of β-cells by the cell-attached configuration of the patch-clamp technique further showed that the density of voltage-dependent Ca^{2+} channels is highest in the plasma membrane area near the region rich in insulin granules (Bokvist et al., 1995). Taken together, these studies suggest that depolarization of the plasma membrane leads to a preferential increase in $[Ca^{2+}]_i$ in the submembrane domain of the cell region that contains the highest density of granules. This organization may constitute the functional basis of polarized insulin secretion (Bonner-Weir, 1988).

Stable expression of the Ca^{2+}-sensitive photoprotein aequorin in mitochondria of INS-1 or MIN-6 cells made it possible to compare the effects of glucose on cytoplasmic and mitochondrial $[Ca^{2+}]_i$. Both were found to vary in a parallel manner (Rutter et al., 1993; Kennedy et al., 1996; Nakazaki et al., 1998). Using indo-1 as Ca^{2+} indicator and confocal fluorescence microscopy, it has been observed that the changes in $[Ca^{2+}]_i$ taking place in the cytoplasm also occurred within the nucleus of β-cells (Brown et al., 1997). It is, therefore, possible that Ca^{2+} also serves as second messenger in glucose-induced metabolic and transcriptional changes.

VI. OSCILLATIONS OF $[Ca^{2+}]_i$ IN PANCREATIC β-CELLS

Microfluorimetric methods have been used to measure $[Ca^{2+}]_i$ in suspensions of islet cells, in single β-cells, in clusters of islet cells, and in intact islets. In all preparations glucose has been found to raise $[Ca^{2+}]_i$, but the characteristics of the response were variable.

A. Cells in Suspension

When large numbers of islet cells are mixed in a cuvette, stimulation with glucose produces an initial reduction in $[Ca^{2+}]_i$ followed by a monotonic, sustained rise. No oscillations are detected, probably because the cells are not synchronized (Gylfe, 1988).

B. Single β-Cells

In single β-cells, glucose induces inconsistent and heterogeneous changes in $[Ca^{2+}]_i$. Some cells do not respond at all. In others, the response is characterized by three phases: an initial, small decrease, a secondary prompt rise and, eventually, oscillations or a sustained elevation of $[Ca^{2+}]_i$ (Herchuelz et al., 1991; Hellman et al., 1992; Wang et al., 1993). The early lowering of $[Ca^{2+}]_i$ coincides with the initial slow depolarization induced by glucose, while the sharp elevation of $[Ca^{2+}]_i$ coincides with the appearance of the electrical activity (Valdeolmillos et al., 1992; Chow et al., 1995). The $[Ca^{2+}]_i$ oscillations, observed in 15% to 85% of the cells according to the studies, usually have a low frequency (0.2 to 0.4 per min) and start from a close to basal level. They are hardly affected by an increase in the glucose concentration, which rather causes a sustained elevation of $[Ca^{2+}]_i$ (Grapengiesser et al., 1992). The links between these oscillations and the electrical activity are not firmly established because the latter has only rarely been recorded in single β-cells at 37°C. In one study using the cell-attached configuration of the patch-clamp technique, bursts of action potentials were found to occur with a similar frequency to that of $[Ca^{2+}]_i$ oscillations (Hellman et al., 1990). It thus seems likely that the pattern of electrical activity in the individual β-cell determines that of the $[Ca^{2+}]_i$ changes. In a minority of single β-cells stimulated by glucose alone, rapid oscillations (2.5/min) superimposed on the slow ones have been observed (Pralong et al., 1990; Herchuelz et al., 1991). The proportion of these cells was increased by raising cAMP in preparations from ob/ob mice (Grapengiesser et al., 1991) but not from normal mice (Fournier et al., 1994).

C. Clusters of β-Cells

When β-cells are aggregated in clusters, the $[Ca^{2+}]_i$ changes induced by glucose are much more reproducible, and typically characterized by large amplitude oscillations (see Figure 3). Their frequency slightly increases with the number of associated cells, but plateaus at about 0.25/min even in clusters of 30 to 50 cells (Jonkers et al., 1998). When the glucose concentration is raised from 3 to 15 mM, a triphasic $[Ca^{2+}]_i$ change occurs, but the possible impact of higher glucose concentrations on the oscillations has not yet been characterized (Miura et al., 1997). That these oscillations correspond to a periodic activation of Ca^{2+} influx is strongly suggested by their abrogation with D600, a blocker of voltage-dependent Ca^{2+} channels (see Figure 3).

D. Intact Islets

Stimulation of intact mouse islets by a sudden rise in the glucose concentration is followed by three phases of $[Ca^{2+}]_i$ changes as in single cells or clusters (see Figure 2A). The initial decrease is best observed when the preparation has been perifused for a longer time with a low-glucose medium, as if the Ca pool of the endoplasmic reticulum was more depleted under these conditions. The subsequent increase in $[Ca^{2+}]_i$ consists in a large sustained rise for 1 to 2 min, followed by oscillations that persist as long as the glucose stimulation is continued. The initial decrease coincides with the slow depolarization to the threshold potential, but no causal link appears to exist between both events. In contrast, there is good evidence that the phases of the $[Ca^{2+}]_i$ rise are directly caused by the influx of Ca^{2+} occurring during the corresponding changes in membrane potential (Santos et al., 1991; Gilon and Henquin, 1992). The essential role of voltage-dependent Ca^{2+} channels is demonstrated by the abolition of the $[Ca^{2+}]_i$ rise when these channels are blocked with pharmacological agents or when their opening is prevented by holding the membrane hyperpolarized with diazoxide.

Two types of $[Ca^{2+}]_i$ oscillations have been recorded in the islets (see Figure 2). Rapid oscillations (2 to 3 per min) are linked to the slow waves of membrane potential in β-cells (Valdeolmillos et al., 1989; Gilon and Henquin, 1992). Slow oscillations (0.2 to 0.3 per min) are also occasionnally observed, either alone or in association with fast oscillations to produce a mixed pattern (Valdeolmillos et al., 1989; Longo et al., 1991; Gilon et al., 1994; Bergsten, 1995). Because mixed oscillations of the membrane potential also exist in certain β-cells (Henquin et al., 1982; Cook, 1983), these complex oscillations of $[Ca^{2+}]_i$ might also entirely depend on the pattern of Ca^{2+} influx. For unknown reasons, slow oscillations are more often observed when the islets are cultured in the presence of 5 rather than 10 mM glucose (Gilon et al., 1994).

VII. LINKS BETWEEN METABOLISM AND $[Ca^{2+}]_i$ OSCILLATIONS

We have already emphasized the crucial role of glucose metabolism for β-cell membrane depolarization. As the latter and the resulting $[Ca^{2+}]_i$ rise oscillate, an obvious question is whether the concentration of metabolic signals also oscillates in β-cells.

In experiments using large numbers of rat islets placed in perifusion chambers, oscillations in O_2 consumption (Longo et al., 1991) and lactate release (Chou et al., 1992) have been observed, but it remains unclear how these isolated islets are synchronized. According to one hypothesis (Tornheim, 1997), intrinsic oscillations of glycolysis lead to oscillations in the ATP/ADP ratio either directly or by providing pyruvate and α-glycerophosphate to mitochondria in a pulsatile manner. Oscillations in the ATP/ADP ratio would then cause oscillations of the activity of K^+-ATP channels, leading to oscillations in membrane potential, Ca^{2+} influx and $[Ca^{2+}]_i$. Fluctuations of the activity of K^+-ATP channels in β-cells perifused with a medium containing 0 or 3 mM glucose (when the membrane potential is high and stable and $[Ca^{2+}]_i$ remains low) indirectly support the existence of intrinsic metabolic oscillations (Dryselius et al., 1994). Whether oscillations of K^+-ATP channel activity are involved in the slow waves of membrane potential remains controversial (Henquin, 1988, 1992; Cook and Ikeuchi, 1989; Smith et al., 1990; Santos et al., 1992; Larsson et al., 1996).

In a small proportion of β-cells (<10%) glucose induced oscillations of NAD(P)H fluorescence with a similar frequency to that of $[Ca^{2+}]_i$ (Pralong et al., 1994). Repetitive depolarizations with high K^+ also induced parallel oscillations of NAD(P)H. These observations are interpreted as evidence for a stimulation of ATP production by Ca^{2+}. Thus, we have already seen that $[Ca^{2+}]_i$ increases both in the cytoplasm and mitochondria (Rutter et al., 1993; Kennedy et al., 1996; Nakazaki et al., 1998), where it may activate Ca^{2+}-sensitive dehydrogenases (Sener et al., 1990; McCormack et al., 1990). According to this hypothesis, the oscillations in ATP production are thus secondary to the rise in $[Ca^{2+}]_i$. Alone, they cannot explain the oscillations of the membrane potential: an additional player must be involved to cause repolarization and eventually lower $[Ca^{2+}]_i$. Testing the hypothesis meets with the difficulty that NAD(P)H oscillations are not observed in intact islets (Panten et al., 1973; Gilon and Henquin, 1992). Although Ca^{2+} increases O_2 consumption (Panten et al., 1986) and NAD(P)H fluorescence (Gilon and Henquin, 1992), the latter remains stable, even when $[Ca^{2+}]_i$ oscillates spontaneously. One plausible explanation is that only a subpopulation of islet cells, perhaps acting as pacemaker cells, display metabolic oscillations, which pass unnoticed in the whole organ. Although no evidence could be obtained for a metabolic heterogeneity of β-cells in situ (Bennett et al., 1996), this question certainly deserves further investigation.

Less attention has been paid to a third way of inducing oscillations of a metabolic signal. Variations in the rate of ATP consumption may lead to oscillations in the ATP/ADP ratio even if the rate of production does not fluctuate. Our recent evidence that Ca^{2+} influx decreases the ATP/ADP ratio in islets (Detimary et al., 1998) may explain how Ca^{2+} exerts a negative feedback on K^+-ATP channels (Henquin, 1990). The rise in $[Ca^{2+}]_i$ leads to ATP consumption and lowering of the ATP/ADP ratio, causing reopening of K^+-ATP channels, arrest of Ca^{2+} influx, and drop in $[Ca^{2+}]_i$. When the ATP/ADP ratio has increased again, K^+-ATP channels close, the membrane depolarizes and a new cycle starts. The final demonstration of such a cycle would require measurements of the changes in the ATP/ADP ratio during spontaneous oscillations of $[Ca^{2+}]_i$, which remains a technical challenge.

VIII. CORRELATIONS AND DISSOCIATIONS BETWEEN $[Ca^{2+}]_i$ AND INSULIN RELEASE

Within an intact islet, β-cells are electrically coupled by gap junctions (Meissner, 1976; Meda, 1996), which explains that the whole organ may behave as a syncytium. Thus, $[Ca^{2+}]_i$ oscillations appear to be synchronous in all regions even when they vary in amplitude (Gilon and Henquin, 1992; Gilon et al., 1994). This synchrony of the triggering signal made it possible to measure simultaneously the changes in $[Ca^{2+}]_i$ and those in insulin secretion from the same islet. We observed an excellent correlation between both events, each $[Ca^{2+}]_i$ oscillation being accompanied by a pulse of insulin secretion (Gilon et al., 1993; Gilon and Henquin, 1995). Similar observations have been made by others when insulin was measured either by immunological (Bergsten et al., 1994; Martin et al., 1995b) or amperometric methods (Barbosa et al., 1996). Although the $[Ca^{2+}]_i$ gradients existing in the cytoplasm of β-cells (see preceding discussion) are not detected by recordings in whole islets, the average changes, which are measured, must adequately reflect the critical changes for secretion. Otherwise insulin and $[Ca^{2+}]_i$ oscillations would not be synchronous.

In addition to the Ca^{2+}-independent pathway discussed previously, a dissociation between $[Ca^{2+}]_i$ and insulin release measured simultaneously has been observed in single rat islets. Glucose only caused an early and brief increase in

$[Ca^{2+}]_i$, while stimulating a normal biphasic release of insulin (Martin et al., 1995a), which might suggest that the second, sustained phase of insulin secretion does not depend on a sustained elevation of $[Ca^{2+}]_i$ in rat β-cells. This is very surprising and calls for a confirmation of these results for several reasons. First, glucose causes a biphasic electrical activity with bursts of Ca^{2+} spikes in rat as in mouse islets (Ikeuchi et al., 1984). Second, others have reported biphasic increases in $[Ca^{2+}]_i$ in single rat islets stimulated with glucose for 20 min (Niwa et al., 1996). Third, addition of a blocker of Ca^{2+} channels or removal of extracellular Ca^{2+} during second phase of glucose-induced release quickly stops insulin secretion from rat islets (Henquin and Lambert, 1975; Malaisse et al., 1976).

The opposite type of dissociation has been observed in single islets from hyperglycemic ob/ob mice. High glucose, and high K^+ in the presence of low glucose, triggered a sustained rise in $[Ca^{2+}]_i$ similar to that reported in other studies, but the secretion of insulin was only transient (Zaitsev et al., 1995). These observations prompted the proposal that $[Ca^{2+}]_i$ has a permissive rather than regulatory role in contrast to protein kinases. We do not believe that the conclusion is warranted because it is based on an atypical pattern of secretion. In other studies with islets from ob/ob or normal mice, a sustained release of insulin is induced by glucose alone (Gilon et al., 1993; Bergsten, 1995) and by high K^+, except when the latter is tested in a glucose-free medium (Gembal et al., 1993). We agree that Ca^{2+} is an insufficient stimulatory signal to produce a full secretory response. Activation of protein kinases and protein phosphorylation certainly increase the efficacy of Ca^{2+} on the secretory machinery (Jones et al., 1985; Tamagawa et al., 1985; Gillis and Misler, 1993; Ammala et al., 1994; Howell et al., 1994), but recent evidence suggests that activation of protein phosphatases may also have a positive effect on secretion (Sato et al., 1998a). Metabolic signals (e.g., ATP) are also important. For instance, when the glucose concentration is raised from 15 to 30 mM, the peak of insulin oscillations increases while that of $[Ca^{2+}]_i$ oscillations is unaffected (Gilon and Henquin, 1995) because of the capacity of glucose to increase the efficacy of Ca^{2+}. However, it would be hard to explain how oscillations of $[Ca^{2+}]_i$ could induce oscillations of secretion if Ca^{2+} did not have a minute-to-minute regulatory role.

The only plausible explanation would be that a metabolic signal oscillating at the same pace as $[Ca^{2+}]_i$ is the actual regulator of secretion. The hypothesis can be tested by measuring insulin release under conditions where $[Ca^{2+}]_i$ and metabolic oscillations are dissociated. Single mouse islets were persistently depolarized with 30 mM K^+ in the presence of diazoxide to clamp β-cell $[Ca^{2+}]_i$ at a stable, elevated level. Insulin secretion from the same islets did not display any oscillations, as should have occurred if an intrinsic metabolic oscillator was governing secretion (Jonas et al., 1998). Our findings, however, conflict with those of two other studies. Pulsatile insulin secretion was recorded from batches of 50 rat islets perifused with a medium containing high K^+ and diazoxide (Cunningham et al., 1996), but it is unclear how isolated islets can be synchronized. Slow oscillations (~ 0.3/min) of insulin release were also measured from single *ob/ob* mouse islets during sustained elevation of $[Ca^{2+}]_i$ by tolbutamide (Westerlund et al., 1997). A second approach was to impose metabolic or $[Ca^{2+}]_i$ oscillations in islets while minimizing the changes in the other parameter. $[Ca^{2+}]_i$ oscillations were clearly more potent than metabolic oscillations in inducing pulsatile insulin secretion (Ravier et al., 1998). Finally, $[Ca^{2+}]_i$ oscillations were induced in permeabilized islet cells in which the cytoplasmic composition is controlled and therefore metabolic oscillations are unlikely to occur. The imposed Ca^{2+} oscillations were sufficient to trigger oscillations of secretion (Jonas et al., 1994).

In conclusion, we favor the hypothesis that the oscillations of $[Ca^{2+}]_i$ play a regulatory role in the secretion of insulin. If metabolic oscillations occur, they may serve to generate oscillations of membrane potential, hence of $[Ca^{2+}]_i$, and to amplify the efficacy of Ca^{2+} on the exocytotic process.

IX. Ca^{2+} SIGNAL IN β-CELLS FROM ANIMAL MODELS OF DIABETES

In Goto-Kakizaki rats (GK-rats), a model of Type 2 diabetes (Goto and Kakizaki, 1981), glucose-stimulation of insulin release is selectively impaired due to abnormal glucose metabolism in β-cells (Östenson et al., 1993). This explains why, although they are intrinsically normal, K^+-ATP channels do not normally close in GK β-cells stimulated by glucose (Tsuura et al., 1993). The result is a poor depolarization of the plasma membrane, which explains the abnormally small rise in $[Ca^{2+}]_i$ that the sugar produces. Ca^{2+} channels cannot be incriminated as their activity is even higher than in normal β-cells (Kato et al., 1996). Another peculiarity of GK β-cells is a marked decrease in the expression of one isoform (SERCA 3) of the endoplasmic Ca^{2+}-ATPase (Váradi et al., 1996b). The impact of this anomaly on Ca^{2+} handling by the β-cells is difficult to evaluate because SERCA 2 is normally expressed, and nothing is known about the respective role of the two isoforms. Contrasting with these abnormalities, the responsiveness of the exocytotic machinery to Ca^{2+} has been found to be normal or increased in GK β-cells (Katayama et al., 1995; Okamoto et al., 1995).

The picture is quite different in β-cells from Zucker diabetic fatty rats (ZDF rats), a model of Type 2 diabetes (Clark et al., 1983) characterized by a mutation in the leptin receptor (Phillips et al., 1996). Compared with control islets, the expression of mRNA encoding α_1 subunits of L-type voltage-dependent Ca^{2+} channels is significantly reduced in islets from ZDF rats (Tokuyama et al., 1995). This correlates well with a reduction in Ca^{2+} currents and in glucose-induced $[Ca^{2+}]_i$ rise, and is accompanied by loss of pulsatile insulin secretion (Roe et al., 1996). A second anomaly is a reduced expression of mRNA encoding Kir 6.2, the pore-forming subunit of K^+-ATP channels; in contrast, mRNA levels for SUR, the other subunit of the channel, are normal (Tokuyama et al., 1996).

The *db/db* syndrome has also been linked to mutations in the leptin receptor (Chen et al., 1996) but it is still unclear whether the progressively aggravating diabetic state of the animals is secondary or involves primary β-cells defects. Several abnormalities of *db/db* β-cells have been identified. They do not show the initial decrease in $[Ca^{2+}]_i$ that normally occurs when the glucose concentration is suddenly raised, an anomaly that can be explained by their lack of SERCA activity (Roe et al., 1994). *db/db* β-Cells also express a K^+ channel that is infrequent in normal β-cells (Kalman et al., 1995), but it is not known whether this channel is responsible for the abnormal electrical activity induced by glucose (Meissner and Schmidt, 1976).

In conclusion, although several defects of $[Ca^{2+}]_i$ signaling have been identified in β cells from animal models of type 2 diabetes, it remains unclear whether they are primitive or secondary to the chronic abnormalities of glucose and lipid metabolism that these animals present.

X. NORMAL AND DIABETIC HUMAN β-CELLS

Glucose stimulation of insulin secretion from human islets correlates well with the rate of glucose oxidation and is abolished by inhibition of Ca^{2+} influx (Harrisson et al., 1985; Misler et al., 1992). Studies of human β-cells with the patch-clamp technique concur in their conclusions that the major effects of glucose are similar to those more extensively characterized in mouse β-cells. In particular, glucose increases the ATP/ADP ratio in human islets (Detimary et al., 1998a). As in rodent β cells, closure of K^+-ATP channels underlies the depolarization leading to activation of voltage-dependent Ca^{2+} channels (Ashcroft et al., 1987; 1989; Falke et al., 1989; Misler et al., 1989, 1992). Whether the electrical activity induced by glucose displays an oscillatory pattern has, however, not yet been convincingly demonstrated. In addition to typical L-type voltage-dependent Ca^{2+} channels (Kelly et al., 1991; Misler et al., 1992), human β-cells possess voltage-insensitive Ca^{2+} channels activated by glucose metabolism (Rojas et al., 1990). Their functional role is still uncertain. The K^+-ATP channel–independent pathway of glucose regulation is also operative in human islets (Straub et al., 1998).

The influence of glucose on β-cell $[Ca^{2+}]_i$ has been evaluated with single cells and whole islets. The results are generally concordant except in one study in which it was concluded that glucose causes a monophasic $[Ca^{2+}]_i$ rise without oscillations, and that this rise is partly due to mobilization of Ca^{2+} from intracellular stores (Rojas et al., 1994). It should be noted that this study was performed at room temperature, which is sufficient to suppress $[Ca^{2+}]_i$ oscillations in mouse islets (Hellman et al., 1990). Other studies evidenced clear $[Ca^{2+}]_i$ oscillations in response to glucose, and concluded that the human β-cell has oscillatory signaling for insulin release similar to that of the mouse β-cell (Hellman et al., 1994, 1997). Similar conclusions were reached by studies using intact human islets (Kindmark et al., 1991, 1994; Martin and Soria, 1996). The three phases of $[Ca^{2+}]_i$ changes induced by glucose, with synchronous oscillations during steady-state stimulation were similar to those first identified in mouse islets.

Only two studies have investigated the biophysical events occurring in β-cells from Type 2 diabetic patients. In one patient, whose diabetes was controlled by diet alone, the properties of β-cell K^+-ATP channels were found to be essentially normal, but the number of functional channels was reduced. Surprisingly, an ATP-activated, sulphonylurea-insensitive K^+ channel was present in the β-cells of this patient (Williams et al., 1993). The impact of these abnormalities on β-cell function is uncertain. In islets from patients with impaired glucose tolerance or diabetes, glucose was found to induce no response, a sustained decrease in $[Ca^{2+}]_i$ or $[Ca^{2+}]_i$ oscillations resembling those observed in islets from control normal subjects (Kindmark et al., 1994). Obviously the limited access to islets of diabetic patients and the heterogeneity of the disease will be two major obstacles in the identification of the pathogenic abnormalities of β-cells. Discovery of mutations in the genes coding for proteins playing a key role in stimulus-secretion coupling is a powerful alternative approach. A number of recent studies have already shown that genetic variations in the SUR-1 and Kir 6.2 genes are not major causative factors in Type 2 diabetes (Zhang et al., 1995; Iwasaki et al., 1996; Sakura et al., 1996; Inoue et al., 1997).

ACKNOWLEDGMENTS

We thank M. Nenquin for editorial help. The authors' work referred to in this chapter has been supported by funds from the Fonds National de la Recherche Scientifique (3.4525.94), the General Direction of Scientific Research of the French Community of Belgium (ARC 95/00-188), and the Federal Office for Scientific, Technical and Cultural Affairs of Belgium (PAI 4/21).

REFERENCES

Aizawa, T., Sato, Y., Ishihara, F., Taguchi, N., Komatsu, M., Suzuki, N., Hashizume, K. and Yamada, T. (1994). ATP-sensitive K^+ channel-independent glucose action in rat pancreatic β-cell. Am. J. Physiol. 266, C622–C627.

Ammala, C., Eliasson, L., Bokvist, K., Berggren, P.O., Honkanen, R.E., Sjöholm, Å. and Rorsman, P. (1994). Activation of protein kinases and inhibition of protein phosphatases play a central role in the regulation of exocytosis in mouse pancreatic β cells. Proc. Natl. Acad. Sci. USA 91, 4343–4347.

Ammala, C., Eliasson, L., Bokvist, K., Larsson, O., Ashcroft, F.M. and Rorsman, P. (1993). Exocytosis elicited by action potentials and voltage-clamp calcium currents in individual mouse pancreatic B-cells. J. Physiol. (Lond.) 472, 665–688.

Ashcroft, F.M., Kakei, M., Gibson, J.S., Gray, D.W. and Sutton, R. (1989). The ATP- and tolbutamide-sensitivity of the ATP-sensitive K-channel from human pancreatic B cells. Diabetologia 32, 591–598.

Ashcroft, F.M., Kakei, M., Kelly, R.P. and Sutton, R. (1987). ATP-sensitive K^+ channels in human isolated pancreatic B-cells. FEBS Lett. 215, 9–12.

Atwater, I., Kukuljan, M. and Perez-Armendariz, E.M. (1994). Molecular biology of the ion channels in the pancreatic β cells. In: *Molecular Biology of Diabetes,* Part I. (Draznin, B. and Leroith, D., Eds.), pp. 303–332. Humana Press, Totowa, NJ.

Barbosa, R.M., Silva, A.M., Tomé, A.R., Stamford, J.A., Santos, R.M. and Rosario, L.M. (1996). Real time electrochemical detection of 5-HT/insulin secretion from single pancreatic islets: Effect of glucose and K^+ depolarization. Biochem. Biophys. Res. Commun. 228, 100–104.

Bennett, B.D., Jetton, T.L., Ying, G.T., Magnuson, M.A. and Piston, D.W. (1996). Quantitative subcellular imaging of glucose metabolism within intact pancreatic islets. J. Biol. Chem. 271, 3647–3651.

Bergsten, P. (1995). Slow and fast oscillations of cytoplasmic Ca^{2+} in pancreatic islets correspond to pulsatile insulin release. Am. J. Physiol. Endocrinol. Metab. 268, E282–E287.

Bergsten, P., Grapengiesser, E., Gylfe, E., Tengholm, A. and Hellman, B. (1994). Synchronous oscillations of cytoplasmic Ca^{2+} and insulin release in glucose-stimulated pancreatic islets. J. Biol. Chem. 269, 8749–8753.

Bertram, R., Smolen, P., Sherman, A., Mears, D., Atwater, I., Martin, F. and Soria, B. (1995). A role for calcium release-activated current (CRAC) in cholinergic modulation of electrical activity in pancreatic β-cells. Biophys. J. 68, 2323–2332.

Best, L., Yates, A.P. and Tomlinson, S. (1992). Stimulation of insulin secretion by glucose in the absence of diminished potassium ($^{86}Rb^+$) permeability. Biochem. Pharmacol. 43, 2483–2485.

Bokvist, K., Eliasson, L., Ammala, C., Renström, E. and Rorsman, P. (1995). Co-localization of L-type Ca^{2+} channels and insulin-containing secretory granules and its significance for the initiation of exocytosis in mouse pancreatic B-cells. EMBO J. 14, 50–57.

Bonner-Weir, S. (1988). Morphological evidence for pancreatic polarity of beta-cell within islets of Langerhans. Diabetes 37, 616–621.

Bozem, M. and Henquin, J.C. (1988). Glucose modulation of spike activity independently from changes in slow waves of membrane potential in mouse B-cells. Pflugers Arch. 413, 147–152.

Brown, G.R., Köhler, M. and Berggren, P.O. (1997). Parallel changes in nuclear and cytosolic calcium in mouse pancreatic β-cells. Biochem. J. 325, 771–778.

Cerasi, E. (1975). Mechanisms of glucose stimulated insulin secretion in health and in diabetes: Some reevaluations and proposals. The Minkowski Award Lecture 1974. Diabetologia 11, 1–14.

Chan, C.B. and MacPhail, R.M. (1996). K_{ATP} channel-dependent and independent pathways of insulin secretion in isolated islets from *fa/fa* Zucker rats. Biochem. Cell Biol. 74, 403–410.

Chen, H., Charlat, O., Tartaglia, L.A., Woolf, E.A., Weng, X., Ellis, S.J., Lakey, N.D., Culpepper, J., Moore, K.J., Breitbart, R.E., Duyk, G.M., Tepper, R.I. and Morgenstern, J.P. (1996). Evidence that the diabetes gene encodes the leptin receptor: Identification of a mutation in the leptin receptor gene in *db/db* mice. Cell 84, 491–495.

Chou, H.-F., Berman, N. and Ipp, E. (1992). Oscillations of lactate released from islets of Langerhans: Evidence for oscillatory glycolysis in β-cells. Am. J. Physiol. Endocrinol. Metab. 262, E800–E805.

Chow, R.H., Lund, P.-E., Löser, S., Panten, U. and Gylfe, E. (1995). Coincidence of early glucose-induced depolarization with lowering of cytoplasmic Ca^{2+} in mouse pancreatic β-cells. J. Physiol. (Lond.) 485, 607–617.

Clark, J.B., Palmer, C.J. and Shaw, W.N. (1983). The diabetic Zucker fatty rat. Proc. Soc. Exp. Biol. Med. 173, 68–75.

Clement, J.P., Kunjilwar, K., Gonzalez, G., Schwanstecher, M., Panten, U., Aguilar-Bryan, L. and Bryan, J. (1997). Association and stoichiometry of K_{ATP} channel subunits. Neuron 18, 827–838.

Cook, D.L. (1983). Isolated islets of Langerhans have slow oscillations of electrical activity. Metabolism 32, 681–685.

Cook, D.L. and Ikeuchi, M. (1989). Tolbutamide as mimic of glucose on beta-cell electrical activity. ATP-sensitive K^+ channels as common pathway for both stimuli. Diabetes 38, 416–421.

Cook, D.L., Satin, L.S. and Hopkins, W.F. (1991). Pancreatic B cells are bursting, but how? Trends Neurosci. 14, 411–414.

Corkey, B.E., Deeney, J.T., Glennon, M.C., Matschinsky, F.M. and Prentki, M. (1988). Regulation of steady-state free Ca^{2+} levels by the ATP/ADP ratio and orthophosphate in permeabilized RINm5F insulinoma cells. J. Biol. Chem. 263, 4247–4253.

Cunningham, B.A., Deeney, J.T., Bliss, C.R., Corkey, B.E. and Tornheim, K. (1996). Glucose-induced oscillatory insulin secretion in perifused rat pancreatic islets and clonal β-cells (HIT). Am. J. Physiol. Endocrinol. Metab. 271, E702–E710.

Detimary, P., Dejonghe, S., Ling, Z., Pipeleers, D., Schuit, F., and Henquin, J.C. (1998a). The changes in adenine nucleotides measured in glucose-stimulated rodent islets occur in β cells but not in α cells and are also observed in human islets. J. Biol. Chem. 273, 33905–33908.

Detimary, P., Gilon, P., and Henquin, J.C. (1998b). Interplay between cytoplasmic Ca^{2+} and the ATP/ADP ratio: A feedback control mechanism in mouse pancreatic islets. Biochem. J. 333, 269–274.

Detimary, P., Gilon, P., Nenquin, M. and Henquin, J.C. (1994). Two sites of glucose control of insulin release with distinct dependence on the energy state in pancreatic B-cells. Biochem. J. 297, 455–461.

Detimary, P., Van den Berghe, G. and Henquin, J.C. (1996). Concentration dependence and time course of the effects of glucose on adenine and guanine nucleotides in mouse pancreatic islets. J. Biol. Chem. 271, 20559–20565.

Dryselius, S., Lund, P.-E., Gylfe, E. and Hellman, B. (1994). Variations in ATP-sensitive K^+ channel activity provide evidence for inherent metabolic oscillations in pancreatic β-cells. Biochem. Biophys. Res. Commun. 205, 880–885.

Dukes, I.D., Roe, M.W., Worley, J.F. III, and Philipson, L.H. (1997). Glucose-induced alterations in β-cell cytoplasmic Ca^{2+} involving the coupling of intracellular Ca^{2+} stores and plasma membrane ion channels. Curr. Opin. Endocrinol. Diab. 4, 262–271.

Efendic, S., Cerasi, E. and Luft, R. (1971). Role of glucose in arginine-induced release in man. Metabolism 20, 568–579.

Eliasson, L., Renström, E., Ding, W.G., Proks, P., and Rorsman, P. (1997). Rapid ATP-dependent priming of secretory granules precedes Ca^{2+}-induced exocytosis in mouse pancreatic β-cells. J. Physiol. 503, 399–412.

Falke, L.C., Gillis, K.D., Pressel, D.M. and Misler, S. (1989). 'Perforated patch recording' allows long-term monitoring of metabolite-induced electrical activity and voltage-dependent Ca^{2+} currents in pancreatic islet B cells. FEBS Lett. 251, 167–172.

Fournier, L., Whitfield, J.F., Schwartz, J.-L. and Bégin-Heick, N. (1994). Cyclic AMP triggers large $[Ca^{2+}]_i$ oscillations in glucose-stimulated β-cells from ob/ob mice. J. Biol. Chem. 269, 1120–1124.

Friel, D.D. (1996). TRP: Its role in phototransduction and store-operated Ca^{2+} entry. Cell 85, 617–619.

Garcia-Barrado, M.J., Gilon, P., Sato, Y., Nenquin, M. and Henquin, J.C. (1996). No evidence for a role of reverse Na^+-Ca^{2+} exchange in insulin release from mouse pancreatic islets. Am. J. Physiol. Endocrinol. Metab. 271,

Gembal, M., Detimary, P., Gilon, P., Gao, Z.Y. and Henquin, J.C. (1993). Mechanisms by which glucose can control insulin release independently from its action on adenosine triphosphate-sensitive K^+ channels in mouse B cells. J. Clin. Invest. 91, 871–880.E426–E433.

Gembal, M., Gilon, P. and Henquin, J.C. (1992). Evidence that glucose can control insulin release independently from its action on ATP-sensitive K^+ channels in mouse B cells. J. Clin. Invest. 89, 1288–1295.

Gillis, K.D. and Misler, S. (1993). Enhancers of cytosolic cAMP augment depolarization-induced exocytosis from pancreatic B-cells: Evidence for effects distal to Ca^{2+} entry. Pflugers Arch. 424, 195–197.

Gilon, P. and Henquin, J.C. (1992). Influence of membrane potential changes on cytoplasmic Ca^{2+} concentration in an electrically excitable cell, the insulin-secreting pancreatic B-cell. J. Biol. Chem. 267, 20713–20720.

Gilon, P. and Henquin, J.C. (1995). Distinct effects of glucose on the synchronous oscillations of insulin release and cytoplasmic Ca^{2+} concentration measured simultaneously in single mouse islets. Endocrinology 136, 5725–5730.

Gilon, P., Jonas, J.C. and Henquin, J.C. (1994). Culture duration and conditions affect the oscillations of cytoplasmic calcium concentration induced by glucose in mouse pancreatic islets. Diabetologia 37, 1007–1014.

Gilon, P., Shepherd, R.M. and Henquin, J.C. (1993). Oscillations of secretion driven by oscillations of cytoplasmic Ca^{2+} as evidenced in single pancreatic islets. J. Biol. Chem. 268, 22265–22268.

Goto, Y. and Kakizaki, M. (1981). The spontaneous-diabetes rat: a model of noninsulin dependent diabetes mellitus. Proc. Jpn. Acad. 57, 381–384.

Grapengiesser, E., Gylfe, E. and Hellman, B. (1991). Cyclic AMP as a determinant for glucose induction of fast Ca^{2+} oscillations in isolated pancreatic β-cells. J. Biol. Chem. 266, 12207–12210.

Grapengiesser, E., Gylfe, E. and Hellman, B. (1992). Glucose sensing of individual pancreatic β-cells involves transitions between steady-state and oscillatory cytoplasmic Ca^{2+}. Cell Calcium 13, 219–226.

Grodsky, G.M. and Bennett, L.L. (1966). Cation requirements for insulin secretion in the isolated perfused pancreas. Diabetes 15, 910–913.

Gromada, J., Frokjaer-Jensen, J. and Dissing, S. (1996). Glucose stimulates voltage- and calcium-dependent inositol trisphosphate production and intracellular calcium mobilization in insulin-secreting βTC3 cells. Biochem. J. 314, 339–345.

Gylfe, E. (1988). Nutrient secretagogues induce bimodal early changes in cytoplasmic calcium of insulin-releasing ob/ob mouse beta-cells. J. Biol. Chem. 263, 13750–13754.

Gylfe, E., Grapengiesser, E. and Hellman, B. (1991). Propagation of cytoplasmic Ca^{2+} oscillations in clusters of pancreatic β-cells exposed to glucose. Cell Calcium 12, 229–240.

Hamakawa, N. and Yada, T. (1995). Interplay of glucose-stimulated Ca^{2+} sequestration and acetylcholine-induced Ca^{2+} release at the endoplasmic reticulum in rat pancreatic β-cells. Cell Calcium 17, 21–31.

Harrison, D.E., Christie, M.R. and Gray, D.W. (1985). Properties of isolated human islets of Langerhans: insulin secretion, glucose oxidation and protein phosphorylation. Diabetologia 28, 99–103.

Hedeskov, C.J. (1980). Mechanism of glucose-induced insulin secretion. Physiol. Rev. 60, 442–509.

Hellman, B., Gylfe, E., Bergsten, P., Grapengiesser, E., Berts, A., Liu, Y.J., Tengholm, A. and Westerlund, J. (1997). Oscillatory signaling and insulin release in human pancreatic β-cells exposed to strontium. Endocrinology 138, 3161–3165.

Hellman, B., Gylfe, E., Bergsten, P., Grapengiesser, E., Lund, P.-E., Berts, A., Tengholm, A., Pipeleers, D.G. and Ling, Z. (1994). Glucose induces oscillatory Ca^{2+} signalling and insulin release in human pancreatic beta cells. Diabetologia 37 Suppl. 2, S11–S20.

Hellman, B., Gylfe, E., Grapengiesser, E., Lund, P.-E. and Berts, A. (1992). Cytoplasmic Ca^{2+} oscillations in pancreatic β-cells. Biochim. Biophys. Acta 1113, 295–305.

Hellman, B., Gylfe, E., Grapengiesser, E., Panten, U., Schwanstecher, C. and Heipel, C. (1990). Glucose induces temperature-dependent oscillations of cytoplasmic Ca^{2+} in single pancreatic beta-cells related to their electrical activity. Cell Calcium 11, 413–418.

Hellman, B., Sehlin, J. and Taljedal, I.B. (1971). Calcium uptake by pancreatic β-cells as measured with the aid of ^{45}Ca and mannitol ^{3}H. Am. J. Physiol. 221, 1795–1801.

Henquin, J.C. (1978). Relative importance of extracellular and intracellular calcium for the two phases of glucose stimulated insulin release: Studies with theophylline. Endocrinology 102, 723–730.

Henquin, J.C. (1980) Tolbutamide stimulation and inhibition of insulin release: Studies of the underlying ionic mechanisms in isolated rat islets. Diabetologia 18, 151–160.

Henquin, J.C. (1988). ATP-sensitive K^+ channels may control glucose-induced electrical activity in pancreatic B-cells. Biochem. Biophys. Res. Commun. 156, 769–775.

Henquin, J.C. (1990). Glucose-induced electrical activity in β-cells: Feedback control of ATP-sensitive K^+ channels by Ca^{2+}. Diabetes 39, 1457–1460.

Henquin, J.C. (1992). Adenosine triphosphate-sensitive K^+ channels may not be the sole regulators of glucose-induced electrical activity in pancreatic B-cells. Endocrinology 131, 127–131.

Henquin, J.C., Gembal, M., Detimary, P., Gao, Z.Y., Warnotte, C. and Gilon, P. (1994). Multisite control of insulin release by glucose. Diabete Metabolisme 20, 132–137.

Henquin, J.C. and Lambert, A.E. (1975). Cobalt inhibition of insulin secretion and calcium uptake by isolated rat islets. Am. J. Physiol. 228, 1669–1677.

Henquin, J.C. and Meissner, H.P. (1984). Significance of ionic fluxes and changes in membrane potential for stimulus-secretion coupling in pancreatic B-cells. Experientia 40, 1043–1052.

Henquin, J.C., Meissner, H.P. and Schmeer, W. (1982). Cyclic variations of glucose-induced electrical activity in pancreatic B-cells. Pflugers Archiv. 393, 322–327.

Herchuelz, A. and Plasman, P.O. (1991). Sodium-calcium exchange in the pancreatic B cell. Ann. NY Acad. Sci. 639, 642–656.

Herchuelz, A., Pochet, R., Pastiels, Ch. and Van Praet, A. (1991). Heterogeneous changes in $[Ca^{2+}]_i$ induced by glucose, tolbutamide and K^+ in single rat pancreatic B cells. Cell Calcium 12, 577–586.

Hermans, M.P., Schmeer, W. and Henquin, J.C. (1987). The permissive effect of glucose, tolbutamide and high K^+ on arginine stimulation of insulin release in isolated mouse islets. Diabetologia 30, 659–665.

Hollander, P.M., Christopher, M.A. and Palmer, J.P. (1982). Glucose modulation of insulin and glucagon secretion in nondiabetic and diabetic man. Diabetes 31, 489–495.

Hoth, M. and Penner, R. (1993). Calcium release-activated calcium current in rat mast cells. J. Physiol. (Lond.) 465, 359–386.

Howell, S.L., Jones, P.M. and Persaud, S.J. (1994). Regulation of insulin secretion: The role of second messengers. Diabetologia 37 (Suppl. 2), S30–S35.

Ihara, Y., Yamada, Y., Fujii, Y., Gonoi, T., Yano, H., Yasuda, K., Inagaki, N., Seino, Y. and Seino, S. (1995). Molecular diversity and functional characterization of voltage-dependent calcium channels (CACN4) expressed in pancreatic β-cells. Mol. Endocrinol. 9, 121–130.

Ikeuchi, M., Fujimoto, W.Y. and Cook, D.L. (1984). Rat islet cells have glucose-dependent periodic electrical activity. Horm. Metab. Res. 16, 125–127.

Inagaki, N., Gonoi, T., Clement, J.P., Namba, N., Inazawa, J., Gonzalez, G., Aguilar-Bryan, L., Seino, S. and Bryan, J. (1995). Reconstitution of I_{KATP}: An inward rectifier subunit plus the sulfonylurea receptor. Science 270, 1166–1170.

Inoue, H., Ferrer, J., Warren-Perry, M., Zhang, Y., Millns, H., Turner, R.C., Elbein, S.C., Hampe, C.L., Suarez, B.K., Inagaki, N., Seino, S. and Permutt, M.A. (1997). Sequence variants in the pancreatic islet β-cell inwardly rectifying K^+ channel Kir 6.2 (Bir) gene—Identification and lack of role in Caucasian patients with NIDDM. Diabetes 46, 502–507.

Iwasaki, N., Kawamura, M., Yamagata, K., Cox, N.J., Karibe, S., Ohgawara, H., Inagaki, N., Seino, S., Bell, G.I. and Omori, Y. (1996). Identification of microsatellite markers near the human genes encoding the β-cell ATP-sensitive K^+ channel and linkage studies with NIDDM, in Japanese. Diabetes 45, 267–269.

Jijakli, H. and Malaisse, W.J. (1997). Cationic events stimulated by D-glucose in depolarized islet cells. Cell. Signal. 9, 283–290.

Jonas, J.C., Gilon, P., and Henquin, J.C. (1998). Temporal and quantitative correlations between insulin secretion and stably elevated or oscillatory cytoplasmic Ca^{2+} in mouse pancreatic β cells. Diabetes 47, 1266–1273.

Jonas, J.C., Li, G., Palmer, M., Weller, U. and Wollheim, C.B. (1994). Dynamics of Ca^{2+} and guanosine 5′-[gamma-thio]triphosphate action on insulin secretion from α-toxin-permeabilized HIT-T15 cells. Biochem. J. 301, 523–529.

Jones, P.M., Stutchfield, J. and Howell, S.L. (1985). Effects of Ca^{2+} and a phorbol ester on insulin secretion from islets of Langerhans permeabilised by high-voltage discharge. FEBS Lett. 191, 102–106.

Jonkers, F.C., Gilon, P., and Henquin, J.C. (1998). The influence of cell number on Ca^{2+} oscillations in clusters of β cells. Diabetologia 41 (Suppl. 1), A144.

Kalman, K., Tseng-Crank, J., Dukes, I.D., Chandy, G., Lancaster, M.E., Spencer, R.H., Aiyar, J., Gutman, G.A. and Chandy, K.G. (1995). Heightened expression of a novel K^+ Shaker-related channel Kvl.7 in diabetic β-cells: Potential role in impaired insulin secretion in diabetes. Biophys. J. 68, A268 [Abstract].

Katayama, N., Hughes, S.J., Persaud, S.J., Jones, P.M. and Howell, S.L. (1995). Insulin secretion from islets of GK rats is not impaired after energy generating steps. Mol. Cell. Endocrinol. 111, 125–128.

Kato, S., Ishida, H., Tsuura, Y., Tsuji, K., Nishimura, M., Horie, M., Taminato, T., Ikehara, S., Odaka, H., Ikeda, H., Okada, Y. and Seino, Y. (1996). Alterations in basal and glucose-stimulated voltage-dependent Ca^{2+} channel activities in pancreatic β cells of non-insulin-dependent diabetes mellitus GK rats. J. Clin. Invest. 97, 2417–2425.

Kelley, G.G., Zawalich, K.C. and Zawalich, W.S. (1994). Calcium and a mitochondrial signal interact to stimulate phosphoinositide hydrolysis and insulin secretion in rat islets. Endocrinology 134, 1648–1654.

Kelly, R.P., Sutton, R. and Ashcroft, F.M. (1991). Voltage-activated calcium and potassium currents in human pancreatic β-cells. J. Physiol. (Lond.) 443, 175–192.

Kennedy, E.D., Rizzuto, R., Theler, J.M., Pralong, W.F., Bastianutto, C., Pozzan, T. and Wollheim, C.B. (1996). Glucose-stimulated insulin secretion correlates with changes in mitochondrial and cytosolic Ca^{2+} in aequorin-expressing INS-1 cells. J. Clin. Invest. 98, 2524–2538.

Kikuchi, M., Wollheim, C.B., Siegel, E.G., Renold, A.E. and Sharp, G.W.G. (1979). Biphasic insulin release in rat islets of Langerhans and the role of intracellular Ca^{++} stores. Endocrinology 105, 1013–1019.

Kindmark, H., Köhler, M., Arkhammar, P., Efendic, S., Larsson, O., Linder, S., Nilsson, T. and Berggren, P.O. (1994). Oscillations in cytoplasmic free calcium concentration in human pancreatic islets from subjects with normal and impaired glucose tolerance. Diabetologia 37, 1121–1131.

Kindmark, H., Köhler, M., Nilsson, T., Arkhammar, P., Wiechel, K.-L., Rorsman, P., Efendic, S. and Berggren, P.O. (1991). Measurements of cytoplasmic free Ca^{2+} concentration in human pancreatic islets and insulinoma cells. FEBS Lett. 291, 310–314.

Komatsu, M., Schermerhorn, T., Aizawa, T. and Sharp, G.W.G. (1995). Glucose stimulation of insulin release in the absence of extracellular Ca^{2+} and in the absence of any increase in intracellular Ca^{2+} in rat pancreatic islets. Proc. Natl. Acad. Sci. USA 92, 10728–10732.

Komatsu, M., Schermerhorn, T., Straub, S.G. and Sharp, G.W.G. (1996). Pituitary adenylate cyclase-activating peptide, carbachol, and glucose stimulate insulin release in the absence of an increase in intracellular Ca^{2+}. Mol. Pharmacol. 50, 1047–1054.

Larsson, O., Kindmark, H., Bränström, R., Fredholm, B. and Berggren, P.O. (1996). Oscillations in K_{ATP} channel activity promote oscillations in cytoplasmic free Ca^{2+} concentration in the pancreatic β cell. Proc. Natl. Acad. Sci. USA 93, 5161–5165.

Leech, C.A., Holz G.G. and Habener, J.F. (1994). Voltage-independent calcium channels mediate slow oscillations of cytosolic calcium that are glucose dependent in pancreatic β-cells. Endocrinology 135, 365–372.

Levin, S.R., Grodsky, G.M., Hagura, R., Smith, D.F. and Forsham, P.H. (1972). Relationships between arginine and glucose in the induction of insulin secretion from the isolated perfused rat pancreas. Endocrinology 90, 624–631.

Liu, Y.J., Grapengiesser, E., Gylfe, E. and Hellman, B. (1996). Crosstalk between the cAMP and inositol trisphosphate-signalling pathways in pancreatic β-cells. Arch. Biochem. Biophys. 334, 295–302.

Longo, E.A., Tornheim, K., Deeney, J.T., Varnum, B.A., Tillotson, D., Prentki, M. and Corkey, B.E. (1991). Oscillations in cytosolic free Ca^{2+}, oxygen consumption, and insulin secretion in glucose-stimulated rat pancreatic islets. J. Biol. Chem. 266, 9314–9319.

Malaisse, W.J., Brisson, G.R. and Baird, L.E. (1973). Stimulus secretion coupling of glucose-induced insulin release. X. Effect of glucose on ^{45}Ca efflux from perifused islets. Am. J. Physiol. 224, 389–394.

Malaisse, W.J., Devis, G., Pipeleers, D.G. and Somers, G. (1976). Calcium-antagonists and islet function. IV. Effect of D600. Diabetologia 12, 77.

Malaisse-Lagae, F. and Malaisse, W.J. (1971). Stimulus-secretion coupling of glucose-induced insulin release. III. Uptake of ^{45}calcium by isolated islets of Langerhans. Endocrinology 88, 72–80.

Martin, F., Reig, J.A. and Soria, B. (1995a). Secretagogue-induced $[Ca^{2+}]_i$ changes in single rat pancreatic islets and correlation with simultaneously measured insulin release. J. Mol. Endocrinol. 15, 177–185.

Martin, F., Ribas, J. and Soria, B. (1997). Cytosolic Ca^{2+} gradients in pancreatic islet-cells stimulated by glucose and carbachol. Biochem. Biophys. Res. Commun. 235, 465–468.

Martin, F., Sanchez-Andres, J.V. and Soria, B. (1995b). Slow $[Ca^{2+}]_i$ oscillations induced by ketoisocaproate in single mouse pancreatic islets. Diabetes 44, 300–305.

Martin, F. and Soria, B. (1996). Glucose-induced $[Ca^{2+}]_i$ oscillations in single human pancreatic islets. Cell Calcium 20, 409–414.

McCormack, J.G., Longo, E.A. and Corkey, B.E. (1990). Glucose-induced activation of pyruvate dehydrogenase in isolated rat pancreatic islets. Biochem. J. 267, 527–530.

Mears, D., Sheppard, N.F. Jr., Atwater, I., Rojas, E., Bertram, R. and Sherman, A. (1997). Evidence that calcium release-activated current mediates the biphasic electrical activity of mouse pancreatic β-cells. J. Membr. Biol. 155, 47–59.

Meda, P. (1996). The role of gap junction membrane channels in secretion and hormonal action. J. Bioenerg. Biomembr. 28, 369–377.

Meissner, H.P. (1976). Electrophysiological evidence for coupling between β cells of pancreatic islets. Nature 262, 502–504.

Meissner, H.P. and Schmidt, H. (1976). The electrical activity of pancreatic β-cells of diabetic mice. FEBS Lett. 67, 371–374.

Meredith, M., Rabaglia, M.E. and Metz, S.A. (1995). Evidence of a role for GTP in the potentiation of Ca^{2+}-induced insulin secretion by glucose in intact rat islets. J. Clin. Invest. 96, 811–821.

Milner, R.D.G. and Hales, C.N. (1967). The role of calcium and magnesium in insulin secretion from rabbit pancreas studied in vitro. Diabetologia 3, 47–49.

Misler, S., Barnett, D.W., Pressel, D.M., Gillis, K.D., Scharp, D.W. and Falke, L.C. (1992). Stimulus-secretion coupling in β-cells of transplantable human islets of Langerhans: Evidence for a critical role for Ca^{2+} entry. Diabetes 41, 662–670.

Misler, S., Gee, W.M., Gillis, K.D., Scharp, D.W. and Falke, L.C. (1989). Metabolite-regulated ATP-sensitive K^+ channel in human pancreatic islet cells. Diabetes 38, 422–427.

Miura,Y., Henquin, J.C. and Gilon, P. (1997) Emptying of intracellular Ca^{2+} pools stimulates two modalities of Ca^{2+} influx in mouse pancreatic β-cells. J. Physiol. (Lond) 503, 387–398.

Nakazaki, M., Ishihara, H., Kakei, M., Inukai, K., Asano, T., Miyazaki, J.I., Tanaka, H., Kikuchi, M., Yada, T., and Oka, Y. (1998). Repetitive mitochondrial Ca^{2+} signals synchronize with cytosolic Ca^{2+} oscillations in the pancreatic beta-cell line, MIN6. Diabetologia 41, 279–286.

Nichols, C.G., Shyng, S.L., Nestorowicz, A., Glaser, B., Clement, J.P., Gonzalez, G., Aguilar-Bryan, L., Permutt, M.A. and Bryan, J. (1996). Adenosine diphosphate as an intracellular regulator of insulin secretion. Science 272, 1785–1787.

Niwa, K., Shibuya, I. and Kanno, T. (1996). Temperature dependence of processes proximal and distal to the glucose-induced $[Ca^{2+}]_i$ rise in stimulus-secretion coupling in rat pancreatic islets. Biol. Signals 5, 30–43.

Okamoto, Y., Ishida, H., Tsuura, Y., Yasuda, K., Kato, S., Matsubara, H., Nishimura, M., Mizuno, N., Ikeda, H. and Seino, Y. (1995). Hyperresponse in calcium-induced insulin release from electrically permeabilized pancreatic islets of diabetic GK rats and its defective augmentation by glucose. Diabetologia 38, 772–778.

Östenson, C.-G., Khan, A., Abdel-Halim, S.M., Guenifi, A., Suzuki, K., Goto, Y. and Efendic, S. (1993). Abnormal insulin secretion and glucose metabolism in pancreatic islets from the spontaneously diabetic GK rat. Diabetologia 36, 3–8.

Panten, U., Christians, J., Kriegstein, E.V., Poser, W. and Hasselblatt, A. (1973). Effect of carbohydrates upon fluorescence of reduced pyridine nucleotides from perifused isolated pancreatic islets. Diabetologia 9, 477–482.

Panten, U., Schwanstecher, M., Wallasch, A. and Lenzen, S. (1988). Glucose both inhibits and stimulates insulin secretion from isolated pancreatic islets exposed to maximally effective concentrations of sulfonylureas. Naunyn Schmiedebergs Arch. Pharmacol. 338, 459–462.

Panten, U., Zunkler, B.J., Scheit, S., Kirchhoff, K. and Lenzen, S. (1986). Regulation of energy metabolism in pancreatic islets by glucose and tolbutamide. Diabetologia 29, 648–654.

Pershadsingh, H.A., McDaniel, M.L., Landt, M., Bry, C.G., Lacy, P.E. and McDonald, J.M. (1980). Ca^{2+}-activated ATPase and ATP-dependent calmodulin-stimulated Ca^{2+} transport in islet cell plasma membrane. Nature 288, 492–494.

Phillips, M.S., Liu, Q., Hammond, H.A., Dugan, V., Hey, P.J., Caskey, C.T., and Hess, J.F. (1996). Leptin receptor missense mutation in the *fatty* Zucker rat. Nature Genetics 13, 18–19.

Plant, T.D. (1988). Properties and calcium-dependent inactivation of calcium currents in cultured mouse pancreatic B-cells. J. Physiol. (Lond.) 404, 731–747.

Pralong, W.F., Bartley, C. and Wollheim, C.B. (1990). Single islet beta-cell stimulation by nutrients: Relationship between pyridine nucleotides, cytosolic Ca^{2+} and secretion. EMBO J. 9, 53–60.

Pralong, W.F., Spät, A. and Wollheim, C.B. (1994). Dynamic pacing of cell metabolism by intacellular Ca^{2+} transients. J. Biol. Chem. 269, 27310–27314.

Prentki, M., Janjic, D., Biden, T.J., Blondel, B. and Wollheim, C.B. (1984). Regulation of Ca^{2+} transport by isolated organelles of a rat insulinoma. Studies with endoplasmic reticulum and secretory granules. J. Biol. Chem. 259, 10118–10123.

Proks, P., Eliasson, L., Ammala, C., Rorsman, P. and Ashcroft, F.M. (1996). Ca^{2+}- and GTP-dependent exocytosis in mouse pancreatic β-cells involves both common and distinct steps. J. Physiol. (Lond.) 496, 255–264.

Putney, J.W. Jr. and Bird, G.St.J. (1993). The inositol phosphate-calcium signaling system in nonexcitable cells. Endocr. Rev. 14, 610–631.

Ravier, M.A., Gilon, P., and Henquin, J.C. (1998). Hierarchy of the signals driving pulsatile insulin secretion. Diabetologia 41 (Suppl. 1), A147.

Regazzi, R. (1999). Mechanism of Insulin Exocytosis. In: *Advances in Molecular and Cell Biology.* (Bittar, E.E. and Howell, S.L., Eds.) Vol. 29, pp. 151–172. JAI Press, Stamford, CT.

Roe, M.W., Lancaster, M.E., Mertz, R.J., Worley, J.F. III and Dukes, I.D. (1993). Voltage-dependent intracellular calcium release from mouse islets stimulated by glucose. J. Biol. Chem. 268, 9953–9956.

Roe, M.W., Philipson, L.H., Frangakis, C.J., Kuznetsov, A., Mertz, R.J., Lancaster, M.E., Spencer, B., Worley, J.F. III and Dukes, I.D. (1994). Defective glucose-dependent endoplasmic reticulum Ca^{2+} sequestration in diabetic mouse islets of Langerhans. J. Biol. Chem. 269, 18279–18282.

Roe, M.W., Worley, J.F. III, Tokuyama, Y., Philipson, L.H., Sturis, J., Tang, J.P., Dukes, I.D., Bell, G.I. and Polonsky, K.S. (1996). NIDDM is associated with loss of pancreatic β-cell L-type Ca^{2+} channel activity. Am. J. Physiol. Endocrinol. Metab. 270, E133–E140.

Rojas, E., Carroll, P.B., Ricordi, C., Boschero, A.C., Stojilkovic, S.S. and Atwater, I. (1994). Control of cytosolic free calcium in cultured human pancreatic β-cells occurs by external calcium-dependent and independent mechanisms. Endocrinology 134, 1771–1781.

Rojas, E., Hidalgo, J., Carroll, P.B., Li, M.X. and Atwater, I. (1990). A new class of calcium channels activated by glucose in human pancreatic beta-cells. FEBS Lett. 261, 265–270.

Rorsman, P. and Trube, G. (1986). Calcium and delayed potassium currents in mouse pancreatic beta-cells under voltage-clamp conditions. J. Physiol. (Lond.) 374, 531–550.

Rutter, G.A., Theler, J.M., Li, G. and Wollheim, C.B. (1994). Ca^{2+} stores in insulin-secreting cells: Lack of effect of cADP ribose. Cell Calcium 16, 71–80.

Rutter, G.A., Theler, J.M., Murgia, M., Wollheim, C.B., Pozzan, T. and Rizzuto, R. (1993). Stimulated Ca^{2+} influx raises mitochondrial free Ca^{2+} to supramicromolar levels in a pancreatic β-cell line. Possible role in glucose and agonist-induced insulin secretion. J. Biol. Chem. 268, 22385–22390.

Sakura, H. and Ashcroft, F.M. (1997). Identification of four *trp*1 gene variants murine pancreatic beta-cells. Diabetologia 40, 528–532.

Sakura, H., Bond, C.T., Warren-Perry, M., Horsley, S., Kearney, L., Tucker, S., Adelman, J., Turner, R. and Ashcroft, F.M. (1995). Characterization and variation of a human inwardly-rectifying K-channel gene (KCNJ6): A putative ATP-sensitive K-channel subunit. FEBS Lett. 367, 193–197.

Sakura, H., Wat, N., Horton, V., Millns, H., Turner, R.C. and Ashcroft, F.M. (1996). Sequence variations in the human Kir 6.2 gene, a subunit of the beta-cell ATP-sensitive K-channel: No association with NIDDM in white Caucasian subjects or evidence of abnormal function when expressed in vitro. Diabetologia 39, 1233–1236.

Santos, R.M., Barbosa, R.M., Silva, A.M., Antunes, C.M. and Rosario, L.M. (1992). High external Ca^{2+} levels trigger membrane potential oscillations in mouse pancreatic β-cells during blockade of K(ATP) channels. Biochem. Biophys. Res. Commun. 187, 872–879.

Santos, R.M., Rosario, L.M., Nadal, A., Garcia-Sancho, J., Soria, B. and Valdeolmillos, M. (1991). Widespread synchronous $[Ca^{2+}]_i$ oscillations due to bursting electrical activity in single pancreatic islets. Pflugers Arch. 418, 417–422.

Satin, L.S. and Smolen, P.D. (1994). Electrical bursting in β-cells of the pancreatic islets of Langerhans. Endocrine 2, 677–687.

Sato, Y., Aizawa, T., Komatsu, M., Okada, N. and Yamada, T. (1992). Dual functional role of membrane depolarization/Ca^{2+} influx in rat pancreatic B-cell. Diabetes 41, 438–443.

Sato, Y. and Henquin, J.C. (1998). The K^+-ATP channel–independent pathway of regulation of insulin secretion by glucose. In search of the underlying mechanism. Diabetes 47, 1713–1721.

Sato, Y., Mariot, P., Detimary, P., Gilon, P. and Henquin, J.C. (1998a). Okadaic acid inhibits insulin secretion by decreasing the magnitude and the efficacy of the Ca^{2+} signal in pancreatic β cells. Brit. J. Pharmacol., 123, 97–105.

Sato, Y., Nenquin, M., and Henquin, J.C. (1998b). Relative contribution of Ca^{2+}-dependent and Ca^{2+}-independent mechanisms to the regulation of insulin secretion by glucose. FEBS Lett. 421, 115–119.

Sener, A., Rasschaert, J. and Malaisse, W.J. (1990). Hexose metabolism in pancreatic islets. Participation of Ca^{2+}-sensitive 2-ketoglutarate dehydrogenase in the regulation of mitochondrial function. Biochim. Biophys. Acta 1019, 42–50.

Silva, A.M., Rosario, L.M. and Santos, R.M. (1994). Background Ca^{2+} influx mediated by a dihydropyridine- and voltage-insensitive channel in pancreatic β-cells. Modulation by Ni^{2+}, diphenylamine-2-carboxylate, and glucose metabolism. J. Biol. Chem. 269, 17095–17103.

Smith, P.A., Ashcroft, F.M. and Rorsman, P. (1990). Simultaneous recordings of glucose dependent electrical activity and ATP-regulated K^+-currents in isolated mouse pancreatic beta-cells. FEBS Lett. 261, 187–190.

Smith, P.A., Rorsman, P. and Ashcroft, F.M. (1989). Modulation of dihydropyridine-sensitive Ca^{2+} channels by glucose metabolism in mouse pancreatic β-cells. Nature 342, 550–553.

Straub, S.G., James, R.F.L., Dunne, M.J., and Sharp, G.W.G. (1998). Glucose activates both K_{ATP} channel-dependent and K_{ATP} channel-independent signaling pathways in human islets. Diabetes 47,758–763.

Takasawa, S., Nata, K., Yonekura, H. and Okamoto, H. (1993). Cyclic ADP-ribose in insulin secretion from pancreatic β cells. Science 259, 370–373.

Tamagawa, T., Niki, H. and Niki, A. (1985). Insulin release independent of a rise in cytosolic free Ca^{2+} by forskolin and phorbol ester. FEBS Lett. 183, 430–432.

Theler, J.M., Mollard, P., Guérineau, N., Vacher, P., Pralong, W.F., Schlegel, W. and Wollheim, C.B. (1992). Video imaging of cytosolic Ca^{2+} in pancreatic β-cells stimulated by glucose, carbachol, and ATP. J. Biol. Chem. 267, 18110–18117.

Tokuyama, Y., Fan, Z., Furuta, H., Makielski, J.C., Polonsky, K.S., Bell, G.I. and Yano, H. (1996). Rat inwardly rectifying potassium channel kir 6.2: Cloning, electrophysiological characterization, and decreased expression in pancreatic islets of male Zucker diabetic fatty rats. Biochem. Biophys. Res. Commun. 220, 532–538.

Tokuyama, Y., Sturis, J., DePaoli, A.M., Takeda, J., Stoffel, M., Tang, J., Sun, X., Polonsky, K.S. and Bell, G.I. (1995). Evolution of β-cell dysfunction in the male Zucker diabetic fatty rat. Diabetes 44, 1447–1457.

Tornheim, K. (1997). Are metabolic oscillations responsible for normal oscillatory insulin secretion? Diabetes 46, 1375–1380.

Tsuura, Y., Ishida, H., Okamoto, Y., Kato, S., Sakamoto, K., Horie, M., Ikeda, H., Okada, Y. and Seino, Y. (1993). Glucose sensitivity of ATP-sensitive K^+ channels is impaired in β-cells of the GK rat: A new genetic model of NIDDM. Diabetes 42, 1446–1453.

Tucker, S.J., Gribble, F.M., Zhao, C., Trapp, S. and Ashcroft, F.M. (1997). Truncation of Kir 6.2 produces ATP-sensitive K^+ channels in the absence of the sulphonylurea receptor. Nature 387, 179–183.

Valdeolmillos, M., Nadal, A., Contreras, D. and Soria, B. (1992). The relationship between glucose-induced K^+_{ATP} channel closure and the rise in $[Ca^{2+}]_i$ in single mouse pancreatic β-cells. J. Physiol. (Lond.) 455, 173–186.

Valdeolmillos, M., Santos, R.M., Contreras, D., Soria, B. and Rosario, L.M. (1989). Glucose-induced oscillations of intracellular Ca^{2+} concentration resembling bursting electrical activity in single mouse islets of Langerhans. FEBS Lett. 259, 19–23.

Vallar, L., Biden, T.J. and Wollheim, C.B. (1987). Guanine nucleotides induce Ca^{2+}-independent insulin secretion from permeabilized RINm5F cells. J. Biol. Chem. 262, 5049–5056.

Van Eylen, F., Lebeau, C., Albuquerque-Silva, J., and Herchuelz, A. (1998). Contribution of Na/Ca exchange to Ca^{2+} outflow and entry in the rat pancreatic β-cell. Studies with antisense oligonucleotides. Diabetes 47, 1873–1880.

Váradi, A., Molnár, E. and Ashcroft, S.J.H. (1996a). A unique combination of plasma membrane Ca^{2+}-ATPase isoforms is expressed in islets of Langerhans and pancreatic β-cell lines. Biochem. J. 314, 663–669.

Váradi, A., Molnár, E., Östenson, C.G. and Ashcroft, S.J.H. (1996b). Isoforms of endoplasmic reticulum Ca^{2+}-ATPase are differentially expressed in normal and diabetic islets of Langerhans. Biochem. J. 319, 521–527.

Wang, J.L., Corbett, J.A., Marshall, C.A. and McDaniel, M.L. (1993). Glucose-induced insulin secretion from purified β-cells. A role for modulation of Ca^{2+} influx by cAMP- and protein kinase C-dependent signal transduction pathways. J. Biol. Chem. 268, 7785–7791.

Ward, W.K., Beard, J.C., Halter, J.B., Pfeifer, M.A. and Porte, D. Jr. (1984). Pathophysiology of insulin secretion in non-insulin-dependent diabetes mellitus. Diabetes Care 7, 491–502.

Webb, D.L., Islam, S., Efanov, A.M., Brown, G., Köhler, M., Larsson, O. and Berggren, P.O. (1996). Insulin exocytosis and glucose-mediated increase in cytoplasmic free Ca^{2+} concentration in the pancreatic β-cell are independent of cyclic ADP-ribose. J. Biol. Chem. 271, 19074–19079.

Westerlund, J., Gylfe, E., and Bergsten, P. (1997). Pulsatile insulin release from pancreatic islets with nonoscillatory elevation of cytoplasmic Ca^{2+}. J. Clin. Invest. 100, 2547–2551.

Williams, B.A., Smith, P.A., Leow, K., Shimizu, S., Gray, D.W. and Ashcroft, F.M. (1993). Two types of potassium channel regulated by ATP in pancreatic B cells isolated from a type-2 diabetic human. Pflugers Arch. 423, 265–273.

Willmott, N.J., Galione, A., and Smith, P.A. (1995). A cADP-ribose antagonist does not inhibit secretagogue-, caffeine-, and nitric oxide-induced Ca^{2+} responses in rat pancreatic β-cells. Cell Calcium 18, 411–419.

Wollheim, C.B., Lang, J. and Regazzi, R. (1996). The exocytotic process of insulin secretion and its regulation by Ca^{2+} and G-proteins. Diabetes Rev. 4, 276–297.

Wollheim, C.B. and Sharp, G.W.G. (1981). Regulation of insulin release by calcium. Physiol. Rev. 61, 914–973.

Worley, J.F. III, McIntyre, M.S., Spencer, B. and Dukes, I.D. (1994a). Depletion of intracellular Ca^{2+} stores activates a maitotoxin-sensitive nonselective cationic current in β-cells. J. Biol. Chem. 269, 32055–32058.

Worley, J.F. III, McIntyre, M.S., Spencer, B., Mertz, R.J., Roe, M.W. and Dukes, I.D. (1994b). Endoplasmic reticulum calcium store regulates membrane potential in mouse islet β-cells. J. Biol. Chem. 269, 14359–14362.

Zaitsev, S.V., Efendic, S., Arkhammar, P., Bertorello, A.M. and Berggren, P.O. (1995). Dissociation between changes in cytoplasmic free Ca^{2+} concentration and insulin secretion as evidenced from measurements in mouse single pancreatic islets. Proc. Natl. Acad. Sci. USA 92, 9712–9716.

Zhang, Y., Warren-Perry, M., Sakura, H., Adelman, J., Stoffel, M., Bell, G.I., Ashcroft, F.M. and Turner, R.C. (1995). No evidence for mutations in a putative β-cell ATP-sensitive K^+ channel subunit in MODY, NIDDM, or GDM. Diabetes 44, 597–600.

ROLES OF PHOSPHOLIPIDS AND PHOSPHOLIPASE ACTIVATION IN β-CELL FUNCTION

Stewart A. Metz

The putative involvement of phospholipid turnover in the function of endocrine glands remains a subject of intense interest but is still of some controversy. The

Advances in Molecular and Cell Biology
Volume 29, pages 277-301.

ISBN: 0-7623-0547-9

objective of this chapter is to review key findings over the past 7 years which shed light on this field. The reader is referred to an earlier article for an overview of the subject prior to 1991 (Metz, 1991). The current overview does not cover emerging evidence of (patho)physiological roles for sphingolipid metabolism in β-cell function.

I. PUTATIVE ROLE(S) OF INTACT PHOSPHOLIPIDS AS CO-FACTORS IN EXOCYTOSIS

In the past, it has been widely assumed that the major functional role of phospholipids in the secretory process is to serve as substrates for phospholipases, thereby generating second messengers that act as effectors of, or second messengers for, secretory agonists. Although this hypothesis has been the subject of the majority of published investigations, it should be noted at the outset that *intact* phospholipids may also have important roles—both structural and dynamic. If so, it is conceivable that it is the abrogation of those effects, rather than the accumulation of second messengers, which mediates some of the effects of phospholipid hydrolysis. Significant degrees of lipid hydrolysis would be expected to change membrane fluidity, and thereby alter the function of associated membranous enzymes. Furthermore, recent work has focused on the role of intact inositol-containing phospholipids (particularly phosphatidylinositol biphosphate [PIP_2] and/or triphosphate [PIP_3]) as direct co-factors in one or more steps in cell signaling. These minor anionic phospholipids may interact with the cytoskeleton, modulate ion channels (Fan and Makielski, 1997), act as co-factors for phospholipase D, regulate phospholipase C (Lomasney et al., 1996; Popova et al., 1997), or modify the activity of various guanosine triphosphate (GTP)-binding proteins (Zheng et al., 1996). Some of these effects are attributable to binding of inositol lipids to the pleckstrin homology (PH) domains intrinsic to several relevant peptides—for example, the β subunit of heterotrimeric GTP-sending proteins (GBPs). It is possible that hydrolysis of inositol lipids would release proteins tethered to the membrane or cytoskeleton by their interaction with phosphatidylinositol biphosphate [$PI(P_2)$] via PH (or other) domains. Recent studies (Hay et al., 1995; Martin et al., 1995) from the laboratory of T. Martin have also focused on inositol-containing lipids and PI transfer proteins as direct mediators of discrete steps in exocytosis, especially the adenosine triphosphate (ATP)–dependent priming of Ca^{2+}-induced secretion. In fact, in those studies, the hydrolysis of inositol phospholipids by exogenous phospholipase C (PLC) resulted in an *inhibition* of calcium-activated exocytosis. Other studies (Schiavo et al., 1996) have focused on interactions of PIP_2 and PIP_3 with proteins at the exocytotic site or scaffold. Furthermore, some isoforms of phosphatidylinositol kinase have been localized to secretory granules in certain secretory cells. The possibility that similar events occur in the islet merits additional study, since glucose modulates the synthesis of phosphoinositides (compare Farese et al., 1986; Rustenback et al., 1994). Interestingly, decreased synthesis *and* degradation of phosphoinositides has been described (Morin et al., 1997) in pancreatic islets from the neonatal streptozotocin-diabetic rat model. Recently, in studying the effects of interleukin-1β (IL-1β) on normal rat islets, we observed (Vadakekalam et al., 1997a) that IL-1β appeared to inhibit both nutrient- and nonnutrient-induced insulin secretion, an effect that could not be readily related to its inhibitory effects on PLC, but rather which seemed to correlate with an impairment of phosphoinositide synthesis.

Additionally, it should be noted that phospholipases may also have important cellular effects that are not mediated via their enzymatic effects (i.e., the breakdown of membrane phospholipid substrates). Recent studies in nonendocrine cells (Ancian et al., 1995) have identified surface receptors for certain phospholipases (particularly phospholipase A_2) and have proposed roles for the activation of these receptors, by phospholipases secreted into the extracellular space, in the process of cell proliferation and/or other nuclear events. Studies of this formulation using pancreatic islets and assessing insulin biosynthesis and/or secretion are needed.

Lastly, although we focus here on the role of phospholipids in acute β-cell function (i.e., in secretion), there is emerging evidence that lipid by-products of phospholipase A_2 (PLA_2) and/or D (PLD) action may also play roles in the mitogenesis of β-cells (Dunlop et al., 1994) and possibly in gene regulation in some cells.

II. PHARMACOLOGICAL MECHANISMS OF ACTION OF PHOSPHOLIPID HYDROLYSIS PRODUCTS

Many of the by-products of phospholipid hydrolysis have well defined biological actions presumed to be relevant to β-cell function. (Vanishingly small amounts of data exist for the function of other intra-islet cells such as α-cells, which are not discussed herein). In part, such effects are inferred from studies designed to increase or decrease cellular levels of the relevant phospholipid hydrolysis products. The former approach has relied on provision of the lipid mediator exogenously, provision of pure phospholipase exogenously (and therefore extracellularly), and/or pharmacological inhibition of the catabolism (clearance) of the endogenous lipid compound. Clearly, any of these approaches *by itself* yields only pharmacological data; however, together, the results suggest working models. Thus, the activation of PLC yields diacylglycerol (DG), as well as inositol phosphates. The former activates protein kinase C (PKC); the latter releases calcium from the endoplasmic reticulum. DG may also have other direct effects that are independent of PKC, including possibly the mobilization of calcium (Thomas and Pek, 1992). Some other recent studies, albeit controverted, suggest that inositol-1,4,5-triphosphate (IP_3) acting probably via the type III receptor (Blondel et al., 1995; Wojcikiewicz, 1995), might also have a role in calcium homeostasis of the secretory granules. According to another recent study (Efanov et al., 1997), inositol hexakisphosphate (IP_6) may directly promote insulin exocytosis.

The activity of PLA_2, working chiefly on cationic or zwitterionic phospholipids, yields a lysophospholipid (lyso-PL), as well as a fatty acid (generally unsaturated) from the *sn*-2 position, of which arachidonic acid (AA) has been the one most studied. Arachidonic acid, in turn, can be oxygenated by any of a number of oxidative enzymes including cyclo-oxygenase, lipoxygenase, or, arguably in islets (Zeldin et al., 1997), mono-oxygenases. The lysophospholipid can also be metabolized (either by a distinct lysophospholipase or by the lysophospholipase activity intrinsic to PLA_2 itself) to yield a fatty acid (often saturated) from the *sn*-1 position, plus a glycerophosphoryl-base. A number of biological actions in β-cells (or other cells) have been tentatively ascribed to lysophospholipids, including mobilization of intracellular calcium (Metz, 1988; Rustenbeck and Lenzen, 1992), and possibly the activation of PKC or the direct promotion of secretory granule exocytosis. It seems curious that these initial observations on the putative effects of lysophospholipids in β-cells have received little follow-up. Recently, however, effects of arachidonic acid, lysophospholipids, and prostaglandins on the β-cell K^+_{ATP} channel have been described by Eddlestone (Eddlestone, 1995); such effects are compatible with their stimulatory effects on insulin release.

Likewise, a number of effects have been attributed to arachidonic acid or its metabolites, including the mobilization of intracellular calcium, the promotion of calcium influx via voltage-activated Ca^{2+} channel (Wolf et al., 1991), modulation of one more protein kinases or phosphatases (Band et al., 1992; Basudev et al., 1992; Gong et al., 1992; Landt et al., 1992), activation of ATP-sensitive K^+ channels (in a PKC-dependent fashion; Müller et al., 1992), and potentially, direct fusogenic effects on the exocytotic event. More recently, a novel mechanism for arachidonic acid (and other biologically active lipid mediators) has been proposed (Kowluru and Metz, 1994a)—namely, modulation of the GTPase activity of (one or more) GTP-binding proteins within specific subcellular compartments of the β-cell. The dose-response curves comparing the effects of arachidonic acid or lysophospholipids upon the function of GTP-binding proteins (GBPs) to their biological efficacy as insulin secretagogues are similar; likewise, the structural specificities of these effects appears to accord quite well. Additionally, at least some arachidonate-derived lipids (such as prostaglandin E_2) modify the activity of one or more GBPs on the β-cell secretory granule membrane (Kowluru and Metz, 1994b).

Activation of a phosphatidylcholine (PC)-directed PLD, recently localized to the pancreatic islet, leads to the generation of phosphatidic acid (PA), plus free choline (which is felt to be inactive). A number of actions have been ascribed to PA or its lyso-derivative including the following: mobilization of intracellular calcium; activation of certain GBPs via an effect on an (apparently) exofascial receptor, activation of mitogenesis and modulation of cyclic adenosine monophosphate (cAMP) formation. Furthermore, PA, like some other anionic phospholipids (see earlier discussion), may directly modulate the activity of a number of enzymes, including one or more PA-activated protein kinases.

Additionally, it is important to keep in mind that various lipids may act indirectly via the generation of other lipid mediators. For example, DG and PA are interconvertible by the action of DG kinase and PA phosphohydrolase, respectively; DG, in turn, can be broken down to release its intrinsic fatty acids.

Having identified such putative effects in β-cells, a more difficult question is to assess the physiological role of endogenous phospholipases.

III. PHOSPHOLIPASES THAT ARE PRESENT IN β-CELLS, AND HOW THEY ARE ACTIVATED

A. Phospholipase A_2

It is beyond the scope of this chapter to review fully the classification of different isoforms of phospholipases that have been identified in various cells. This is an intensely complicated and rapidly growing subject. However, it is established that several isoforms of PLA_2 do exist in the pancreatic β-cell, including calcium-sensitive and calcium-"insensitive" forms (Hazen and Gross, 1993; Jolly et al., 1993; Ramanadham et al., 1994; Loweth et al., 1995; Chen et al., 1996; Parker et al., 1996); forms that may be activated by ATP (Ramanadham et al., 1994); and forms of type I or II secretory PLA_2 isoforms (Loweth et al., 1995; Chen et al., 1996) similar to those found in the exocrine pancreas (and which therefore might serve an extracellular role in the islet, potentially via by a receptor-mediated mechanism; see earlier discussion). The gene for a type I PLA_2 has been detected; its cognate protein was not detected in another study in rat insulinoma (RIN) cells, although it was detected in human and rat islets (Loweth et al., 1995). Type II $sPLA_2$ was found in hamster insulinoma (HIT) cells but not in islets in one study (Loweth et al., 1995). Two other isoforms, however, have generated the most recent attention. One, cytosolic ($cPLA_2$), typically can be shown to be translocated from soluble to particulate fractions (the site of its phospholipid substrates) by the action of calcium. However, it is not directly activated by calcium; rather, it may be activated by various protein kinases including MAP (mitogen-activated protein) kinases and possibly PKC. This isoform of PLA_2 is often selective for the release of arachidonic acid. These attributes make it of potential interest to intracellular signaling in the β-cell and, indeed, multiple reports (Loweth et al., 1995; Chen et al., 1996; Parker et al., 1996; and S. Seavey, A. Kowluru, and S. Metz, unpublished) have identified the presence of this enzyme within islets and β-cells. Although it is clear that substantial amounts of arachidonic acid are released by physiologic agonists such as glucose in the β-cell, it remains to be demonstrated that such agonists specifically activate the $cPLA_2$ isoform in a physiological fashion. Carbachol and glucose together (but not either alone) increased activities of

Ca^{2+}-dependent (but not Ca^{2+}-independent) PLA_2 (Jolly et al., 1993) in islets whereas glucose alone may activate Ca^{2+}-independent AA release (Wolf et al., 1991; Ramanadham et al., 1994). (Glucose-induced release of AA has also been ascribed to PLC activation with generation of AA-enriched DG species.) Glucose alone (in physiological concentrations) has not been shown to generate a rapid increase in content of endogenous lysophospholipids when measured in whole islets; however, some increase in lysophospholipid content occurs using very high glucose levels and delayed (30 min) time points (Konrad et al., 1992a). The failure to see an earlier rise may be attributable in part to the rapid metabolism of these minor polar phospholipids, to their selective accumulation only in certain subcellular organelles (Rustenbeck et al., 1994), or to the transient nature of the activation. Recently mouse islets stimulated by glucose were noted (Capito et al., 1996) to show the accumulation of glycerophosphorylcholine (GPC), the water-soluble metabolite derived from the sequential de-esterification of lysophosphatidylcholine; lyso-PC itself was not increased. This observation may provide further evidence for the transient activation of PLA_2 in intact islets. However, an increased generation of GPC induced by physiological nutrients was not previously noted in studies of neonatal rat islet cells and conceivably could be due to de novo synthesis of GPC by the Infante pathway (Infante, 1985). Carbachol alone does induce a rapid accumulation of lysophospholipids (and release of AA) in islets (Konrad et al., 1992a), an effect potentiated by glucose.

Turk and co-workers in St. Louis have recently identified a calcium-independent, ATP-stimulatable PLA_2 that selectively cleaves ethanolamine-containing plasmalogens. This sensitivity to ATP (and adenosine diphosphate [ADP]) is of particular relevance to islet function, as are recent studies suggesting that the catalytic subunit of this isoform of PLA_2 can form a macromolecular complex with an 85-KDa isoform of phosphofructokinase in β-cells (Hazen and Gross, 1993; Ramanadham et al., 1994). The latter implies a potential link between glycolysis and phospholipase activation. Furthermore, a semiselective inhibitor of this enzyme blocks eicosanoid release and physiologic insulin release (Hazen and Gross, 1993); however, more recent studies (Balsinde and Dennis, 1996) have cast some degree of doubt on the specificity of the inhibitor of this enzyme used in the aforementioned studies. Interestingly, the ATP sensitivity of this enzyme seemed to be lost in RINm5F tumoral β-cells (Gross et al., 1993). A second (but possibly related) Ca^{2+}-independent PLA_2 was recently isolated from islets (Ma et al., 1997). Of interest is the homology of this isoform to domains of ankyrin which bind tubulin, since the latter may transfer GTP to subunits of G-proteins in some cells (Popova et al., 1977).

B. Phospholipase D

One or more isoforms of PLD also exist in pancreatic islets, including one directed against phosphatidylcholine (Dunlop et al., 1994) and another directed against glycosyl-PI (Metz et al., 1991; Deeg and Verchere, 1997). The former may be activated by physiological agonists, such as glyceraldehyde, and by phorbol ester activators of PKC, at least in neonatal islet cells. However the apparent lack of effect of glucose or carbachol alone, in studies to date of choline release (Wolf et al., 1989; Capito et al., 1996; Vadakekalam and Metz, in press), is noteworthy. Glucose may potentiate the activation of the enzyme when it is triggered via G-protein–mediated agonists, such as cholinergic agonists or NaF (Capito et al., 1996). We observed (Vadakekalam and Metz, in press), when studying the activation of PLD in intact islets via the release of choline into the extracellular space, that even the combined presence of a high concentration of glucose and the cholinergic agonist carbachol activated PLD only in a delayed fashion (>30 min). This delay was not attributable to a slow release of the ^{3}H-choline into the extracellular space, since a phorbol ester induced a rapid release (Vadakekalam and Metz, in press). Also, in neonatal β-cells, the activation of PLD by the glucose metabolite glyceraldehyde is apparently consequent to, and therefore temporally follows, the rapid activation of PLC. These findings together suggest that PLD might be an unlikely candidate to play a role in the acute first phase secretory responses to agonists, although the possibility remains of roles in later phases of secretion or in other effects, such as mitogenesis (especially in neonatal islet cells; Dunlop et al., 1994). Note also that PA can be synthesized de novo rapidly from glucose in islets (Farese et al., 1986; Rustenbeck et al., 1994), which may partially explain the very rapid generation of PA induced by glucose alone, or synergistically with carbachol (Konrad et al., 1991).

Prolonged secretion into the surrounding medium (or presumably into serum) of glycosyl-PI-directed PLD is stimulated by phorbol ester and indeed, this enzyme co-localizes with insulin in β-cells (Deeg and Verchere, 1997). The significance of this finding is unclear.

C. Phospholipase C

PLC has been much more extensively studied in β-cells (for review, see Zawalich, 1996); nonetheless, its exact physiological role is still not, in the opinion of this reviewer, firmly established. With respect to the assay of PLC, it should be noted the majority of the studies in the β-cell literature have been carried out using islets prelabeled with radioactive inositol under conditions that likely did not achieve isotopic equilibrium. Furthermore, even if isotopic steady state was achieved, this presumably would have been disrupted in some studies by the inclusion in the medium of unlabeled inositol (designed to prevent the reincorporation of any released label). Other studies in the literature (Li et al., 1992; Vadakekalam et al., 1996), however, do support the formulation that glucose increases the islet content of biologically active isomers of IP_3 and IP_4, as assessed using radioreceptor mass assays. Nonetheless, the catabolism of IPs itself is a regulated phenomenon (Biden et al., 1988); therefore, it would be attractive to supplement measurements of intracellular IPs with another index of PLC activation—ideally, the stoichiometric generation of diglyceride (DG). However, measurements of DG accumulation are problematic. First, DG can be synthesized by a number of routes in addition to its direct formation via PLC activation. Second, any DG generated is likely to be rapidly metabolized either to monoglycerides or to phosphatidic acid, or reincorporated into phospholipids. Third, as illustrated in the studies of several investigators, "basal" levels of DG in β-cells appear to be very high, even approaching 1 mM. It is not yet clear whether these apparently high levels represent an artifact of cell preparation, or whether in fact there is some degree of basal or "constitutive" activation of PLC even at "basal" glucose levels.

The physiological relevance (if any) of "basal" or stimulated levels of DG cannot be readily interpreted without knowing the molecular species of DG involved, since only some species activate PKC. Furthermore, there may be compartmentalization of phosphoinositides in discrete structures in the plasma membrane named caveoli (Pike and Casey, 1996); such pools of substrate may be tightly coupled to degradation by G-protein–activated PLC, which has also been reported to be present in caveolae. Thus, it is possible that such localized accumulations of DG would be obscured in measurements of total DG content. Indeed, glucose appears to generate little if any rise in total intracellular DG content (Peter-Riesch et al., 1988; Wolf et al., 1989; Rustenbeck et al., 1994). In contrast, several studies provide strong evidence that various receptor-mediated agonists (such as carbachol) generate detectable amounts of DG via the activation of PLC (Peter-Riesch et al., 1988; Hughes et al., 1990).

Efflux of inositol into the medium during perifusion has also been used to monitor PLC activation (Zawalich et al., 1995b; Zawalich, 1996; Zawalich and Zawalich, 1996; Vadakekalam et al., 1997a, 1997b); however, care must be exercised to exclude base exchange mechanisms as the cause of some of the inositol efflux, or activation of the aforementioned glycosylphosphatidyl-directed PLD. Note that culture conditions can markedly alter subsequent responsivity of islet PLC (Zawalich et al., 1995a; Zawalich and Zawalich, 1996). Thus, interpretation of studies of PLC activation is not always straightforward.

Using immunoblots, a number of isoforms of PLC have been identified in β-cells and rat islets (Kelley et al., 1995; Zawalich et al., 1995a; S. Seavey, A. Kowluru, S. Metz, unpublished data); these include PLC-β_1, PLC-β_3, PLC-γ_1 (S. Seavey, A. Kowluru, S. Metz, unpublished data), and PLC-δ. The content of PLC-δ and PLC-β is greatly reduced in mouse islets (Kelley et al., 1995; Zawalich et al., 1995a). Recently a novel isoform PLC-δ coupled to G_h (a unique large molecular mass G-protein) has been described in nonislet cells. G_h contains intrinsic transglutaminase activity; the G-protein activity in the molecule seems to be favored by the presence of adequate GTP, whereas the tissue transglutaminase function is favored in the presence of GTP deficit. Such interrelationships could help to explain the inhibition of PLC by GTP depletion in the pancreatic islet (Vadakekalam et al., 1996), and indeed, we have recently observed that GTP depletion augments tissue transglutaminase activity in β cells *in situ* (Li and Metz, unpublished observations). The various isoforms of PLC may be differentially distributed in subcellular compartments, but this also remains to be studied systematically in the islet.

It is classically stated that β isoforms of PLC are activated by GTP-binding proteins, whereas PLC-γ is activated by tyrosine kinases (Öberg et al., 1994; Jonas and Henquin, 1996). However, it is now clear that this distinction is not absolute, given the cross-talk between these two signaling systems. Furthermore, G-proteins may modulate synthesis of phosphatidylinositol-derived PLC substrates (Zhang et al., 1996). Stable analogs of GTP, provided exogenously, activate total PLC activity in β-cells, suggesting the involvement of one or more G-proteins. Furthermore, adequate amounts of endogenous GTP appear to be obligatory for the activation of PLC by nutrients or by increments in cytosolic calcium concentrations (Vadakekalam et al., 1996; Vadakekalam et al., 1997b). This effect of GTP has been tentatively ascribed to activation of low molecular mass G-proteins, which might potentiate the PLC response to increments in $[Ca^{2+}]_i$. Cdc42 is the leading candidate (Kowluru et al., 1996) for the putative G-protein involved in islets. (In contradistinction, in other cells, studies have focused on the role of small GBPs in the activation of PLD rather than PLC, although recently cdc42 has been coupled to PLC in mast cells as well; Illenberger et al., 1998.) There is also evidence that membrane potential can modulate PLC activity in β-cells even independently of any associated changes in cytosolic free calcium levels (Biden et al., 1993). In contrast, receptor-mediated agonists may activate PLC to a large degree independently of extracellular calcium influx (in contradistinction to the activation induced by nutrients); this effect may rely on activation of the α subunits of a heterotrimeric GTP-binding protein, presumably of the G_q/G_{11} family. Alternatively, however, the activation in some cells of PLC by high concentrations of agonists coupled to the pertussis toxin–sensitive G-proteins G_i and/or G_o has suggested a role for the released βγ subunits of trimeric G-proteins in the regulation of PLC.

It is clear that physiological agonists activate PLC in the islet; however, it has been difficult to show an increase in DG content of glucose-stimulated islets (see earlier discussion). Even if total DG increases, it will be necessary to show that biologically relevant molecular species of DG are involved, since most or all of the DG is synthesized de novo, a route that may yield DG species inefficient as PKC activators. Likewise, the majority of IP_3 generated by glucose is the (1,3,4) isomer, which is biologically inactive. Thus, although activation of PLC follows closely a large number of biological events in the pancreatic islet *pari passu*, there is actually little *direct* evidence *causally* linking the two phenomenon. Under certain conditions (Vadakekalam et al., 1977b), we have been able to demonstrate that the activation of PLC is (at least in part) an epiphenomenon, rather than the event "driving" the concurrent insulin release. Additionally, in a number of conditions, the activation of PLC is neither required (e.g., depolarizing concentrations of K^+) nor sufficient (receptor-mediated agonists, when provided at low glucose levels) to induce exocytosis. Protein kinase A, protein kinase C, and calmodulin-activated protein kinase are negative feedback modulators of (at least certain) isoforms of PLC (Hedeskov et al., 1991; Zawalich et al., 1991), although they potentiate insulin secretion.

Calcium has a direct, albeit limited, effect to activate PLC in β-cells; this action is potentiated by other effects of glucose, possibly including an increase in the GTP content of the β-cell (Vadakekalam et al., 1996). Glucose may interact synergistically with neurohumoral agonists (including cholinergic agonists, bombesin, arginine vasopressin, and extracellular ATP; Li et al., 1991); it has been suggested that such synergistic interactions (Kelley et al., 1995) might represent the concurrent activation of distinct isozymes of PLC. It has also been suggested that the consequent activation of PKC plays a role in time-dependent potentiation of β-cell function (Zawalich, 1996), and possibly as well in the desensitization induced by long-term exposure of islets to high glucose levels. The truncated second phase of insulin secretion seen in mouse (compared to rat or human) islets also has been attributed (Zawalich et al., 1995b) to deficient "second phase" activation of PLC in murine models. Likewise, glucose may not adequately stimulate PLC activity in HIT-T15 cells (Gonzalez and Rana, 1993), a fact that could be related to the failure of glucose to translocate PKC in these cells (Wolf et al., 1989), in contrast to normal islets (Ganesan et al., 1990). (Note, however, that the significant secretory response to glucose present in HIT-T15 cells underscores the lack of a clear requirement for PLC activation in order for glucose to initiate exocytosis). In other transformed cell lines, PLC appears to be constitutively activated (Giaever et al., 1993), which may obfuscate the delineation of any further response to physiological agonists such as glucose. In neonatal-streptozotocin diabetic rats, PLC activation was said (Morin et al., 1997) to be impaired, although the content of substrates for PLC (PIP and PIP_2) was also severely reduced. However, in the Goto-Kakizaki rat islet, we have been unable to find any intrinsic defect in PLC activation in response to glucose or 40 mM K^+ (J. Vadakekalam, M, Rabaglia, S. Metz, unpublished data).

Unfortunately there are no specific inhibitors of PLC with which to test the physiological importance of PLC activation. Conclusions derived via the use of nonspecific PLC inhibitors (e.g., mepacrine) or PKC inhibitors (e.g., staurosporine) have severe intrinsic limitations (Persaud et al., 1993); nonetheless, they have not always inhibited the insulin response to glucose (compare Wolf et al., 1990). Neomycin acts as a nonspecific inhibitor of PLC "activity," principally by binding to substrate, not the enzyme itself. U-73122 had been suggested to inhibit G-protein–activated phospholipase C in some cells. However, it did not have a *specific* effect in islets or β-TC3 lines (Alter et al., 1994). This failure was tentatively attributed by Hsu and colleagues to the metabolic instability of the drug (Chen and Hsu, 1995); however, in preliminary studies taking the latter variable into effect, we have not found this drug to be useful as a specific and selective inhibitor of PLC activity in islets (J. Vadakekalam, M. Rabaglia, S. Metz, unpublished data). Recently, this drug has been suggested to have direct effects on calcium fluxes, a fact that could explain its nonspecificity.

IV. ROLES OF PROTEIN KINASE C

Since previous overviews of the role of PKC in pancreatic islet function, the picture has become only moderately more clear. We now know that "protein kinase C" is a large gene family (for review, see Dekker et al., 1995), which can be subdivided into conventional (c) forms (α, β, and γ), the Ca^{2+}-unresponsive novel or "n" subfamily, the atypical or "a" family, and the μ-like or "m" group. These isoforms display differential responsiveness to calcium, DG, phosphatidylserine, arachidonic acid, and phorbol esters; conversely, only some can be inhibited by using various pharmacological probes or by "downregulation" using prolonged exposure to high concentrations of phorbol ester. Thus, since some isoforms of PKC found in insulin-secreting cells are not responsive to activation by DG, their status may not relate at all to the activation of PLC. This complexity makes it extremely difficult to make dogmatic statements about the "role of PKC" in islet function, especially if one assumes that various isoforms have different biological roles.

Several PKC isoforms have been identified in rodent islets or β-cells (Table 1). Age may modulate the expression of isoforms (Fletcher and Ways, 1991). In a number of studies (Ganesan et al., 1990; Zaitsev et al., 1995; Deeney et al., 1996; Yedovitzky et al., 1997), physiological agonists such as glucose have led to the translocation of one or more isoforms of PKC from the cytosol to the plasma membrane or to perinuclear sites; to this translocation has been ascribed a role in

Table 1. PKC Isoforms in Islet Cells

Isoforms Found	*Isoforms Not Found*	*Cells*	*Author/Year*[a]
PKC-α	PKC-β	Adult rat β-cells	Fletcher (1991), Ganesan (1990)
PKC-γ	PKC-β	Adult rat α-cells	Fletcher (1991)
PKC-ε	PKC-β	Adult rat δ-cells	Fletcher (1991)
	PKC-α, -γ	Fetal rat islets	Fletcher (1991)
PKC-α		Rat islet β-cell	Ganesan (1990)
PKC-α, -ε, -β_1		Mouse *ob/ob* islets	Zaitsev (1995), Arkhammar (1994)
PKC-δ, -α, -β, -ε, -ζ	PKC-γ	Insulinoma, islet β-cells	Knutson (1996), Knutson (1994)
PKC-α, -βII, -ε, -ζ, -λ, -μ		MIN6 cells	Tian (1996)
PKC-δ		RINm5F cells	Tian (1996)
Type III (PKC-α)	Type I, II (PKC-γ, -β)	Rat, rabbit β-cells; insulinoma cells	Onoda (1990)
Type II (PKC-β)	Type I (PKC-γ)	Rat, rabbit α-cells	Onoda (1990)
PKC-βII		Rat islet β-cells	Ito (1989)
	PKC-βII	Rat insulinoma cells	Ito (1989)
PKC-ι		RINm5F	Selbie (1993)
PKC-α, -δ, -ζ, -ε		Rat islets	Yedovitsky (1997)
PKC-α, -β_1		Mouse β-cells	Arkhammar (1994)

Note: [a]Author cited refers to first author, only.

Table 2. Some Sites of Action or Effects on Intracellular Signalling Inferred for PKC in Insulin-secreting Cells[a]

Action on (direction of effect)[b]	*Probe*[c]	*PKC Isoform(s) Putatively Involved*	*Cell Type*	*Author, Year*[d]
1. Apoptosis (↓)	Staurosporine	?	RINm5F	Sanchez-Margalet (1993)
2. cAMP accumulation (↑)	PE	PE-sensitive	Mouse islets	Thams (1988)
3. Ca^{2+} channels (↑)	PE	PE-sensitive	RINm5F	Platano (1996)
4. PLC (↓)	PE	PE-sensitive	Rat islets	Zawalich (1991)
	PE	PE-sensitive	RINm5F	Li (1990)
5. MAP kinase (↑)	PE	PE-sensitive	INS-1	Frödin (1995)
6. Depolarization, rise in $[Ca_i]$ (↓)(→)	PE	PE-sensitive	HIT cells	Hughes (1988, 1992)
(↑)	PE	PE-sensitive	RINm5F	Yada (1989)
7. Glucose- or vasopressin-increase induced in $[Ca^{2+}]_i$ (↓)	PE	PE-sensitive	HIT	Hughes (1992)
8. Capacitative Ca^{2+} entry (↑) (see number 7)	PE	PE-sensitive	RINm5F	Bode (1994)
9. Glucose-induced oscillations in $[Ca^{2+}]_i$ (convert to sustained response by inhibiting slow oscillations)	PE	PE-sensitive	*ob/ob* β-cells	Kindmark (1992)
10. K^+_{ATP} channels (↓ then ↑)	PE	PE-sensitive	RINm5F	Dunne (1994)
(↑)	PE, PKC inhibitors	PE-sensitive	HIT, RIN (excised patches)	Ribalet (1995)
11. Insulin biosynthesis (↑)	PE	PE-sensitive	Rat islets	Gwilliam (1993)
12. Glucagon gene expression (↑)	H-7	?	Rat islets	Yamato (1990)

Notes: [a] From studies since 1988.
[b] Arrows in parenthesis indicate that the deduced or presumed effect of PKC activation is to increase (↑), decrease (↓), or not change (→) the indicated biological activity.
[c] PE, phorbol ester.
[d] Author cited refers to first author, only.

mediating the effect of agonists or alternatively in inducing desensitization to agonists. In one study, PKC-δ was found to associate with the β-cell cytoskeleton, an effect reduced by fatty acid (Knutson and Hoenig, 1996) but an effect of glucose or carbachol was not readily apparent (Knutson and Hoenig, 1996).

The purported effects ascribed to the activation or inactivation of "PKC" are extremely diffuse. Some of this diffuseness may be attributed to the use of high concentrations of long-lasting (and therefore nonphysiological) activators of various isoforms of PKC, such as phorbol esters; use of the latter has been the typical screening maneuver to look for roles of PKC. Some of these putative effects of PKC described over the past 8 years are listed in Table 2; they should be viewed as unproven, but possibly physiological, actions of (one or more isoforms of) protein kinase C, whose roles will require additional study.

Recently various inhibitors, most of them pharmacological, have been described that provide increased specificity to earlier studies in which highly unspecific agents (such as staurosporine or H-7) were used to inhibit PKC. Using synthetic pseudosubstrate peptide inhibitors of *conventional* isoforms of PKC (Harris et al., 1996b), workers at King's College in London were able to detect a role for protein kinase C in cholinergic, but not in glucose-stimulated, insulin secretion. This finding would accord with a number of other studies (Persaud et al., 1989; Weng et al., 1993; Gao et al., 1994; Tian et al., 1996), which convincingly show that carbachol (or acetylcholine) increases DG content, and activates PKC, both effects that presumably contribute to the transient acute rise in cytosolic free calcium level as well as the delayed phase of insulin secretion. In contrast, Yedovitsky et al. (1997) suggested a role for PKC-α (a conventional isoform) in glucose-induced secretion, using peptide inhibitors of its translocation. Using other pharmacological inhibitors with differential sensitivities for various PKC isoforms, Jones's group was able to suggest (Harris et al., 1996a; Persaud and Jones, 1996) that glucose-induced insulin secretion may be dependent, at least in part, upon the activation of an *atypical*, possibly Ca^{2+}-independent isoform of PKC. This finding would support other data from the same laboratory (Harris et al., 1996a) and that of others (Yedovitsky et al., 1997), suggesting that phorbol ester-*insensitive* isoforms of PKC also may be involved in glucose-stimulated insulin release, and would accord with earlier data suggesting that the downregulation of phorbol ester-sensitive isoforms of PKC does not materially affect glucose-stimulated insulin release (at least during static incubations of rat islets). However, an alternative explanation for earlier negative studies is that such downregulation may differentially affect both phases of insulin release as seen during *perifusions* of mouse islets (Thams et al., 1990)—that is, it potentiates the first phase and inhibits the second phase of insulin release (Thams et al., 1990). These effects might cancel each other out in static incubations.

It is widely believed that a principal role of PKC activation, and its concomitant calcium-dependent translocation to the plasma membrane, may be the potentiation of calcium-induced exocytosis. In one of the most interesting studies on this question (Jones et al., 1989; Jones et al., 1992), Jones and his group reported that the secretory response to persistent elevations in intracellular calcium levels (studied in electrically permeabilized islets) was only transient; however, the addition of phorbol ester not only increased the maximal response but maintained elevated rates of secretion. Therefore, PKC may well play a role in potentiating the second phase of nutrient-induced insulin secretion, as well as in maintaining its duration, perhaps thereby obviating the "third phase" or desensitization phase of prolonged islet stimulation, as described by Grodsky and colleagues.

As indicated above, certain physiological and pharmacological stimuli to the β-cell do increase the intracellular translocation of various isoforms of PKC; additionally, insulin secretagogues are known to increase the phosphorylation of appropriate PKC substrates (Easom et al., 1990; Calle et al., 1992; Gao et al., 1994). However, divergent data have been reported (Thomas et al., 1989; Persaud et al, 1991; Persaud and Jones, 1996) when insulin secretion is stimulated by glyceraldehyde, a triose metabolite of glucose. Therefore, the case for a requirement for PKC in *nutrient-induced* secretion remains incomplete. Note also that the activation of PKC (at least as judged by its translocation) tends to be transient; this finding accords with other findings in endocrine cells, suggesting that there is a pool of DG (specifically the stearoyl-arachidonyl molecular species) that is derived by PLC activation, the effects of which are rapidly terminated. However, if this formulation is true, how can it accord with the findings that PKC activation appears to have stimulatory effects on insulin secretion in a *delayed and persistent* fashion (even after the activator has disappeared from the cells; Deeney et al., 1996)? To resolve this conundrum, more studies are needed to identify the specific substrates for the phosphorylation induced by the various PKC subspecies, and to determine whether the reversal of PKC translocation and/or the rapid termination of DG production is accompanied by reversal of the phosphorylation of these substrates.

If PKC activation does have a delayed effect to potentiate insulin release, this finding must still be put in a physiological framework. The delayed effects of glucose on insulin secretion appear to be complex and are comprised of more than one phenomenon; namely, the second phase of insulin secretion (which is rapidly reversible upon removal of glucose) and a time-dependent potentiation (TDP) or priming (Zawalich and Zawalich, 1996) of subsequent insulin release (an effect that is more persistent). It remains unclear how these effects are related to the depolarization-independent "calcium-potentiating" effects of glucose in the pancreatic islet, which were originally described by groups in Belgium and Japan (see refs. in Vadakekalam et al., 1997). As long as these phenomena are only operationally defined and their molecular basis remains moot, the involvement of PKC in any of them will likewise remain unsettled. Two groups have suggested that the calcium-potentiating effect of glucose are dependent upon PKC, largely based on the use of nonspecific inhibitors of PKC. However, Henquin's group reached different conclusions. My colleagues and I recently reexamined this question (Vadakekalam et al., 1997b). We found that under certain circumstances, the calcium-potentiating effects of glucose *were* accompanied by evidence of activation of PLC (and presumably the activation of PKC); however, the activation of PLC could be shown to be an epiphenomenon, since under other conditions this effect of glucose on insulin release could be observed in the total absence of PLC activation. To further complicate matters, there is now evidence that the calcium-potentiating effects of glucose can be subdivided into two effects, one of which is dependent upon GTP and is seen in glucose-primed islets, and the other, apparently independent of GTP and seen in the unprimed state (Vadakekalam et al., 1997b). Very recently, potentiating effects of glucose have been described that do not require the presence of concomitant elevations in $[Ca^{2+}]_i$. Obviously more work is needed in these areas.

More studies are also needed to identify other physiological endogenous activators of various PKC isoforms in β-cells, in addition to DG. We previously suggested (based on a limited number of direct and indirect studies) that arachidonic acid may be one such activator in the β-cell (as it is in other cells); however, these conclusions were not confirmed by Howell, Jones and colleagues, albeit using a slightly different protocol (Band et al., 1992; Basudev et al., 1992). More recent studies (Thams et al., 1993) suggest that arachidonic acid does indeed have important effects on PKC in the islet and imply that the actions of this fatty acid should be reexamined in this context. In fact, it has recently been suggested (Owada et al., 1999) that arachidonic acid (presumably released via Ca^{2+}-independent phospholipase A_2) inhibits Na^+, K^+-ATPase in the beta cells of *ob/ob* mice; this effect, which seemingly involved metabolism of the arachidonate by 12-lipoxygenase, was mediated by activation of PKC.

V. ARACHIDONIC ACID RELEASE, LYSOPHOSPHOLIPIDS, AND PHOSPHOLIPASE A_2 ACTIVATION

Arachidonic acid (AA) can be released by mechanisms other than the direct activation of PLA_2. These include the activation of PLC (with the subsequent hydrolysis of DG), as well as the activation of PLD (with the subsequent hydrolysis of phosphatidic acid; PA). For the ease of discussion, in the following presentation I will ignore: (1) the possibility of multiple enzymatic sources of its release; and (2) recent data suggesting that even when cellular release of arachidonic acid derives primarily via PLA_2 activation, the AA that is liberated may be compartmentalized into "pools," which may have differing cellular effects. The different "pools" may arise as a result of being released by different PLA_2 isoforms, and may be differentially coupled to pathways of the oxidative metabolism of AA (i.e., cyclo-oxygenase I vs. cyclo-oxygenase II vs. lipoxygenase; see Marshall et al., 1997; Reddy and Herschman, 1997). Also, some "pools" of AA are retained intracellularly whereas others are released into the extracellular space; some are released rapidly, but others only in a delayed fashion. Thus, it is not at all surprising that measurements of *total* AA *release* into *the medium* have revealed some discrepancies with rates of exocytosis.

As summarized previously, a growing number of isoforms of PLA_2 have been localized to β-cells or pancreatic islets. Although it is widely held that it is the intracellular forms that should play a predominant role in stimulus-secretion coupling, it is premature to exclude secretory forms of PLA_2 as being of relevance to the pancreatic islet (see earlier discussion). One interesting hypothesis is that PLA_2 may have a specific role in exocytosis due to the potential localization of one or more isoforms of PLA_2 in the secretory granules themselves, thereby placing them in a strategic location to provide high local concentrations of arachidonic acid (and lysophospholipids) to participate in granule fusion, fission, or other events central to hormone release. Preparation of an adequately pure secretory granule fraction to study this issue is not facile, and the conclusion that PLA_2 may be associated with secretory granules in certain cells has been controverted and remains uncertain in β-cells.

In passing, it is noted that a PLA_2-activating protein (PLAP) has recently been described. Its possible presence and/or effects in the β-cell have not been studied. However, it is intriguing that there is sequence homology between PLAP and the β subunits of G-proteins, raising the interesting possibility that some of the reported effects of G-protein β subunits might be due to an ability to activate PLA_2 directly via homology to PLAP. In pancreatic acinar tissue, PLAP has been linked to calcium oscillations and amylase secretion in conjunction with release of arachidonate (Tsunoda and Owyang, 1994).

In the islet as in some other cells, the rate of reuptake and resterification of arachidonic acid (and lysophospholipids) may be a rate-limiting step in their intracellular availability, in addition to the rate of phospholipase activity. It has recently been suggested (Ma et al., 1996) that interleukin-1, acting via generation of nitric oxide, enhances the release of arachidonic acid and/or its metabolites in part by impairing the esterification of arachidonate into islet phospholipids, thereby mimicking the activation of PLA_2. However, there are as yet no data unequivocally demonstrating a role for the released arachidonic acid in the β-cell death associated with prolonged exposure to interleukin. Likewise, IL-1β inhibits PLC in islets (Vadakekalam et al., 1997) but has additional effects to block insulin secretion which cannot be easily explained by phospholipase inhibition.

A number of recent studies in various nonislet cells have examined the relationship between the release of arachidonate, and exocytosis (Churcher et al., 1990; Morgan and Burgoyne, 1990; Cockcroft, 1991; Vitale et al., 1994; Coorssen, 1996). In several of these studies, discrepancies between arachidonate release and hormone secretion were noted at the high and low ends of the dose-response curves; the authors therefore concluded that no requisite role exists for arachidonate release in the process of exocytotic secretion. However, some readers might be more impressed by the *similarities* in these studies in the dose-response characteristics of the two phenomena; the disparities at the high and low ends of the dose-response curves might well be explicable by threshold and saturation kinetics, as well as by the fact (as noted earlier) that measurement of *total arachidonic acid release* provides very little information on specific compartments and pools and, in fact, may fail to detect significant release of free arachidonic acid from phospholipids in β-cells. In a recent publication (Ray et al., 1993), it was noted that botulinum toxin, which inhibits acetylcholine release from PC12 cells, also inhibited arachidonic acid release; the effect of the toxin on neurotransmitter secretion could be prevented by increasing the availability of arachidonic acid.

Thus, a modulatory role (and perhaps a mediatory role) for arachidonic acid in late steps of stimulus-secretion coupling remains plausible.

There are a few recent pharmacological studies designed to test this possibility in insulin-secreting cells. A purportedly "specific" inhibitor of $cPLA_2$ has been demonstrated (Loweth et al., 1996) to inhibit glucose-induced insulin secretion from isolated rat islets, suggesting that this enzyme might play a role in physiological glucose-induced insulin release; however, the specificity of this inhibitor has been questioned (Thams and Capito, 1997). Other studies using an inhibitor of the plasmalogen-selective PLA_2 have suggested a role for this phospholipase as well; as often happens, the specificity of this inhibitor as well has been recently questioned (Balsinde and Dennis, 1996). Another pharmacological inhibitor of PLA_2 ([P-amylcinnamoyl] anthranilic acid) also inhibits nutrient-induced (Konrad et al., 1992b), but not receptor-mediated insulin secretion (Chen et al., 1994); the exact specificity of this agent is still uncertain, but it appears not to inhibit PLC or glucose oxidation (Konrad et al., 1992b).

Recent studies (Kowluru and Metz, 1994a) showing that arachidonic acid potentially can increase the activation of GTP-binding proteins by increasing the binding of GTP as well as impeding its hydrolysis, suggest a new approach to unraveling the potential mechanisms whereby arachidonic acid may act. A number of other interesting effects of arachidonic acid as well as of lysophospholipids have been described. In liver cells or islet cells, either product can release calcium from mitochondria or other intracellular sites (Metz, 1988; Rustenbeck and Lenzen, 1992); interestingly, lysophospholipids may selectively accumulate in mitochondria (Rustenbeck et al., 1994). These effects could potentially contribute to a rise in cytosolic free calcium levels. However, such an effect on mitochondria would presumably be associated with a *reduction* in intramitochondrial calcium concentration. Since the latter has been suggested to be a coupling signal linking glycolysis to oxidative metabolism of glucose, an effect of AA to release mitochondrial Ca^{2+} stores might actually be viewed as a negative modulatory event.

It should be noted that in many studies, arachidonic acid was added exogenously to cells already having a normal (i.e., replete) complement of endogenous arachidonic acid; it may be that the further addition of exogenous AA led to supraphysiological levels. This may explain some of the reported inhibitory effects (e.g., Gong et al., 1992) of exogenous arachidonate. The putative role of arachidonic acid in activating PKC remains controversial (see earlier discussion). The direct effects of arachidonic acid to promote insulin secretion are also seen with several other polyunsaturated fatty acids (Konrad et al., 1996). Such acute *stimulatory* effects are probably unrelated to the acute stimulatory effects tentatively ascribed to certain acyl-CoA esters, or to the *suppressive* effects of a more subacute or chronic exposure to high fatty acid levels (Elks, 1993; Milburn et al., 1995).

VI. POTENTIAL ROLE OF PHOSPHOLIPASE D

Since the subject was recently reviewed, little progress has been made in understanding the potential role of PLD in the islet, although the combination of glucose and a cholinergic agonist (but not either one alone), can activate the PC-directed PLD in β-cells (see earlier discussion). Isoforms of PLD are just now being described in nonendocrine cells and no data exist on various subtypes in β-cells.

VII. CONCLUSION

The burgeoning literature on phospholipases continues to bewitch students of insulin secretion, but also to discourage them as the complexity of this field increases. It seems likely that one or more phospholipases is intimately involved in the initiation or potentiation of insulin release induced by certain receptor agonists (especially cholinergic agents) and might play roles in the potentiating effects of glucose on such secretion. However, an unequivocal role in the direct (initiating) effects of glucose has not been defined. It is worth bearing in mind, as one interprets the extant literature, that duplicative or overlapping pathways culminating in insulin release probably exist in the β-cell. This might make it difficult to unmask a critical role for a single signal transduction pathway, even one that is widely accepted. For example, cAMP is widely accepted to be a physiological modulator of insulin release, yet one study using inhibitors of cAMP-directed protein kinase failed to find an effect on secretion; Harris et al., 1997). In all likelihood, definitive answers to the question of the physiological role(s) of phospholipase activation will require new molecular approaches (including antisense "knock-out," and transient expression of particular PLA_2 isoforms in models of phospholipase deficiency or as part of attempts to "bioengineer" β-cells using nonendocrine cells). Cautious and thoughtful interpretation of the findings will be needed to reconcile all the results in this difficult field.

ACKNOWLEDGMENTS

The studies by the author were funded by the Veterans Administration and the National Institutes Health (DK37312).

REFERENCES

Alter, C.A., Amagasu, M., Shah, K., Jolly, Y.C., Major, C. and Wolf, B.A. (1994). U-73122 does not specifically inhibit phospholipase C in rat pancreatic islets and insulin-secreting β-cell lines. Life Sci. 54, 107–112.

Ancian, P., Lambeau, G., Mattéi, M.-G. and Lazdunski, M. (1995). The human 180-kDa receptor for secretory phospholipases A_2. J. Biol. Chem. 270, 8963–8970.

Arkhammar, P., Juntti-Berggren, L., Larsson, O., Welsh, M., Nånberg, E., Sjöholm, A., Köhler, M., Berggren, P.-O. (1994). Protein kinase C modulates the insulin secretory process by maintaining a proper function of the β-cell voltage-activated Ca^{2+} channels. J. Biol. Chem. 269, 2743–2749.

Balsinde, J. and Dennis, E.A. (1996). Bromoenol lactone inhibits magnesium-dependent phosphatidate phosphohydrolase and blocks triacylglycerol biosynthesis in mouse $P388D_1$ macrophages. J. Biol. Chem. 271, 31937–31941.

Band, A.M., Jones, P.M. and Howell, S.L. (1992). Arachidonic acid-induced insulin secretion from rat islets of Langerhans. J. Mol. Endocrinol. 8, 95–101.

Basudev, H., Jones, P.M., Persaud, S.J. and Howell, S.L. (1992). Arachidonic acid induces phosphorylation of an 18 kDa protein in electrically permeabilised rat islets of Langerhans. FEBS Lett. 296, 69–72.

Biden, T.J., Davison, A.G.M. and Prugue, M.L. (1993) Regulation by membrane potential of phosphatidylinositol hydrolysis in pancreatic islets. J. Biol. Chem. 268, 11065-11072.

Biden, T.J., Vallar, L. and Wollheim, C.B. (1988). Regulation of inositol 1,4,5-trisphosphate metabolism in insulin-secreting RINm5F cells. Biochem. J. 251, 435–440.

Blondel, O., Takeda, J., Janssen, H., Seino, S. and Bell, G.I. (1995). Sequence and functional characterization of a third inositol trisphosphate receptor subtype, IP3R-3, expressed in pancreatic islets, kidney, gastrointestinal tract, and other tissues. J. Biol. Chem. 268, 11356–11363.

Calle, R., Ganesan, S., Smallwood, J.I. and Rasmussen, H. (1992). Glucose-induced phosphorylation of myristoylated alanine-rich C kinase substrate (MARCKS) in isolated rat pancreatic islets. J. Biol. Chem. 267, 18723–18727.

Capito, K., Hansen, S.E. and Thams, P. (1996). Production of [^{3}H]choline-labelled metabolites from endogenously ^{3}H-labelled phosphatidylcholine in mouse pancreatic islets. J. Mol. Endocrinol. 17, 101–107.

Chen, M., Yang, Z.-D., Naji, A. and Wolf, B.A. (1996). Identification of calcium-dependent phospholipase A_2 isoforms in human and rat pancreatic islets and insulin secreting β-cell lines. Endocrinology 137, 2901–2909.

Chen, T.-H. and Hsu, W.H. (1995). U-73122 inhibits carbachol-induced increases in $[Ca^{2+}]_i$, IP_3, and insulin release in β-TC3 cells. Life Sci. 56, 103–108.

Chen, T.-H., Lee, B. and Hsu, W.H. (1994). Arginine vasopressin-stimulated insulin secretion and elevation of intracellular Ca^{++} concentration in rat insulinoma cells: Influences of a phospholipase C inhibitor 1-[6-[[17β-methoxyestra-1,3,5(10)-trien-17-yl]amino]hexyl]-1H-pyrrol-2,5-dione (U-73122) and a phospholipase A_2 inhibitor N-(ρ-Amylcinnamoyl)anthranilic acid. J. Pharmacol. Exp. Ther. 270, 900–904.

Churcher, Y., Allan, D. and Gomperts, B.D. (1990). Relationship between arachidonate generation and exocytosis in permeabilized mast cells. Biochem. J. 266, 157–163.

Cockcroft, S. (1991). Relationship between arachidonate release and exocytosis in permeabilized human neutrophils stimulated with formylmethionyl-leucyl-phenylalanine (fMetLeuPhe), guanosine 5′-[γ-thio]triphosphate (GTP[S]) and Ca^{2+}. Biochem. J. 275, 127–131.

Coorssen, J.R. (1996). Phospholipase activation and secretion: Evidence that PLA_2, PLC, and PLD are not essential to exocytosis. Am. J. Physiol. (Cell. Physiol. 39) 270, C1153–C1163.

Deeg, M.A. and Verchere, C.B. (1997). Regulation of glycosylphosphatidylinositol-specific phospholipase D secretion from βTC3 cells. Endocrinology 138, 819–826.

Deeney, J.T., Cunningham, B.A., Chheda, S., Bokvist, K., Juntti-Berggren, L., Lam, K., Korchak, H.M., Corkey, B.E. and Berggren, P.-O. (1996). Reversible Ca^{2+}-dependent translocation of protein kinase C and glucose-induced insulin release. J. Biol. Chem. 271, 18154–18160.

Dekker, L.V., Palmer, R.H. and Parker, P.J. (1995). The protein kinase C and protein kinase C related gene families. Curr. Opin. Struct. Biol. 5, 396–402.

Dunlop, M., Metz, S. and Clark, S. (1994). Guanine nucleotide-dependent β-cell replication by phospholipase D and A_2 products. In: *Diabetes*. (Baba, S. and Kaneko, T., Eds.), pp. 198–202. Elsevier Science B.V., Amsterdam.

Easom, R.A., Landt, M., Colca, J.R., Hughes, J.H., Turk, J. and McDaniel, M. (1990). Effects of insulin secretagogues on protein kinase C-catalyzed phosphorylation of an endogenous substrate in isolated pancreatic islets. J. Biol. Chem. 265, 14938–14946.

Eddlestone, G.T. (1995). ATP-sensitive K channel modulation by products of PLA_2 action in the insulin-secreting HIT cell line. Am. J. Physiol. 268 (Cell Physiol 37), C181–190.

Efanov, A.M., Zaitsev, S.V. and Berggren, P.-O. (1997). Inositol hexakisphosphate stimulates non-Ca^{2+}-mediated and primes Ca^{2+}-mediated exocytosis of insulin by activation of protein kinase C. Proc. Natl. Acad Sci. USA Vol: 94, 4435–4439.

Elks, M.L. (1993). Chronic perifusion of rat islets with palmitate suppresses glucose-stimulated insulin release. Endocrinology 133, 208–214.

Fan, Z. and Makielski, J.C. (1997). Anionic phospholipids activate ATP-sensitive potassium channels. J. Biol. Chem. 272, 5388–5395.

Farese, R.V., DiMarcom P.E., Barnesm D.E., Sabir, M.A., Larson, R.E., Davis, J.S. and Morrison, A.D. (1986). Rapid glucose-dependent increases in phosphatidic acid and phosphoinositides in rat pancreatic islets. Endocrinology 118, 1498–1503.

Fletcher, D.J. and Ways, D.K. (1991). Age-dependent expression of protein kinase C isoforms in rat islets. Diabetes 40, 1496–1503.

Ganesan, S., Calle, R., Zawalich, K., Smallwood, J.I., Zawalich, W.S. and Rasmussen, H. (1990). Glucose-induced translocation of protein kinase C in rat pancreatic islets. Proc. Natl. Acad. Sci. USA 87,9893–9897.

Gao, Z.-Y., Gilon, P. and Henquin, J.-C. (1994). The role of protein kinase-C in signal transduction through vasopressin and acetylcholine receptors in pancreatic B-cells from normal mouse. Endocrinology 135, 191–199.

Giæver, A.K., Haukland, H.H., Bertinussen, A., Vonen, B., Malm, D., Huseby, N.E. and Florholmen, J. (1993) Phosphoinositide metabolism in a polyoma-BK-virus-transformed pancreatic islet cell line: Evidence for constitutively activated phospholipase C. Int. J. Cancer 53, 80–86.

Gong, M.C., Fuglsang, A., Alessi, D., Kobayashi, S., Cohen, P., Somlyo, A.V. and Somlyo, A. P. (1992). Arachidonic acid inhibits myosin light chain phosphatase and sensitizes smooth muscle to calcium. J. Biol. Chem. 267, 21492–21498.

Gonzalez, R. and Rana, R.S. (1993). Vasopressin and oxytocin but not glucose stimulate hydrolysis of phosphatidylinositol and phosphatidyl choline in a hamster insulinoma. Life Sci. 53, 1179–1183.

Gross, R.W., Ramanadham, S., Kruszka, K.K., Han, X., and Turk, J. (1993). Rat and human pancreatic islet cells contain a calciumion independent phospholipase A_2 activity selective for hydrolysis of arachidonate which is stimulated by adenosine triphosphate and is specifically localized to islet β-cells. Biochemistry 32, 327–336.

Harris, T.E., Persaud, S.J. and Jones, P.M. (1997). Pseudosubstrate inhibition of cyclic AMP-dependent protein kinase in intact pancreatic islets: Effects on cyclic AMP-dependent and glucose-dependent insulin secretion. Biochem. Biophys. Res. Commun. 232, 648–651.

Harris, T.E., Persaud, S.J. and Jones, P.M. (1996a). Atypical isoforms of PKC and insulin secretion from pancreatic β-cells: Evidence using Gö 6976 and Ro 31-8220 as PKC inhibitors. Biochem. Biophys. Res. Commun. 227, 672–676.

Harris, T.E., Persaud, S.J., Saermark, T. and Jones, P.M. (1996b). A myristoylated pseudosubstrate peptide inhibitor of protein kinase C: effects on glucose- and carbachol-induced insulin secretion. Mol. Cell. Endocrinol. 121, 133–141.

Hay, J.C., Fisette, P.L., Jenkins, G.H., Fukami, K., Takenawa, T., Anderson, R.A. and Martin, T.F.J. (1995). ATP-dependent inositide phosphorylation required for Ca^{2+}-activated secretion. Nature 374, 173–177.

Hazen, S.L. and Gross, R.W. (1993). The specific association of a phosphofructokinase isoform with myocardial calcium-independent phospholipase A_2. J. Biol. Chem. 268, 9892–9900.

Hedeskov, C.J., Thams, P., Gembal, M., Malik, T. and Capito, K. (1991). Ca^{2+}-and ATP-dependent reversible inactivation of pancreatic islet phosphoinositide-specific phospholipase C activity. Mol. Cell. Endocrinol. 82, 81–88.

Hughes, S.J., Chalk, J.G. and Ashcroft, S.J.H. (1990). The role of cytosolic free Ca^{2+} and protein kinase C in acetylcholine-induced insulin release in the clonal β-cell line, HIT-T15. Biochem. J. 267, 227–232.

Illenberger, D., Schwald, F., Pimmer, D., Binder, W., Maier, G., Dietrich, A., and Gierschik, P. (1998). Stimulation of phospholipase C-beta2 by the Rho GTPases Cdc42Hs and Rac1. EMBO J. 17, 6241–6249.

Infante, J.P. (1985). Defective synthesis of glycerophosphorylcholine in murine muscular dystrophy; the primary molecular lesion? FEBS Lett. 186, 205–210.

Ito, A., Saito, N., Taniguchi, H., Chiba, T., Kikkawa, U., Saitoh, Y. and Tanaka, C. (1989). Localization of βII subspecies of protein kinase C in β-cells. Diabetes 38, 1005-1011.

Jolly, Y.C., Major, C., Wolf, B.A. (1993). Transient activation of calcium-dependent phospholipase A_2 by insulin secretogogues in isolated pancreatic islets. Biochemistry 32, 12209–12217.

Jonas, J.-C. and Henquin, J.-C. (1996). Possible involvement of a tyrosine kinase-dependent pathway in the regulation of phosphoinositide metabolism by vanadate in normal mouse islets. Biochem. J. 315, 49–55.

Jones, P.M., Persaud, S.J. and Howell, S.L. (1989). Time-course of Ca^{2+}-induced insulin secretion from perifused, electrically permeabilised islets of Langerhans: Effects of cAMP and phorbol ester. Biochem. Biophys. Res. Commun. 3, 998–1003.

Jones, P.M., Persaud, S.J. and Howell, S.L. (1992). Ca^{2+}-induced insulin secretion from electrically permeabilized islets. Biochem. J. 285, 973–978.

Kelley, G.G., Zawalich, K.C. and Zawalich, W.S. (1995). Synergistic interaction of glucose and neurohumoral agonists to stimulate islet phosphoinositide hydrolysis. Am. J. Physiol. 269, E575–E582.

Knutson, K. and Hoenig, M. (1994). Identification and subcellular characterization of protein kinase-C isoforms in insulinoma β-cells and whole islets. Endocrinology 135, 881–886.

Knutson, K. and Hoenig, M. (1996). Regulation of distinct pools of protein kinase C δ in beta cells. J. Cell. Biochem. 60,130–138.

Konrad, R.J., Jolly, Y.C., Major, C. and Wolf, B.A. (1992a). Carbachol stimulation of phospholipase A_2 and insulin secretion in pancreatic islets. Biochem. J. 287, 283–290.

Konrad, R.J., Jolly, Y.C., Major, C. and Wolf, B.A. (1992b). Inhibition of phospholipase A_2 and insulin secretion in pancreatic islets. Biochim. Biophys. Acta 1135, 215–220.

Konrad, R.J., Jolly, C. and Wolf, B.A. (1991). Glucose and carbachol synergistically stimulate phosphatidic acid accumulation in pancreatic islets. Biochem. Biophys. Res. Commun. 180, 960–966.

Konrad, R.J., Stoller, J.Z., Gao, Z.-Y. and Wolf, B.A. (1996).Eicosapentaenoic acid (C20:5) augments glucose-induced insulin secretion from β-TC3 insulinoma cells. Pancreas 13, 253–258.

Kowluru, A. and Metz, S.A. (1994a). Regulation of guanine–nucleotide binding proteins in islet subcellular fractions by phospholipase-derived lipid mediators of insulin secretion. Biochim. Biophys. Acta 1222, 360–368.

Kowluru, A. and Metz, S.A. (1994b). Stimulation by prostaglandin E_2 of a high-affinity GTPase in the secretory granules of normal rat and human pancreatic islets. Biochem. J. 297, 399–406.

Kowluru, A., Seavey, S.E., Li, G., Sorenson, R.L., Weinhaus, A.J., Nesher, R., Rabaglia, M.E., Vadakekalam, J., Metz, S.A. (1996). Glucose- and GTP-dependent stimulation of the carboxyl methylation of CDC42 in rodent and human pancreatic islets and pure β cells. J. Clin. Invest. 98, 540–555.

Landt, M., Easom, R.A., Colca, J.R., Wolf, B.A., Turk, J., Mills, L.A. and McDaniel, M.L. (1992). Parallel effects of arachidonic acid on insulin secretion, calmodulin-dependent protein kinase activity and protein kinase C activity in pancreatic islets. Cell. Calcium 13, 163–172.

Li, G., Milani, D., Dunne, M.J., Pralong, W.-F., Theler, J.-M., Petersen, O.H. and Wollheim, C.B. (1991). Extracellular ATP causes Ca^{2+}-dependent and -independent insulin secretion in RINm5F cells. J. Biol. Chem. 266, 3449–3457.

Li, G., Pralong, W.-F., Pittet, D., Mayr, G.W., Schlegel, W. and Wollheim, C.B. (1992). Inositol tetrakisphosphate isomers and elevation of cytolic Ca^{2+} in vasopressin-stimulate insulin-secreting RINm5F cells. J. Biol. Chem. 267, 4349–4356.

Lomasney, J.W., Cheng, H.-F., Wang, L.-P., Kuan, Y.-S., Liu, S.-M., Fesik, S.W., King, K. (1996). Phosphatidylinositol 4,5-bisphosphate binding to the pleckstrin homology domain of phospholipase C-δ1 enhances enzyme activity. J. Biol. Chem. 271, 25310–25326.

Loweth, A.C., Scarpello, J.H.B. and Morgan, N.G. (1995). Phospholipase A_2 expression in human and rodent insulin-secreting cells. Mol. Cell. Endocrinol. 112, 177–183.

Loweth, A.C., Scarpello, J.H.B. and Morgan, N.G. (1996). A specific inhibitor of cytosolic phospholipase A_2 activity, $AACOCF_3$, inhibits glucose-induced insulin secretion from isolated rat islets. Biochem. Biophys. Res. Commun. 218, 423–427.

Ma, Z., Ramanadham, S., Corbett, J.A., Bohrer, A., Gross, R.W., McDaniel, M.L. and Turk, J. (1996). Interleukin-1 enhances pancreatic islet arachidonic acid 12-lipoxygenase product generation by increasing substrate availability through a nitric oxide-dependent mechanism. J. Biol. Chem. 271, 1029–1042.

Ma, Z., Ramanadham, S., Kempe, K., Chi, X. S., Ladenson, J. and Turk, J. (1997). Pancreatic islets express a Ca^{2+}-independent phospholipase A_2 enzyme that contains a repeated structural motif homologous to the integral membrane protein binding domain of ankyrin. J. Biol. Chem. 272, 11118–11127.

Marshall, L. A., Bolognese, B., Winkler, J.D. and Roshak, A. (1997). Depletion of human monocyte 85-kDa phospholipase A_2 does not alter leukotriene formation. J. Biol. Chem. 272, 759–765.

Martin, T.F.J., Hay, J.C., Bannerjee, A., Barry, V.A., Ann, K., Yom, H.-C., Porter, B.W. Kowalchyk, J.A. (1995). Late ATP-dependent and Ca^{++}-activated steps of dense core granule exocytosis. Cold Spring Harb. Symp. Quant. Biol. Vol LX, 197–204.

Metz, C.N., Zhang, Y., Guo, Y., Tsang, T.C., Kochan, J.P., Altszuler, N. and Davitz, M.A. (1991). Production of the glycosylphosphatidylinositol-specific phospholipase D by the islets of Langerhans. J. Biol. Chem. 266, 17733–17736.

Metz, S.A. (1988). Mobilization of cellular Ca^{2+} by lysophospholipids in rat islets of Langerhans. Biochim. Biophys. Acta 968, 239–252.

Metz, S.A. (1991). The pancreatic islet as Rubik's Cube. Diabetes 40, 1565–1573.

Milburn, J.L. Jr., Hirose, H., Lee, Y.H., Nagasawa, Y., Ogawa, A., Ohneda, M., BeltrandelRio, H., Newgard, C.B., Johnson, J.H. and Unger, R.H. (1995). Pancreatic β-cells in obesity. J. Biol. Chem. 270, 1295–1299.

Morgan, A. and Burgoyne, R.D. (1990). Stimulation of Ca^{2+}-independent catecholamine secretion from digitonin-permeabilized bovine adrenal chromaffin cells by guanine nucleotide analogues. Biochem. J. 269, 521–526.

Morin, L., Giroix, M.-H., Gangnerau, M.-N., Bailbé and Portha, B. (1997).Impaired phosphoinositide metabolism in glucose-incompetent islets of neonatally streptozotocin-diabetic rats. Am. J. Physiol. (Endocrinol Metab 35) 272, E737–E745.

Müller, M., Szewczyk, A., De Weille, J.R. and Lazdunski, M. (1992). ATP-sensitive K^+ channels in insulinoma cells are activated by nonesterified fatty acids. Biochemistry 31, 4656–5661.

Öberg, C., Waltenberger, J., Claesson-Welsh, L. and Welsh, M. (1994). Expression of protein tyrosine kinases in islet cells: Possible role of the Flk-1 receptor for β-cell maturation from duct cells. Growth Factors 10, 115–126.

Onoda, K., Hagiwara, M., Hachiya, T., Usuda, N., Nagata, T. and Hidaka, H. (1990). Different expression of protein kinase C isozymes in pancreatic islet cells. Endocrinology 126, 1235–1240.

Owada, S., Larsson, O., Arkhammar, P., Katz, A.I., Chibalin, A.V., Berggren, P.-O., and Bertorello, A.M. (1999). Glucose decreases Na^+, K^+-ATPase activity in pancreatic β-cells. J. Biol. Chem. 274, 2000–2008.

Parker, K.J., Jones, P.M., Hunton, C.H., Persaud, S.J., Taylor, C.G. and Howell, S.L. (1996). Identification and localisation of a type IV cytosolic phospholipase A_2 in rat pancreatic β-cells. J. Mol. Endocrinol. 17, 31–43.

Persaud, S.J. and Jones, P.M. (1996). Inhibition of glucose-stimulated insulin secretion by Ro 31-8220, a protein kinase C inhibitor. Endocrine 3, 285–289.

Persaud, S.J., Jones, P.M. and Howell, S.L. (1991). Activation of protein kinase C is not required for glyceraldehyde-stimulated insulin secretion from rat islets. Biochim. Biophys. Acta 1095, 183–185.

Persaud, S.J., Jones, P.M. and Howell, S.L. (1993). Staurosporine inhibits protein kinases activated by Ca^{2+} and cyclic AMP in addition to inhibiting protein kinase C in rat islets of Langerhans. Mol. Cell. Endocrinol. 94, 55–60.

Persaud, S.J., Jones, P.M., Sugden, D. and Howell, S.L. (1989). The role of protein kinase C in cholinergic stimulation of insulin secretion form rat islets of Langerhans. Biochem. J. 264, 753–758.

Peter-Riesch, B., Fathi, M., Schlegel, W. and Wollheim, C.B. (1988). Glucose and carbachol generate 1,2-diacylglycerols by different mechanisms in pancreatic islets. J. Clin. Invest. 81, 1154–1161.

Pike, L.J. and Casey, L. (1996). Localization and turnover of phosphatidylinositol 4,5-bisphosphate in caveolin-enriched membrane domains. J. Biol. Chem. 271, 26453–26456.

Popova, J.S., Garrison, J.C., Rhee, S.G. and Rasenick, M.M. (1997). Tubulin, G_q, and phosphatidylinositol 4,5-bisphosphate interact to regulate phospholipase $C\beta_1$ signaling. J. Biol. Chem. 272, 6760–6765.

Ramanadham, S., Wolf, M.J., Jett, P.A., Gross, R.W. and Turk, J. (1994). Characterization of an ATP-stimulatable Ca^{2+}-independent phospholipase A_2 from clonal insulin-secreting HIT cells and rat pancreatic islets: a possible molecular component of the β-cell fuel sensor. Biochemistry 33, 7442–7452.

Ray, P., Berman, J.D., Middleton, W. and Brendle, J. (1993). Botulinum toxin inhibits arachidonic acid release associated with acetylcholine release from PC12 cells. J. Biol. Chem. 268, 11057–11064.

Reddy, S.T. and Herschman, H.R. (1997). Prostaglandin synthase-1 and prostaglandin synthase-2 are coupled to distinct phospholipases for the generation of prostaglandin D_2 in activated mast cells. J. Biol. Chem. 272, 3231–3237.

Rustenbeck, I. and Lenzen, S. (1992).Effect of lysophospholipids, arachidonic acid and other fatty acids on regulation of Ca^{2+} transport in permeabilized pancreatic islets. Cell. Calcium 13, 193–202.

Rustenbeck, I., Matthies, A. and Lenzen, S. (1994). Lipid composition of glucose-stimulated pancreatic islets and insulin-secreting tumor cells. Lipids 29, 685–692.

Schiavo, G., Gu, Q.-M., Prestwich, G.D., Söllner, T.H. and Rothman, J.E. (1996). Calcium-dependent switching of the specificity of phosphoinositide binding to synaptotagmin. Proc. Natl. Acad. Sci. USA 93, 13327–1332.

Selbie, L.A., Schmitz-Peiffer, C., Sheng, Y. and Biden, T.J. (1993). Molecular cloning and characterization of PKCι, an atypical isoform of protein kinase C derived from insulin-secreting cells. J. Biol. Chem. 268, 24296–24302.

Thams, P. and Capito, K. (1997). Inhibition of glucose-induced insulin secretion by the diacylgycerol lipase inhibitor RHC 80267 and the phospholipase A_2 inhibitor ACA through stimulation of K^+ permeability without diminution by exogenous arachidonic acid. Biochem. Pharmacol. 53, 1077–1086.

Thams, P., Capito, K., Hedeskov, C.J. and Kofod, H. (1990). Phorbol-ester-induced down-regulation of protein kinase C in mouse pancreatic islets. Biochem. J. 265, 777–787.

Thams, P., Hedeskov, C.J. and Capito, K. (1993). Exogenous arachidonic acid inactivates protein kinase C in mouse pancreatic islets. Acta Physiol. Scand. 149, 227–235.

Thomas, T.P., Ellis, T.R. and Pek, S.B. (1989). Insulin release in RINm5F cells and glyceraldehyde activation of protein kinase C. Diabetes 38, 1371–1376.

Thomas, T.P and Pek, S.B. (1992). Diacylglycerol inhibits potassium-induced calcium influx and insulin release by a protein kinase-C-independent mechanism in HIT T-15 islet cells. Endocrinology 131, 1985–1992.

Tian, Y.-M., Urquidi, V. and Ashcroft, S.J.H. (1996). Protein kinase C in beta-cells: Expression of multiple isoforms and involvement in cholinergic stimulation of insulin secretion. Mol. Cell. Endocrinol. 119, 185–193.

Tsunoda, Y. and Owyang, C. (1994). A newly cloned phospholipase A_2-activating protein elicits Ca^{2+} oscillations and pancreatic amylase secretion via mediation of G protein beta/phospholipase A_2/arachidonic acid cascades. Biochem. Biophys. Res. Commun. 203, 1716–1724.

Vadakekalam, J. and Metz, S. (1999). Isotopic efflux studies as indices of phospholipase activation, In: *Methods in Molecular Biology.* (Bird, I., Ed.), Vol. 105, pp. 175–185. Humana Press, Totowa, NJ. (In press.)

Vadakekalam, J., Rabaglia, M.E., Chen, Q.-H. and Metz, S.A. (1996). Role for GTP in glucose-induced phospholipase C activation in pancreatic islets. Am. J. Physiol. 271, E85–E95.

Vadakekalam, J., Rabaglia, M.E. and Metz, S.A. (1997a). Interleukin-1β inhibits phospholipase C and insulin secretion at sites apart from the K^+ (ATP) channel. Am. J. Physiol. 273, E942–E950.

Vadakekalam, J., Rabaglia, M. E. and Metz, S.A. (1997b). Roles of GTP and phospholipase C in the potentiation of Ca^{2+}-induced insulin secretion by glucose in rat pancreatic islets. J. Endocrinol. 153, 61–71.

Vitale, N., Thiersé, D., Aunis, D. and Bader, M.-F. (1994). Exocytosis in chromaffin cells: Evidence for a MgATP-independent step that requires a pertussis toxin-sensitive GTP-binding protein. Biochem. J. 300, 217–227.

Weng, L., Davies, M. and Ashcroft, S.J.H. (1993). Effects of cholinergic agonists on diacylglycerol and intracellular calcium levels in pancreatic β-cells. Cellular Signalling 5, 777–786.

Wojcikiewicz, R.J.H. (1995). Type I, II, and III inositol 1,4,5-trisphosphate receptors are unequally susceptible to down-regulation and are expressed in markedly different proportions in different cell types. J. Biol. Chem. 270, 11678–11683.

Wolf, B.A., Easom, R.A., Hughes, J.H., McDaniel, M.L. and Turk, J. (1989). Secretagogue-induced diacylglycerol accumulation in isolated pancreatic islets. Mass spectrometric characterization of the fatty acyl content indicates multiple mechanisms of generation. Biochemistry 28, 4291–4301.

Wolf, B.A., Easom, R.A., McDaniel, M.L. and Turk, J. (1990). Diacylglycerol synthesis de novo from glucose by pancreatic islets isolated from rats and humans. J. Clin. Invest. 85, 482–490.

Wolf, B.A., Pasquale, S.M. and Turk, J. (1991). Free fatty acid accumulation in secretagogue-stimulated pancreatic islets and effects of arachidonate on depolarization-induced insulin secretion. Biochemistry 30, 6372–6379.

Yedovitsky, M., Mochly-Rosen, D., Johnson, J. A., Gray, M. O., Ron, D., Abramovitch, E., Cerasi, E. and Nesher, R. (1997). Translocation inhibitors define specificity of protein kinase C isoenzymes in pancreatic β-cells. J. Biol. Chem. 272, 1417–1420.

Zaitsev, S.V., Efendić, S., Arkhammar, P., Bertorello, A.M. and Berggren, P.-O. (1995). Dissociation between changes in cytoplasmic free Ca^{2+} concentration and insulin secretion as evidenced from measurements in mouse single pancreatic islets. Proc. Natl. Acad. Sci. USA 92, 9712–9716.

Zawalich, W.S. (1996). Regulation of insulin secretion by phosphoinositide-specific phospholipase C and protein kinase C activation. Diabetes Rev. 4, 160.

Zawalich, W.S. and Zawalich, K.C. (1996). Species differences in the induction of time-dependent potentiation of insulin secretion. Endocrinology 137, 1664–1669.

Zawalich, W.S., Zawalich, K.C., Ganesan, S., Calle, R. and Rasmussen, H. (1991). Effects of the phorbol ester phorbol 12-myristate 13-acetate (PMA) on islet-cell responsiveness. Biochem. J. 278, 49–56.

Zawalich, W.S., Zawalich, K.C. and Kelley, G.G. (1995a). Effects of short-term culturing on islet phosphoinositide and insulin secretory responses to glucose and carbachol. Acta Diabetol. 32, 158–164.

Zawalich, W.S., Zawalich, K.C. and Kelley, G.G. (1995b). Regulation of insulin release by phospholipase C activation in mouse islets: Differential effects of glucose and neurohumoral stimulation. Endocrinology 136, 4903–4909.

Zeldin, D.C., Foley, J., Boyle, J.E., Moomaw, C.R., Tomer, K.B., Parker, C., Steenbergen, C. and Wu, S. (1997). Predominant expression of an arachidonate epoxygenase in islets of Langerhans cells in human and rat pancreas. Endocrinology 138, 1338–1346.

Zhang, C., Schmidt, M., Von Eichel-Streiber, C., Jakobs, K.H. (1996). Inhibition by toxin B of inositol phosphate formation induced by G protein-coupled and tyrosine kinase receptors in N1E-115 neuroblastoma cells: Involvement of Rho proteins. Mol. Pharmacol. 50, 864–869.

Zheng, Y., Glaven, J.A., Wu, W.J., and Cerione, R.A. (1996). Phosphatidylinositol 4,5-bisphosphate provides an alternative to guanine nucleotide exchange factors by stimulating the dissociation of GDP from Cdc42Hs. J. Biol. Chem. 271, 23815–23819.

SIGNAL TRANSDUCTION
PROTEIN PHOSPHORYLATION AND THE REGULATION OF INSULIN SECRETION

Peter M. Jones, Shanta J. Persaud, and Tracey E. Harris

I. INTRODUCTION

The mechanisms through which β-cells recognize external signals are becoming increasingly well understood, and it is apparent that a variety of disparate extracellular signals are transduced into a limited number of intracellular regulators, the most studied of which are Ca^{2+}, cyclic adenosine monophosphate (cAMP) and

Advances in Molecular and Cell Biology
Volume 29, pages 303-334.

ISBN: 0-7623-0547-9

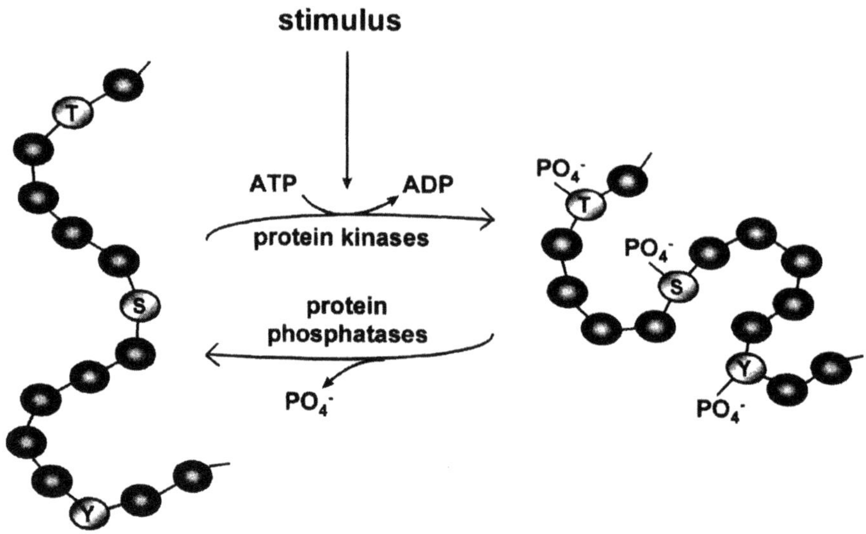

Figure 1. Protein kinases and phosphoprotein phosphatases. Protein kinase enzymes catalyse the transfer of the γ-phosphate group from adenosine triphosphate (ATP) into threonine (T), serine (S), or tyrosine (Y) amino acid residues in their substrate proteins, thus changing the structure and/or activity of the substrate protein. The reverse dephosphorylation reaction is catalyzed by the phosphoprotein phosphatase enzymes.

products of phospholipid hydrolysis such as diacylglycerol (DAG), inositol *tris*-phosphate (IP_3), and arachidonic acid (AA). These regulators have been implicated in the control of insulin secretion, but the mechanisms through which changes in their concentrations produce a controlled insulin secretory response are not yet well understood. Although they differ greatly in chemical structure, these regulatory molecules share a common ability to activate specific classes of protein kinase enzymes within β-cells. In this chapter, we consider whether the activation of protein kinases is a common transduction mechanism for the diverse physiological signals that regulate insulin secretion.

Protein kinases catalyze the transfer of phosphate from adenosine triphosphate (ATP) to specific protein substrates, which alters the function of the substrate protein. For example, if the substrate protein is an enzyme, the phosphorylation may enhance or inhibit its catalytic activity. The phosphorylation process is reversible by another class of enzymes known as phosphoprotein phosphatases, so phosphorylation offers a selective, reversible means of regulating cellular function at the protein level (Figure 1). The protein kinases that are thought to be important in β-cells can be subdivided into those which phosphorylate their substrates on serine and/or threonine residues (protein serine/threonine kinases), and those which incorporate phosphate into tyrosine residues (protein tyrosine kinases). Of these, much more is known about the protein serine/threonine kinases in β-cells since these are the enzymes whose activities are regulated by Ca^{2+}, cyclic nucleotides, and phospholipid derivatives. Accordingly, we focus most of our attention here to the regulated protein serine/threonine kinases, which may be involved in insulin secretion, but we also briefly consider whether protein tyrosine kinases and phosphoprotein phosphatases play important roles in β-cell responses.

II. PROTEIN PHOSPHORYLATION AND INSULIN SECRETION

Little is yet known about the precise substrate-specificity of protein kinases in β-cells, so it is common (and useful) to classify them on the basis of their activators, rather than on the more biochemically correct basis of their substrate specificity. Thus, several distinct classes of protein serine/threonine kinases whose activities can be modified by Ca^{2+}, cyclic nucleotides, and products of phospholipid hydrolysis have been identified both in islets of Langerhans and in insulin-secreting cell lines. Protein kinases exist as families of similar isoforms which may differ in their mode of activation, their substrate specificity, or their intracellular location, and those expressed in β-cells are summarized in Table 1. For historical reasons, the cAMP-dependent protein kinases are often referred to as protein kinase A (PKA), while the Ca^{2+}/phospholipid-dependent protein kinases are collectively known as protein kinase C (PKC). The kinase activities that have been identified in β-cells to date are similar (or identical) to enzymes that have been studied in much more detail in other tissues, and the expression, structure, and regulation of these enzymes have been reviewed in detail elsewhere (Colbran et al., 1989; Taylor et al., 1990; Scott, 1991; Hug and Sarre, 1993; Hardie and Hanks, 1995). This chapter, therefore, concentrates on the occurrence and characteristics of protein kinase activities in islets of Langerhans and insulin-secreting cells, and the physiological relevance of these enzymes to the control of insulin secretion.

Much of the evidence implicating particular protein kinases in physiologically relevant insulin secretory responses is inconclusive, confusing, and often contradictory. In general, most studies of the role(s) of protein kinases in β-cells have concentrated on three experimental approaches: (1) determining whether insulin secretagogues activate the kinase; (2) measuring the consequences of kinase activation on protein phosphorylation and insulin secretion; and (3) measuring the consequences of kinase inhibition. Of these, the use of kinase inhibitors is by far the most common experimental approach, although much of the confusion in the literature can be attributed to the use of protein kinase inhibitors of uncertain specificity.

Given the importance of kinase inhibitors in shaping the consensus view of β-cell function, it is worth considering briefly the different types of protein kinase

Table 1. Protein Kinases and Phosphoprotein Phosphatases Expressed in Islets of Langerhans and in Insulin-secreting Cells[a]

	Isoform/Subtype		
	Islets	*β-Cell Lines*	*References*
Kinase			
CaMKII	$\alpha, \beta_3, \gamma_B, \gamma_E, \gamma_{SRP}, \delta_2$	$\beta_3, \gamma, \delta_2, \delta_6$	Takasawa et al. (1995), Urquidi and Ashcroft (1995), Breen and Ashcroft (1997a, 1997b), Mohlig et al. (1997)
PKA	Types I and II	unknown	Corbin et al. (1975), Sugden et al. (1979)
PKC	$\alpha, \beta, \delta, \varepsilon, \zeta, \iota$	$\alpha, \beta, \delta, \varepsilon, \zeta, \iota, \mu$	Ito et al. (1989), Ganesan et al. (1990, 1992), Onoda et al. (1990), Fletcher and Ways (1991), Calle et al. (1992), Wetsel et al. (1992), Selbie et al. (1993), Knutson and Hoenig (1994, 1996), Zaitsev et al. (1995), King et al. (1996), Tian et al. (1996) Yedovitsky et al. (1997)
MAPK	p42, p44	p38, p42, p44	Frödin et al. (1995), Persaud et al. (1996), Burns et al. (1997), Khoo et al. (1997), Macfarlane et al. (1997)
PTK	BSK, FGFR-4, Flk-1, IGF-IR, IRβ, JAK2, MEK,[b] TrkA	BSK, Flk-1, IRβ, MEK,[b] TrkC	Öberg et al. (1994), Kanaka-Gantenbein et al. (1995), Rothenberg et al. (1995), Velloso et al. (1995), Persaud et al. (1996), Tazi et al. (1996)
Phosphatase			
Tyrosine	IA-2, phogrin[c]	IA-2, phogrin[c]	Lan et al. (1994), Wasmeier et al. (1996), Hatfield et al. (1997)
Serine/threonine	PP1, PP2A, PP2B	PP1, PP2A, PP2B	Gagliardino et al. (1991, 1997), Sjöholm et al. (1993, 1995), Ammala et al. (1994), Ebihara et al. (1996), Kowluru et al. (1996), Murphy and Jones (1996)

Notes: BSK, β-cell src-homology tyrosine kinase; CaMK II, Ca^{2+}/calmodulin-dependent protein kinase II; FGFR-4, fibroblast growth factor receptor-4; Flk-1, fetal liver kinase-1; IGF-IR, insulinlike growth factor-I receptor; IRβ, β-subunit of insulin receptor; JAK2, Janus kinase 2; MAPK, mitogen-activated protein kinase; MEK, MAP kinase kinase; PKA, cAMP-dependent protein kinase; PKC, Ca^{2+}/phospholipid-dependent protein kinase; PP, phosphoprotein phosphatase; PTK, protein tyrosine kinase; TrkA/C, neurotrophin tyrosine kinase receptor A/C.

[a] The table lists the major protein kinase and protein phosphatase subtypes that have been identified in islets and in insulin-secreting β-cell lines.

[b] MEK is a dual specificity tyrosine/threonine kinase.

[c] IA-2 and phogrin have domains with high sequence homology to tyrosine phosphatase catalytic subunits, but it is unclear whether they express tyrosine phosphatase activity.

inhibition that have been used in pancreatic β-cells. The most commonly used inhibitors interact with the ATP-binding site of the enzyme. Older inhibitors of this type tend to be poorly selective since the ATP-binding sites of the common regulated serine/threonine protein kinases are very similar (Hanks and Hunter, 1995). More recently developed ATP-binding site inhibitors show a higher degree of selectivity (Davies et al., 1989; Martiny-Baron et al., 1993), although it is uncertain whether this approach will permit specific inhibition of individual kinase isoforms. Inhibitors that act by interfering with the binding of the intracellular activators (e.g., Ca^{2+}, cAMP, DAG) offer potentially more selectivity between classes of regulated kinases (de Wit et al., 1984; Kobayashi et al., 1989), although these inhibitors may not differentiate between kinase isoforms, and they may also inhibit binding of the activators to response elements other than protein kinases. Synthetic peptide sequences that bind to and inhibit the substrate site of the kinases offer potentially specific inhibitors (Kemp et al., 1991), but the unmodified peptide sequences are poorly membrane-permeant which has largely limited their use to single cell microinjection or patch-clamp preparations (Ammala et al., 1993b) and to populations of permeabilized cells (Basudev et al., 1995). Chemical modification of the peptides can make them more membrane-permeant, and this approach has been applied to peptide inhibitors of PKC and PKA in β-cells (Harris et al., 1996b, 1997).

The problems of inhibitor specificity and delivery can be circumvented by generating β-cells deficient in the enzyme of interest. One useful pharmacological approach, that of downregulation of the DAG-sensitive conventional and novel isoforms of PKC by prolonged treatment with phorbol esters, has been widely applied to study the roles of PKC in β-cells, as described in the section on *Ca^{2+}/phospholipid-dependent protein kinases* later in this chapter. Transfection of β-cells with antisense nucleotide sequences to prevent the expression of particular protein kinases is feasible (Persaud and Jones, 1994), but it is difficult to introduce sufficient amounts of the antisense sequence into populations of cells in a reproducible manner. Stable antisense transfects to prevent kinase expression have not yet been reported for β-cells, but this may offer an effective means of selectively depleting cells of kinase isoforms. Conversely, stable transfection of RINm5F cells with a viral vector coding for the δ isoform of CaMK II has been reported (Möhlig et al., 1995), and this overexpression approach may provide useful information complementary to inhibitor studies. Transgenic animal models are now becoming more widely available, and several protein kinase knockout animals have been reported (Silva et al., 1992; Abeliovich et al., 1993), although information is not yet available about β-cell function in these models.

Rather than relying solely on kinase inhibitors, an accurate evaluation of the importance of a particular kinase requires several different experimental approaches. Thus, a protein kinase can be considered to be involved in the secretory process if: (1) physiologically relevant stimuli increase the phosphorylating activity of the kinase in situ; (2) physiological and/or pharmacological activation of the kinase increases insulin secretion; and (3) kinase inhibition causes defective secretory responses. In the rest of this chapter, we consider to what extent the individual classes of regulated protein kinases fulfill these criteria.

A. Cyclic Nucleotide-Dependent Protein Kinases

It is well documented that physiological regulators of insulin secretion can induce changes in the concentration of cAMP within β-cells, and there have been numerous reports over the past two decades that experimentally induced elevations in β-cell cAMP can enhance insulin secretion. The reported effects are remarkably consistent, whether β-cell cAMP is elevated by activating adenylate cyclase using forskolin or cholera toxin; through inhibition of cAMP degradation by phosphodiesterases (Hughes and Ashcroft, 1992); by introducing cAMP in permeabilized β-cells (Jones et al., 1986); or by using membrane-permeant analogues of cAMP in intact β-cells (Morgan and Montague, 1985).

The major, if not the only, β-cell response element for changes in cAMP is the cAMP-dependent protein kinase (PKA). PKA was first isolated from rat and guinea-pig islets of Langerhans over 25 years ago (Montague and Howell, 1972) and shown to consist of a catalytic subunit and a regulatory subunit, which dissociates from the catalytic subunit in the presence of cAMP. Subsequent studies demonstrated that rat islets contain two distinct isoforms of PKA, which can be distinguished by their kinetic characteristics (Sugden et al., 1979). It therefore seems likely that, in common with many other cell types, pancreatic β-cells contain two isoforms of PKA, each of which comprises a holoenzyme of two regulatory (R) subunits and two catalytic (C) subunits. Under resting conditions, the phosphorylating activity of the catalytic subunits is inhibited by association with the regulatory subunits. Increased availability of cAMP results in its binding to the regulatory subunits, causing a dissociation of the holoenzyme into a $R_2.cAMP_4$ dimer and two free active catalytic subunits. A subsequent decrease in the concentration of cAMP will favor dissociation of cAMP from the regulatory subunits, the reassociation of regulatory and catalytic subunits, and the resultant inhibition of catalytic activity.

The PKA holoenzyme can be formed from at least four different types of regulatory subunit (RIα, RIβ, RIIα, RIIβ) and three catalytic subunits (Cα, Cβ, Cγ), which are expressed in a tissue-specific manner. To date, there is no information about which isoforms of the regulatory and catalytic subunits are expressed in β-cells, although it seems likely that they will express RIα, RIIα, and Cα since these are found in all tissues examined (Scott and Carr, 1992). The subcellular localization of PKA is dependent upon the type of R subunit in the holoenzyme. Both types of RI subunit are primarily cytosolic, so type I PKA is a cytosolic enzyme. In contrast, RII subunits can interact with specific binding proteins located on intracellular membranes, and type II PKA can thus be associated with plasma, nuclear, and secretory granule membranes.

Increases in islet cAMP induced by nutrients (Christie and Ashcroft, 1984) or by the receptor-operated nonnutrients, such as glucagon (Suzuki et al., 1983), result in increased phosphorylation of endogenous PKA substrates. In common with many other protein kinases, the activity of PKA may itself be regulated by phosphorylation. The RII regulatory subunit of type II PKA can be phosphorylated by the catalytic subunit, and it has been suggested that a 54-kDa islet phosphoprotein is an autophosphorylated regulatory subunit of PKA (Kowluru et al., 1984). A number of other PKA substrates have been identified in β-cells and islets, and modulation of their functions by phosphorylation may directly affect the insulin secretory process. For example, although glucose transport across the β-cell plasma membrane is not considered to be rate-limiting for insulin secretion, it has been demonstrated that PKA activation by forskolin results in rapid serine/threonine phosphorylation of the β-cell glucose transporter, type 2 (GLUT-2) (Thorens et al., 1996). Closure of the ATP-sensitive K^+ channel is certainly important for nutrient-induced insulin secretion, and both the K_{ATP} channel and its associated sulphonylurea receptor have several potential phosphorylation sites for PKA (Inagaki et al., 1995), suggesting that K^+ channel activity may be modulated by PKA-dependent phosphorylation. The activation of PKA has also been shown to increase the phosphorylation of the α_1 subunit of the voltage-dependent Ca^{2+} channel VDCC in β-cells, suggesting a role for PKA in the regulation of Ca^{2+} influx (Leiser and Fleischer, 1996). Future studies using channels mutated at potential phosphorylation sites will provide insights into the physiological relevance of channel phosphorylation in the regulation of insulin secretion.

Until recently, studies of the physiological functions of PKA were hampered by the lack of specific, membrane-permeant inhibitors of the enzyme. In an early report, 2-deoxyadenosine, an inhibitor of islet adenylate cyclase, blocked the PKA-mediated phosphorylation of a 15-kDa islet protein, with a concomitant reduction in insulin secretion (Christie and Ashcroft, 1984), but the effects of this inhibitor on secretory responses to physiologically relevant stimuli were not investigated. More recently, membrane-permeant competitive antagonists of cAMP have been developed and shown to inhibit PKA-mediated events in islets (Persaud et al., 1990), single β-cells (Ammala et al., 1993a), and α-cells (Ding et al., 1997). Rp-cAMPS has been used to block protein phosphorylation and insulin secretory responses induced by cAMP, forskolin, and GIP (Persaud et al., 1990; Ammala et al., 1993a; Ding and Gromada, 1997). These observations lend support to the notion that the activation of PKA is the major route through which nonnutrient, receptor-operated, adenylate cyclase-coupled secretagogues modulate insulin secretion, as depicted schematically in Figure 2.

There is general agreement that agents which elevate β-cell cAMP enhance nutrient-induced insulin secretion, but do not initiate secretory responses from intact β-cells at substimulatory glucose concentrations, suggesting that the activation of PKA acts as a potentiator rather than a primary trigger for insulin secretion. Although PKA activation is normally not sufficient to initiate insulin secretion, it

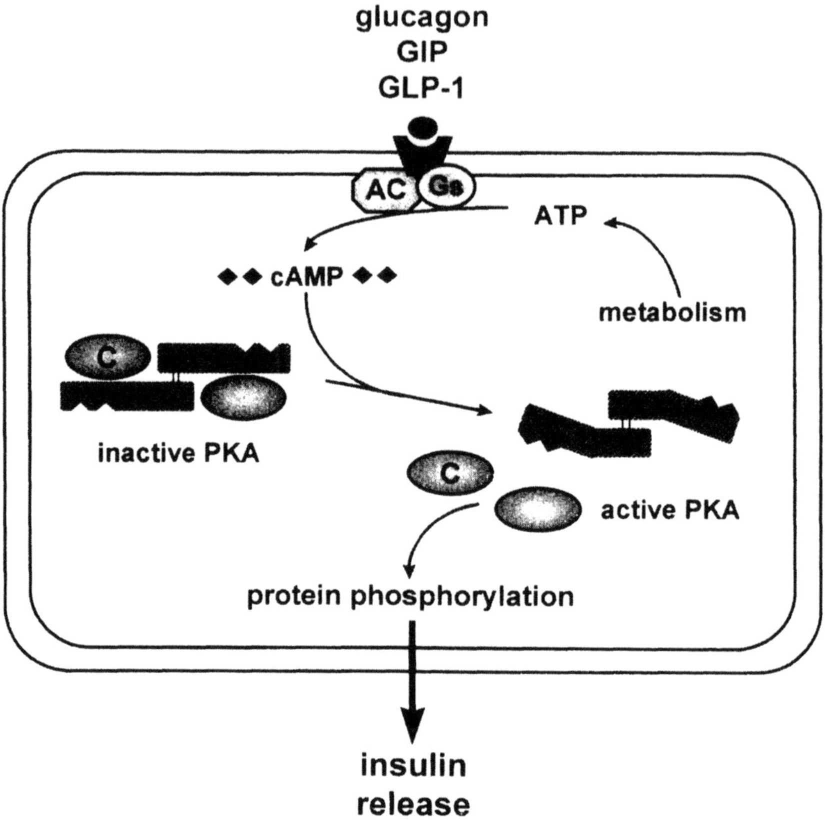

Figure 2. Protein Kinase A (PKA) activation transduces signals from cell surface receptors on β-cells. Under resting conditions cyclic adenosine monophosphate (cAMP)-dependent protein kinase (PKA) exists as an inactive tetramer comprising two regulatory (R) subunits and two catalytic (C) subunits. Agonists such as glucagon, GIP, or glucagonlike peptide-1 (GLP-1) bind to cell surface receptors which are coupled via the heterotrimeric G-protein, G_s, to adenylate cyclase (AC). Receptor occupancy activates AC, with the consequent generation of cAMP (cAMP ♦) from ATP. The cAMP binds to two sites on the regulatory subunits, causing the inactive tetramer to dissociate, liberating active catalytic subunits (C), which stimulate insulin secretion through the phosphorylation of protein substrates.

remains a possibility that PKA activation is required for secretory responses to nutrients. There have been a number of reports that poor or absent β-cell responses to glucose can be remedied by agonist-induced or pharmacological elevations in intracellular cAMP in isolated patch-clamped β-cells (Holz et al., 1993), in purified populations of β-cells (Pipeleers et al., 1985), and in freshly isolated islets

(Barnett et al., 1994), implying that the β-cell content of cAMP, and thus the activation state of PKA, is a vital component in nutrient-induced insulin secretion. In contrast to this, glucose-induced insulin secretion from freshly isolated islets was not significantly inhibited by Rp-cAMPS (Persaud et al., 1990), nor by a membrane-permeant, myristoylated peptide inhibitor of PKA (Harris et al., 1997), although both inhibitors were shown to inhibit phosphorylation and secretory responses to cAMP and forskolin. There are not yet available any experimental models in which the expression of β-cell PKA is suppressed. Such models could determine decisively whether the activation of PKA is obligatory for nutrient-induced insulin secretion, and the extent to which nonnutrients modify β-cell responses by PKA-dependent protein phosphorylation.

A cyclic guanosine monophosphate (GMP)–dependent protein kinase activity (PKG) has also been identified in rat islets (Kuo, 1974), but the regulation of guanylate cyclase by insulin secretagogues and the importance of cyclic GMP and PKG in the control of insulin secretion is uncertain. Nutrients stimulate small increases in islet cyclic GMP content, although with a reported time-course which varies from 2 to 3 minutes (Schmidt et al., 1992) to at least 30 minutes (Jones et al., 1992a), perhaps suggesting that it is not related to insulin secretion. There is also some confusion in the literature about the effects of experimental elevations in β-cell cyclic GMP, which are reported to be stimulatory (Laychock, 1987), inhibitory (Vara and Tamarit-Rodriguez, 1991) or without effect (Jones et al., 1992) on insulin secretion. Some cytokines increase islet cyclic GMP by a nitric oxide–mediated activation of soluble guanylate cyclase (Southern et al., 1990), but the changes in cyclic GMP can be dissociated from the inhibitory effects of cytokines on glucose-induced insulin secretion (Cunningham et al., 1994). The activation of PKG therefore seems unlikely to play a major role in the regulation of β-cell secretory responses to metabolic stimuli or to receptor-operated agonists. However, a recent report has implicated cyclic GMP, and by implication PKG, in the induction of apoptotic cell death in β-cells (Loweth et al., 1997), which may open up many new and clinically important areas of research in the future.

B. Ca^{2+}/Calmodulin-Dependent Protein Kinases

Changes in intracellular Ca^{2+} are known to play a vital role in the regulation of insulin secretion by both nutrient and nonnutrient secretagogues, and experiments using permeabilized islets (Jones et al., 1985) or patch-clamp measurements in single β-cells (Ammala et al., 1993b) have demonstrated that increasing intracellular Ca^{2+} is alone a sufficient stimulus to induce an ATP-dependent stimulation of insulin secretion.

One mechanism by which β-cells respond to changes in intracellular Ca^{2+} is to alter the activity of several protein kinases. A number of protein serine/threonine kinases are dependent upon the presence of Ca^{2+} and the Ca^{2+}-binding protein, calmodulin (CaM), for their activation and β-cells are known to express at least three types of Ca^{2+}/CaM-dependent protein kinase: the Ca^{2+}/CaM-dependent protein kinases CaMK II, CaMK III, and myosin light chain kinase (MLCK) have been identified in rat and hamster insulinoma cells and in rat and human islets of Langerhans (see Harrison et al., 1984). Most investigations of Ca^{2+}-dependent protein kinases in β-cells have focused on the multifunctional CaMK II rather than on CaMK III or MLCK, since CaMK II has been implicated in the regulation of secretion in other tissues. There is no evidence that CaMK III is involved in the regulation of insulin secretion, although the identification of elongation factor 2 as the major CaMK III substrate in islets suggests a role in the regulation of β-cell protein synthesis (Hughes et al., 1993). In contrast, there is some evidence that MLCK may be involved in the secretory process, presumably through the phosphorylation of its preferred substrate, myosin light chains. A 20-kDa β-cell phosphoprotein detected in several studies is almost certainly myosin light chain (Penn et al., 1982), and glucose-induced insulin secretion can be inhibited to some extent using the MLCK inhibitors ML-9 or wortmannin (Niki et al., 1993; Hagiwara et al., 1995).

The evidence that CaMK II is important in regulating insulin secretion in response to Ca^{2+}-mobilizing signals is more plentiful and more convincing than for the other Ca^{2+}/CaM-dependent kinases. β-Cell CaMK II probably exists as a holoenzyme of approximately 550-kDa molecular mass, consisting of a number (8 to 10) of subunits, each of which contain both regulatory and catalytic activity. Four independent genes coding for α, β, γ, and δ isoforms of CaMK II, and corresponding splice variants, have been identified in a variety of tissues. All four isoforms (α, β_3, γ_B, γ_E, and δ_2) have been detected in islets (Urquidi and Ashcroft 1995; Breen and Ashcroft 1997a), with the α isoform reported to be localized to rat islet microsomes (Takasawa et al., 1995) and the δ_2 isoform to β-cell secretory granules (Möhlig et al., 1997). Human islets may express a truncated γ isoform of CaMK II (Breen and Ashcroft 1997b), but the functional significance of the truncation has not been established. When activated by increased intracellular Ca^{2+}, CaMK II autophosphorylates on several serine and threonine residues which reduces its Ca^{2+}-dependency and produces a Ca^{2+}-independent form of the enzyme. CaMK II is dephosphorylated and deactivated in β-cells by the combined actions of a number of phosphoprotein phosphatases (Easom et al., 1998).

Ca^{2+}-mobilizing stimuli that enhance insulin secretion also activate CaMK II and induce Ca^{2+}-dependent phosphorylation of CaMK II substrates. For example, Ca^{2+}-induced insulin secretion from permeabilized islets is accompanied by Ca^{2+}/CaM-dependent phosphorylation of endogenous substrates in situ (Jones et al., 1988), while glucose, carbachol, and depolarizing concentrations of K^+ activate β-cell CaMK II as a consequence of the entry of extracellular Ca^{2+} or the mobilization of intracellular Ca^{2+} (Wenham et al., 1994; Babb et al., 1996). The endogenous β-cell substrates for CaMK II have not yet been identified with any certainty, and it is therefore uncertain how increased CaMK II activity leads to a controlled secretory response. It seems probable that the β-cell proteins of molecular masses 53 to 55 kDa, which have been reported by many groups to be substrates for CaMK II, are autophosphorylated subunits of CaMK II itself, although other possible identities for CaMK substrates have also been reported. For example, a 57-kDa substrate for CaMK was reported to react with antipyruvate kinase antibodies (MacDonald et al., 1985), although this identification has not been confirmed. In a separate study, the 57-kDa and 54-kDa islet substrates for Ca^{2+}-dependent phosphorylation were identified as the α and β subunits of tubulin on the basis of their electrophoretic mobilities and immunoreactivity with antitubulin antibody (Colca et al., 1983), although this identification was later disputed on the basis of the isoelectric point of the 57-kDa protein (Kowluru and MacDonald, 1984). It is possible that the Ca^{2+}-regulated step in insulin secretion may be at the level of exocytosis, and phosphoproteins may be involved in that process. Annexin-I, a protein implicated in membrane fusion events, is present in insulin secretory granules and is phosphorylated in a glucose-dependent manner (Ohnishi et al., 1995) and an 84-kDa secretory granule–associated synapsin protein is phosphorylated by CaMK II in glucose-stimulated mouse insulinoma MIN6 β-cells (Matsumoto et al., 1995).

Studies using inhibitors of CaMK II are also generally supportive of a role for the enzyme in the regulation of insulin secretion. Insulin secretory responses to nutrients and other Ca^{2+}-mobilizing stimuli are inhibited by a number of CaMK II inhibitors, including trifluoperazine (Schubart et al., 1980), dehydrouramil (Harrison et al., 1986), a synthetic CaMK II inhibitory peptide (Ammala et al., 1993b), and KN-62 (Wenham et al., 1992). However, the inhibitory effects of KN-62 on insulin secretion from an insulinoma cell line were attributed to an inhibition of Ca^{2+} entry rather than an inhibition of CaMK II activity, and KN-62 did not inhibit Ca^{2+}-induced insulin release from permeabilized β-cells (Li et al., 1992). The inhibitory effect of KN-62 on Ca^{2+} current has also been observed in single mouse β-cells, but in this model KN-62 also caused a decrease in depolarization-induced membrane capacitance, indicative of an inhibition of exocytosis, suggesting that the activation of CaMK II is required for the exocytotic release of insulin (Ammala et al., 1993b).

In summary, insulin secretagogues increase intracellular Ca^{2+} in β-cells, and secretion correlates well with increased CaMK II activity and the phosphorylation of endogenous CaMK II substrates; experimental elevations in β-cell cytosolic Ca^{2+} stimulate phosphorylation and insulin secretion; and a number of CaMK inhibitors inhibit insulin secretion. CaMK II may, therefore, be responsible for the initiation of an insulin secretory response to nutrient secretagogues, as depicted schematically in Figure 3.

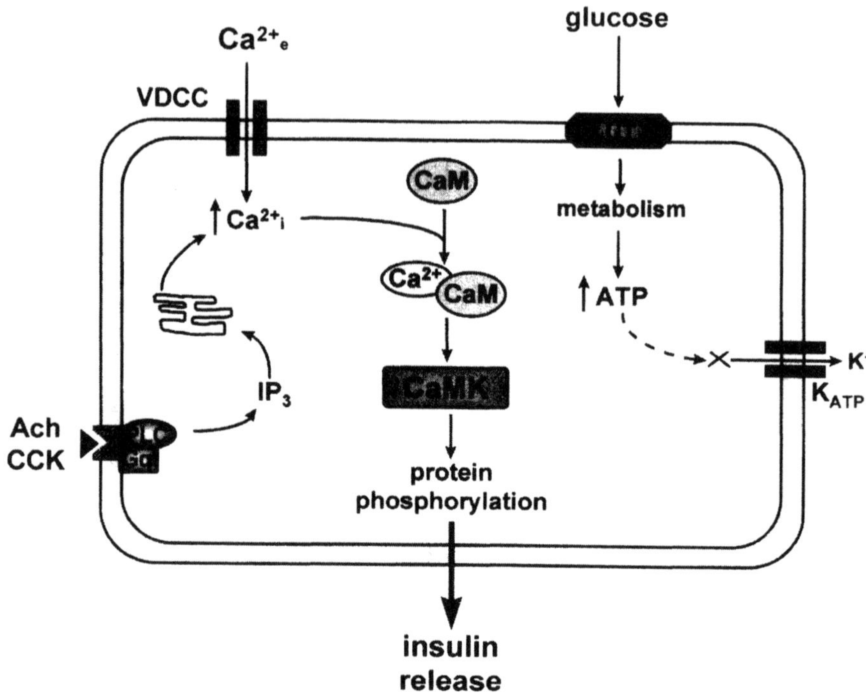

Figure 3. CaMK as a Ca^{2+} sensor in β-cell responses to nutrient and nonnutrient stimuli. The schematic diagram shows how CaMK II may be involved in transducing signals from nutrients (e.g., glucose) and nonnutrients (e.g., acetylcholine, ACh, cholecystokinin, CCK). *Nutrients:* Glucose enters the β-cell on the GLUT-2 transporter and is metabolized with a consequent generation of ATP and closure of plasma membrane K_{ATP} channels. The decreased efflux of K^+ leads to depolarization of the β-cell with the consequent opening of VDCCs and an influx of extracellular Ca^{2+} (Ca^{2+}_e) down its concentration gradient. Once inside the β-cell Ca^{2+} (Ca^{2+}_i) binds to calmodulin (CaM) and the Ca^{2+}/CaM complex binds to and activates Ca^{2+}/calmodulin-dependent protein kinase II (CaMK). CaMK phosphorylates β-cell substrate proteins, and these phosphorylation events stimulate the secretory process. *Nonnutrients:* Agonists bind to cell-surface receptors, which are coupled via the heterotrimeric guanosine triphosphate (GTP)–binding protein G_q to phospholipase C (PLC). Receptor occupancy activates PLC with the consequent generation of inositol *tris*phosphate (IP_3) by the hydrolysis of membrane inositol phospholipids. IP_3 releases stored Ca^{2+} from the endoplasmic reticulum and the elevation in intracellular Ca^{2+} (Ca^{2+}_i) activates CaMK.

C. Ca^{2+}/Phospholipid-Dependent Protein Kinases

The Ca^{2+}/phospholipid-dependent protein kinases known as protein kinase C (PKC) can be classified into three groups of isoforms: (1) those that are Ca^{2+}- and DAG-dependent (conventional isoforms: α, β, γ); (2) those that are Ca^{2+}-independent but DAG-dependent (novel isoforms: δ, ε, η, θ); and (3) those that are Ca^{2+}-independent and do not bind DAG or phorbol esters (atypical isoforms: ζ, ι/λ, μ) (reviewed by Hug and Sarre, 1993). The expression and functions of PKC in β-cells and islets have been the subjects of numerous studies over the past decade, but there is still some disagreement about the role of this class of protein serine/threonine kinase in the regulation of insulin secretion. This area of β-cell research has been reviewed extensively elsewhere (Wollheim and Regazzi, 1990; Persaud et al., 1992; Zawalich, 1996) and readers are referred to these reviews for detailed comment on this complex area.

There have been many, often contradictory, reports of PKC isoform expression in β-cells. A number of studies suggest that pancreatic β-cells and islets contain the α isoform and the βII isoform, but not the βI isoform nor the γ isoform, whereas rat insulinoma (RIN) cells contain the α isoform, but not the β isoforms (Ito et al., 1989; Onoda et al., 1990; Fletcher and Ways, 1991; Ganesan et al., 1992; Wetsel et al., 1992). However, there is some debate about which endocrine cells within the islets express particular PKC isoforms. Thus, the α isoform has been localized in either the β-cells (Onoda et al., 1990; Fletcher and Ways, 1991; Ganesan et al., 1992), or in peripheral islet cells (Wetsel et al., 1992), while a β isoform was found in β-cells (Ito et al., 1989), or in peripheral islet cells but not in β-cells (Ganesan et al., 1992; Wetsel et al., 1992), or was not detectable in any islet cells (Fletcher and Ways, 1991). There is general agreement that the γ isoform of PKC is not found in β-cells, although there is a report that the γ isoform is detectable in glucagon-containing α-cells (Fletcher and Ways, 1991). It has been reported that the ε-isoform of PKC is found only in somatostatin-containing δ-cells (Fletcher and Ways, 1991), but other studies have indicated that it was present in a radiation-induced transplantable insulinoma (Knutson and Hoenig, 1994), in MIN6 β-cells (Tian et al., 1996) and in islets from *ob/ob* mice which contain >90% β-cells (Zaitsev et al., 1995). Observations that the δ isoform was present in pancreatic islets but not localized to a particular endocrine cell type (Wetsel et al., 1992), have been extended by the report that PKC-δ was the predominant isoform in both insulinoma-derived β-cells and whole islets and was associated with the cytoskeleton (Knutson and Hoenig, 1994). Two subsequent studies have indicated that MIN6, RINm5F, and βTC3 β-cell lines express the μ isoform (King et al., 1996; Tian et al., 1996), but it is not yet clear whether this particular PKC species is expressed in β-cells of normal islets. The ζ isoform of PKC has been detected in islets and β-cell lines (Wetsel et al., 1992; Selbie et al., 1993; Knutson and Hoenig, 1994; King et al., 1996; Tian et al., 1996), as has a novel

PKC isoform, ι, which shows approximately 70% sequence homology with the ζ isoform (Selbie et al., 1993; King et al., 1996; Tian et al., 1996).

There are a number of possible reasons for the confusion about the expression of PKC isoforms. First, identification of PKC isoforms has relied to a large extent on immunological measurements, and some of the discrepancies may reflect non-specificity of the antisera used. Second, the expression of PKC isoforms differs in islets from fetal, neonatal, and adult rats (Fletcher and Ways, 1991), so some of the confusion may stem from developmental changes in the different experimental models employed. Finally, much of the reported work on isoform expression in β-cells has used transformed cell lines, which may not offer an accurate reflection of PKC isoform expression in authentic β-cells.

A number of β-cell substrates for PKC have been identified whose identities are consistent with an involvement in the insulin secretory process. The sulphonylurea receptor associated with the K^+_{ATP} channel has potential phosphorylation sites for PKC (Inagaki et al., 1995), and electrophysiological studies have suggested that the K^+_{ATP} channels in rat insulinoma cells are substrates for PKC (Wollheim et al., 1988). The subunits of the VDCC contain consensus phosphorylation sites for PKC, as well as for other kinases (Seino, 1995), and it has been suggested that PKC may regulate Ca^{2+} influx into β-cells by changing the phosphorylation state of the L-type VDCC (Arkhammar et al., 1994), although there has, as yet, been no direct demonstration of PKC-mediated phosphorylation of β-cell VDCC subunit(s). PKC activation may also regulate other transduction systems within β-cells. For example, the phosphorylation of G_i by PKC may modulate β-cell AC activity, and a PKC substrate of the appropriate molecular mass (~40 kDa) for the α subunit of G_i has been reported in several studies (Thams et al., 1988; Persaud et al., 1993a). In addition, cytosolic phospholipase A_2 ($cPLA_2$) may be involved in the Ca^{2+}-dependent generation of arachidonic acid in β-cells, and the 100-kDa $cPLA_2$ is a substrate for PKC in neonatal rat islets (Dunlop and Clarke, 1995). A protein of 80-kDa molecular weight has been reported as a PKC substrate in β-cells in several studies, and it has been suggested that it is autophosphorylated PKC (Dunlop and Larkins, 1986). It seems more likely that the 80-kDa protein is the myristoylated alanine-rich C kinase substrate (MARCKS) protein, a filamentous actin cross-linking protein whose function is modified by PKC-mediated phosphorylation (Hartwig et al., 1992). Enhanced phosphorylation of MARCKS in response to PKC-activating phorbol esters, to cholinergic agonists, and to glucose has been reported in rat and mouse islets (Easom et al., 1990; Calle et al., 1992).

Although there may not be complete agreement on the precise role for PKC in the regulation of insulin secretion, there is little doubt that PKC is involved. Physiological regulators of insulin secretion can generate activators of PKC within β-cells, notably DAG and AA (Peter-Riesch et al., 1988; Regazzi et al., 1990; Konrad et al., 1993), and activation of DAG-sensitive isoforms of PKC within β-cells stimulates insulin secretion (reviewed by Persaud et al., 1992). Under unstimulated conditions, DAG/Ca^{2+}-sensitive PKC activity is mainly associated with a cytosolic fraction in islets and β-cells, but translocates to a membrane-associated fraction upon activation (e.g., Persaud et al., 1989b; Ganesan et al., 1992; Deeney et al., 1996; Knutson and Hoenig, 1996). This property of PKC has been widely used to assess its activation in response to insulin secretagogues, either by measuring enzyme activity in cytosolic and membrane fractions, or by immunodetection of the distribution of particular PKC isoforms. The activation and translocation of PKC is followed by its proteolytic cleavage by membrane-associated, Ca^{2+}-activated neutral proteases (Suzuki et al., 1987). Prolonged exposure to phorbol esters therefore produces cells that are deficient in some DAG-sensitive isoforms, although the atypical PKC isoforms (ζ, ι/λ, and μ) are unaffected (Tian et al., 1996). Such PKC-depleted islets and β-cells have also been widely used to determine the requirement for DAG-sensitive isoforms of PKC in secretory responses to nutrients and nonnutrients.

There is general agreement that the activation of DAG-sensitive isoforms of PKC is required for normal secretory responses to PLC-coupled nonnutrient secretagogues such as acetylcholine or cholecystokinin (CCK): these secretagogues promote the translocation of PKC between subcellular fractions, whether measured by phosphorylation of an exogenous substrate or by quantitative immunoblotting (e.g., Persaud et al., 1989b; Ganesan et al., 1990; Regazzi et al., 1990); PKC depleted islets or β-cells show reduced secretory responses to cholinergic agonists (Persaud et al., 1989a, 1991a; Hughes et al., 1990; Gao et al., 1994), to CCK (Karlsson and Ahrén, 1991; Persaud et al., 1993c), and to AVP (Gao et al., 1994); and secretory responses to these nonnutrients are reduced by a variety of PKC inhibitors, including H-7, clomiphene, polymyxin B, and staurosporine (reviewed by Persaud et al., 1992).

The importance of DAG-sensitive PKC isoforms for nutrient-induced insulin secretion is less clear. Most studies agree that nutrients do not stimulate the translocation of PKC enzyme activity in islets or β-cells (Easom et al., 1989; Persaud et al., 1989b, 1991b; Regazzi et al., 1990; Deeney et al., 1996), but immunoblotting and immunocytochemical approaches have shown nutrient-induced translocation of the α and ε isoforms of PKC (Ganesan et al., 1990, 1992; Calle et al., 1992; Zaitsev et al., 1995). Similarly, PKC-depleted islets have not produced totally consistent results. Thus, some groups find that nutrient-induced secretion is not affected by PKC-depletion (Metz, 1988; Persaud et al., 1989a, 1991a, 1991b, 1993c; Jones et al., 1992b), but downregulation of PKC has been reported to reduce the second phase of glucose-induced secretion from mouse islets (Thams et al., 1990). Confusing effects of nutrients have also been reported in studies using the glucose-insensitive RINm5F cell line, in which glyceraldehyde-stimulated insulin secretion has been reported to be either decreased (Thomas et al., 1989) or increased (Li et al., 1990) in PKC-depleted cells. Studies using pharmacological inhibitors of PKC have not produced a clear answer on whether or not PKC is required for a secretory response to nutrients. The PKC inhibitors H-7, clomiphene, polymyxin B, staurosporine, and Ro31-8220 all inhibit nutrient-induced insulin secretion to some extent (Malaisse et al., 1985; Stutchfield et al., 1986; Easom et al., 1989; Zawalich et al., 1991; Persaud et al., 1993c; Persaud and Jones, 1995; Harris et al., 1996a), but glucose-induced insulin secretion is not inhibited by a myristoylated peptide inhibitor of the α and β isoforms of PKC (Harris et al., 1996b), nor by the indolocarbazole Gö 6976, which shows selectivity for the conventional isoforms of PKC (α, β, γ) (Harris et al., 1996a). These seemingly disparate results are compatible with the model shown in Figure 4, in which the activation of a DAG-insensitive atypical PKC isoform(s) is required for nutrient-induced insulin secretion. Thus, the older, nonisoform-specific PKC inhibitors could inhibit insulin secretion by inhibiting all PKC isoforms, including the atypical PKC isoform(s), whereas the observed lack of effects of Gö 6976 and the peptide inhibitors could reflect their selectivity for inhibition of conventional isoforms.

In summary, the available experimental observations are consistent with nutrient secretagogues operating, at least in part, through the activation of one or more DAG-insensitive atypical isoform(s) of PKC (ζ, ι/λ, and/or μ), whereas the potentiation of nutrient-induced secretion by PLC-activating nonnutrients requires the activation of one or more of the DAG-sensitive isoforms of PKC that are expressed in β-cells (α, β, δ, and/or ε).

D. Mitogen-Activated Protein Kinases

The mitogen-activated protein kinases (MAPK) are a distinct group of protein serine/threonine kinases, which are activated by dual phosphorylation on threonine and tyrosine residues. They appear to be ubiquitous in mammalian cells, and include p42/44 MAP kinases, p38 reactivating kinase (RK), and stress-activated protein kinases (SAP kinases). There have been many recent reviews on the structure and regulation of the MAPK family of kinases and readers are referred to these for detailed comment (e.g., Marshall, 1994; Cano and Mahadaven, 1995; Seger and Krebs, 1995).

At least two forms of MAP kinase (42/44 kDa) are expressed in islets and β-cell lines (Frödin et al., 1995; Persaud et al., 1996; Burns et al., 1997). MAP kinases are activated by phosphorylation on tyrosine and threonine residues by MAP kinase kinase (MEK; 45 to 46 kDa), which is itself phosphorylated and activated by the upstream protein serine/threonine kinases, MEK kinase and Raf-1 (reviewed by Marshall, 1994). All of the elements of this cascade are expressed in β-cells, but the identities and functions of β-cell MAP kinase substrates have not been established. The limited available information suggests that MAP kinase activation is neither sufficient nor essential for regulated insulin secretion. Thus, glucose and PKC activation increase the activity of the 44-kDa MAP kinase in β-cells (Frödin et al., 1995), but MAP kinase activation is not associated with changes in insulin secretion (Frödin et al., 1995; Persaud et al., 1996; Burns et al.,

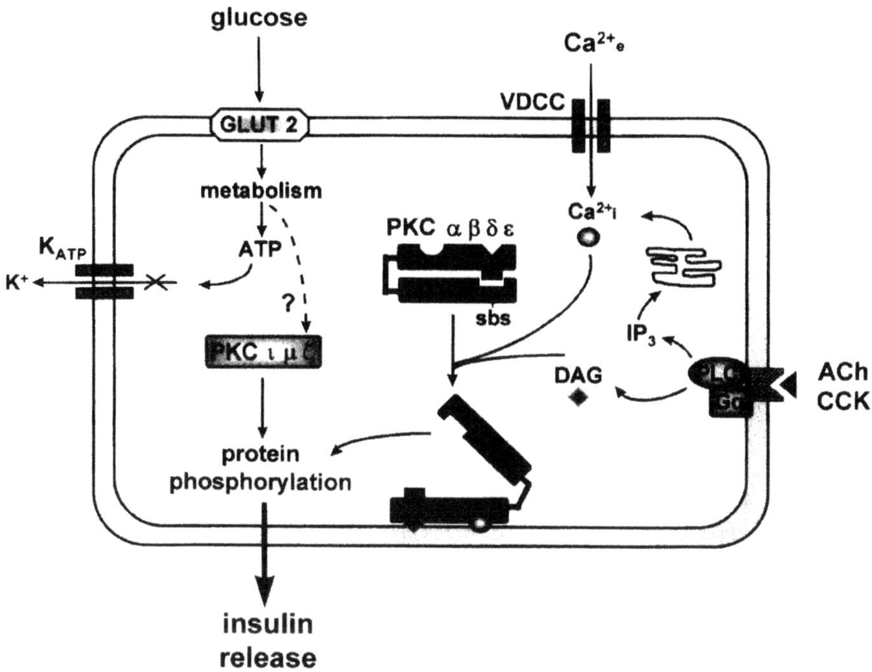

Figure 4. Multiple roles for protein kinase C (PKC) isoforms in the regulation of insulin secretion. A number of PKC isoforms are expressed in β-cells, including the diacylglycerol (DAG)–dependent α, β, δ, and ε isoforms, and the DAG-independent ζ, ι, and *μ* isoforms. The schematic diagram shows how these classes of isoforms may be involved in the regulation of insulin secretion, although little information is yet available about individual isoforms. *Nutrients:* Glucose enters the β-cell and induces depolarization and Ca^{2+} entry, as described in the legend to Figure 3. Some unknown consequence (?) of glucose metabolism may activate one or more of the atypical isoforms of PKC (ζ, ι, *μ*), leading to phosphorylation of protein substrates and insulin secretion. *Nonnutrients:* Under unstimulated conditions, PKC is autoinhibited by the binding of a pseudosubstrate region of PKC to its substrate-binding site (sbs). Activators induce conformational changes in the PKC molecule, which remove the autoinhibition and expose the substrate-binding site. Agonists such as acetylcholine (ACh) or cholecystokinin (CCK) bind to cell-surface receptors, which are coupled via the heterotrimeric G-protein, G_q, to phospholipase C (PLC). Receptor occupancy activates PLC with the consequent generation of inositol *tris*phosphate (IP_3) and diacylglycerol (DAG) by the hydrolysis of membrane inositol phospholipids. Ca^{2+} released from the endoplasmic reticulum by IP_3 may be important for activation of the conventional α and β isoforms, and the DAG can activate the conventional (α, β) and novel (δ, ε) isoforms of PKC.

1997). Similarly, MAP kinase inhibitors do not affect nutrient-induced insulin secretion (Burns et al., 1997; Khoo and Cobb, 1997). MAP kinases may be involved in other β-cell responses, and MAP kinase activation is associated with the induction of early response genes (*jun*B, *nur*77, and *zif*268) in β-cells (Frödin et al., 1995), perhaps suggestive of a role for MAP kinases in proliferative responses.

Even less is known of the role(s) of RK and SAP kinases in β-cells. Stimulatory concentrations of glucose increase the activity of RK in MIN6 β-cells (Macfarlane et al., 1997) and, to a lesser extent, in INS-1 β-cells (Khoo and Cobb, 1997). Activation of this stress-activated pathway may be involved in insulin gene transcription through the phosphorylation and activation of insulin upstream factor-1 (IUF1), a β-cell transcription factor (Macfarlane et al., 1997). The physiological roles of MAP kinases, RK, and SAP kinases in β-cells are potentially rewarding areas to which future effort should be directed.

E. Protein Tyrosine Kinases

Protein tyrosine kinases (PTKs) are important in cellular regulation, and several families of PTK have been identified in eukaryotic cells (see Hunter, 1991), but relatively little information is yet available about the identities and possible functions of PTKs and their substrates in the endocrine pancreas. The use of antiphosphotyrosine antibodies has demonstrated that a wide range of β-cell proteins are substrates for PTKs, and that these proteins are also substrates for very active phosphotyrosine phosphatase activities, suggesting rapid phosphorylation/dephosphorylation cycling of the substrates (Rothenberg et al., 1995; Velloso et al., 1995; Jonas and Henquin, 1996; Konrad et al., 1996; Persaud et al., 1996; Tazi et al., 1996).

A number of PTK activities have already been identified in β-cells, including a novel member of the *src* family of cytoplasmic PTKs (Öberg-Welsh et al., 1995); the nonreceptor tyrosine kinase, JAK2 (Sorenson and Stout, 1995); the nerve growth factor receptor PTK, TrkA (Kanaka-Gantenbein et al., 1995); and the neurotrophin receptor PTK, TrkC (Tazi et al., 1996). The insulin receptor is probably one of the most studied mammalian PTKs, and the recent detection of insulin receptor mRNA and of the insulin receptor substrate (IRS-1) protein in β-cells implies that insulin may modulate β-cell function by autocrine interactions (Harbeck et al., 1996). Glucose and insulinlike growth factor-I (IGF-I) stimulate the tyrosine phosphorylation of the 97-kDa insulin receptor β-subunit and of the 170/180-kDa insulin receptor substrates IRS-1 and IRS-2 (Rothenberg et al., 1995; Velloso et al., 1995), and glucose is also reported to enhance tyrosine phosphorylation of an unidentified 125-kDa β-cell substrate (Konrad et al., 1996). Some PTK substrates (e.g., MAP kinase, PLC-γ, phosphatidylinositol-3-kinase, TrkC) have been identified in islets and β-cell lines based on their increased tyrosine phosphorylation in response to inhibition of phosphotyrosine phosphatases or activation of receptor PTKs (Öberg-Welsh et al., 1995; Jonas and Henquin, 1996; Persaud et al., 1996; Tazi et al., 1996).

The consequences to the secretory process of increased changes in the tyrosine phosphorylation state of islet proteins are uncertain. Receptor-mediated activation of PTKs either stimulates (Brelje et al., 1994) or has no effect on (Sekine et al., 1996; Tazi et al., 1996) insulin release from cell lines and islets. An alternative approach is to enhance tyrosine phosphorylation of a wide range of islet proteins by inhibiting tyrosine phosphatases with vanadate or tungstate (Zhang et al., 1991; Persaud et al., 1996). In a recent study using mouse islets, this was accompanied by increased production of inositol phosphates and inositol phospholipids, effects which were attributed to increased tyrosine phosphorylation and activation of PLC-γ and PI 3-kinase, respectively (Jonas and Henquin, 1996). However, the vanadate-induced inhibition of tyrosine phosphatases can be dissociated from insulin secretion. Under conditions where vanadate caused marked increases in the tyrosine phosphorylation state of a number of islet proteins, it had no effect on basal insulin secretion (Zhang et al., 1991; Persaud et al., 1996), although higher concentrations of vanadate did cause some enhancement of glucose-induced insulin secretion, an effect that was attributed to alterations in membrane potential, PI metabolism, and Ca^{2+}-handling in the β-cells (Zhang et al., 1991).

Studies based on PTK inhibitors have produced conflicting results. The PTK inhibitor genistein stimulates insulin secretion from β-cell lines and islets suggesting that PTKs exert a tonic inhibitory effect on the secretory process (Sorenson et al., 1994; Jonas et al., 1995; Verspohl et al., 1995; Hisatomi et al., 1997), but other inhibitors have been reported to have no effect (Hisatomi et al., 1997), or to inhibit insulin secretion in response to nutrients and nonnutrients (Konrad et al., 1996; Hisatomi et al., 1997; Zhou and Egan, 1997; Persaud et al., 1999). PTK inhibitors have diverse effects on metabolic processes unrelated to the inhibition of PTK activity (Young et al., 1993), and it has been suggested that the stimulation of insulin release by genistein is unrelated to inhibition of PTKs (Jonas et al., 1995). As for the protein serine/threonine kinases, interpretations of the physiological functions of PTKs based solely on the use of inhibitors should be viewed with caution.

F. Phosphoprotein Phosphatases

The phosphorylation state of a protein is dependent upon the balance between the rate of its phosphorylation by protein kinases and its dephosphorylation by phosphoprotein phosphatases (PPs). The activity of PPs was originally thought to be unregulated, but it is now clear that some classes of PP are controlled by extracellular signals and are thus positioned to play key roles in the regulation of cellular processes. Many excellent reviews are available on the structure and function of PPs (e.g., Cohen, 1991; Hunter, 1995), so here we merely give a brief overview of what is known about these enzymes in β-cells.

Islets and β-cell lines express several types of protein serine/threonine phosphatase, including the divalent cation-independent PP1 and PP2A and the Ca^{2+}-dependent PP2B, which is also known as calcineurin (Sjöholm et al., 1993; Ammala et al., 1994; Murphy and Jones, 1996). It is not yet known whether the Mg^{2+}-dependent PP2C is expressed in β-cells. There have been a number of reports of insulin secretagogues affecting β-cell PP activity. For example, glucose causes a transient increase in PP1/2A activities in intact islets (Gagliardino et al., 1994; Ebihara et al., 1996), although arginine and KCl decrease PP1/2A activities in RINm5F β-cells (Sjöholm et al., 1993), and sulphonylureas are reported to increase PP1/2A activities in islet homogenates (Gagliardino et al., 1997).

Most of the evidence implicating PPs in β-cell secretory responses has been obtained using PP inhibitors, and the same caution should be applied to these studies as to studies using inhibitors of protein kinases. In addition, PPs have broad substrate specificities and, as we have seen, there are many potential sites of action of protein phosphorylation in the secretory pathway, so global inhibition of PP activity may affect many stages of the stimulus-secretion pathway.

The PP1/2A inhibitor, okadaic acid (OA), stimulates Ca^{2+}influx through voltage-activated L-type Ca^{2+} channels in RINm5F cells, which may account for its stimulatory effects on insulin secretion from these cells (Haby et al., 1994; Mayer et al., 1994). However, OA also stimulates insulin secretion from permeabilized islets maintained at a substimulatory Ca^{2+} concentration (Ratcliff and Jones 1993), and enhances exocytosis in single β-cells with minimal effects on Ca^{2+} current (Ammala et al., 1994), suggesting that it has effects on the secretory system other than enhancing Ca^{2+} influx. The actions of OA are further complicated since OA does not affect basal insulin release from intact islets, but inhibits glucose-stimulated insulin release (Ratcliff and Jones, 1993; Murphy and Jones 1996), suggesting that PP1 and/or PP2A activities are required for stimulated insulin secretion. These results could best be explained by assuming that OA is inhibiting the actions of PP1/2A at a number of sites in β-cells, including the following: (1) An early excitatory/permissive effect in secretory responses of whole cells to nutrients, the inhibition of which by OA would result in the inhibition of nutrient-induced secretion. (2) An inhibitory effect close to the point of depolarization-induced Ca^{2+} entry, the blocking of which would enhance Ca^{2+} influx and insulin release. This could be a direct effect on a VDCC subunit. (3) An inhibitory effect at a more distal site(s) in the stimulus-secretion coupling pathway, the blocking of which would result in enhanced basal and stimulated secretory responses. This could be via one or more of the proteins involved in secretory granule docking and fusion.

The role of PP2B in insulin secretion is also uncertain. The PP2B inhibitor, cyclosporin A (CsA), enhances secretory responses of MIN6 and HIT-T15 β-cells to glucose, tolbutamide, and KCl (Ebihara et al., 1996), but has no effect on either basal or glucose-stimulated insulin secretion from rat islets (Murphy and Jones, 1996). PP2B has recently been implicated in the regulation of gene transcription in β-cells. Cyclic AMP response element–mediated gene transcription in HIT cells was inhibited by CsA and by another PP2B inhibitor, FK506 (Schwaninger et al., 1993, 1995), and arginine vasopressin (AVP) has been reported to induce CsA- and FK506-sensitive phosphorylation and activation of the CRE-binding protein in these cells (Eckert et al., 1996).

The phosphorylating actions of protein tyrosine kinases are reversed by phosphotyrosine phosphatases (PTP), but very little information is currently available about these enzymes in β-cells. Two closely related proteins, IA-2 and phogrin (IA-2β), with sequence homology to PTPs, have been identified in a variety of insulinoma cell lines and in islets, and phogrin has been localized to the membranes of insulin secretory granules (Lan et al., 1994; Wasmeier and Hutton, 1996; Hatfield et al., 1997). The importance of PTPs in regulating β-cell secretory responses may become more apparent when more is known about the role of protein tyrosine kinases in this process.

It is worth remembering that for protein phosphorylation to be a useful regulatory mechanism within cells, it needs to be rapidly reversible, so there is no question that PPs are essential for normal β-cell function. However, the important, and as yet largely unanswered, question is whether the PPs themselves offer a point of regulation. Future studies directed toward whether, and through what mechanisms, physiologically relevant insulin secretagogues modulate the activities of β-cell PPs should go a long way toward answering this question.

III. SUMMARY AND CONCLUSIONS

Studies of pancreatic β-cell physiology are often driven, albeit indirectly, by Type 2 diabetes mellitus. This is because we need a full understanding of how normal β-cells recognize and respond to external signals before we can understand the β-cell defects that cause diabetes. Furthermore, we need to know which pathways are important for normal function before we can target new therapies at defective β-cells by enhancing or inhibiting existing intracellular pathways. The numerous studies quoted in this chapter demonstrate that protein kinases play vital roles in the regulation of β-cell function, and are intimately involved in transducing extracellular signals into physiologically appropriate insulin secretory responses. It is now clear that β-cells use metabolic processes and a variety of cell surface receptors to recognize a wide range of external signals, and that these numerous and disparate signals can be transduced into the activation of a few regulated protein kinase enzymes. A deeper knowledge of how these enzymes regulate insulin secretion may be of great value in devising new pharmacological approaches to Type 2 diabetes, and in designing genetically engineered replacement cells for people with type I diabetes.

Although there are some inconsistencies in the existing literature, most of the published studies concur with the experimental model for insulin secretion shown

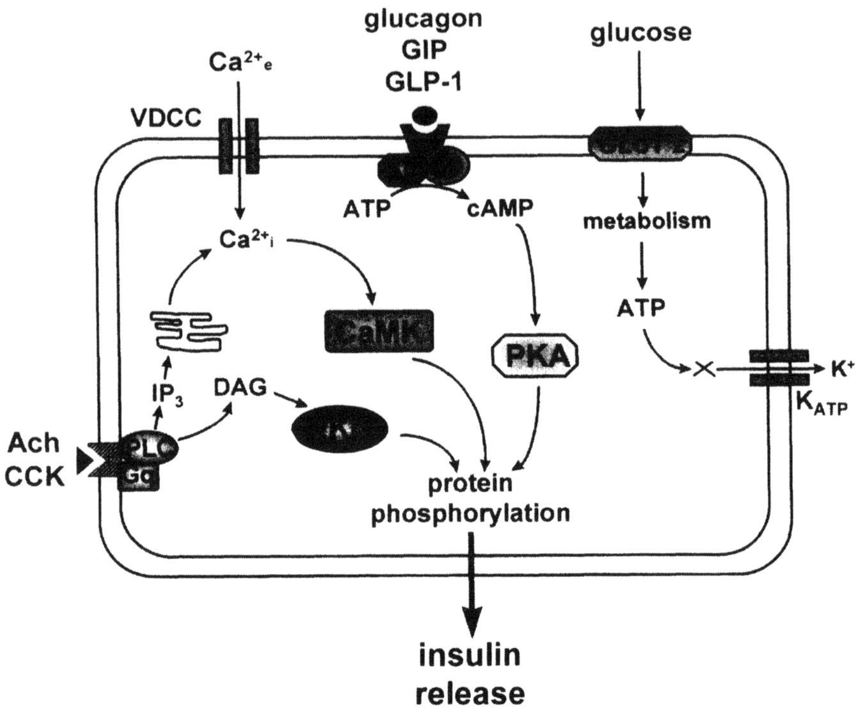

Figure 5. Protein phosphorylation and the regulation of insulin secretion: a consensus model. The schematic diagram shows a consensual model of the involvement of protein kinases in the regulation of insulin secretion from pancreatic β-cells, incorporating elements from Figures 2 through 4. Nutrient secretagogues influence the secretory process through the activation of CaMK II as a consequence of an influx of extracellular Ca^{2+} to increase $Ca^{2+}{}_i$. Receptor-operated nonnutrients which are coupled to phospholipase C (PLC) may activate CaMK by the IP_3-induced release of Ca^{2+} from intracellular stores, and activate conventional and/or novel isoforms of PKC by the generation of diacyl glycerol (DAG). Other receptor-operated nonnutrients, which are coupled to adenylate cyclase (AC) activate protein kinase A (PKA) by increasing intracellular concentrations of cyclic adenosine monophosphate (cAMP).

in Figure 5. In this consensus model, Ca^{2+}-regulated protein kinases, particularly CaMK II, are involved in the initiation of insulin secretory responses by the physiologically important nutrient secretagogues, although other Ca^{2+}-sensitive response elements may also be required for a full secretory response. The involvement of PKA and PKC in nutrient-induced insulin secretion is still a matter of some debate, although there is no doubt that kinase isoforms from the PKA and PKC families are required for the modulation of nutrient-induced secretory responses by receptor-operated, nonnutrient secretagogues acting via adenylate cyclase or phospholipase C, respectively. There is ample evidence that β-cells express other kinase activities, including the MAP kinase cascades and numerous tyrosine kinases, but there is currently insufficient information to determine whether these transduction pathways are directly involved in the regulation of insulin secretion. It also seems likely that at least some of these elements will be involved in the regulation of β-cell growth and proliferation, and the ability to manipulate these parameters may also have important long-term implications for treatment of diabetes.

ACKNOWLEDGMENTS

The research into β-cell function performed in our laboratories at King's College, London, is supported by the British Diabetic Association and the Medical Research Council.

REFERENCES

Abeliovich, A., Paylor, R., Chen, C., Kim, J.J., Wehner, J.M. and Tonegawa, S. (1993). PKC-gamma mutant mice exhibit mild deficits in spatial and contextual learning. Cell 75, 1263–1271.

Ammala, C., Ashcroft, F.M. and Rorsman, P. (1993a). Calcium-independent potentiation of insulin release by cyclic AMP in single β-cells. Nature 363, 356–358.

Ammala, C., Eliasson, L., Bokvist, K., Berggren, P.-O., Honkanen, R.E., Sjöholm, Å. and Rorsman, P. (1994). Activation of protein kinases and inhibition of protein phosphatases play a central role in the regulation of exocytosis in mouse pancreatic β cells. Proc. Natl. Acad. Sci. USA 91, 4343–4347.

Ammala, C., Eliasson, L., Bokvist, K., Larsson, O., Ashcroft, F.M. and Rorsman, P. (1993b). Exocytosis elicited by action potentials and voltage-clamp calcium currents in individual mouse pancreatic B-cells. J. Physiol. 472, 665–688.

Arkhammar, P., Juntti-Berggren, L., Larsson, O., Welsh, M., Nanberg, E., Sjoholm, Å., Kohler, M. and Berggren, P.-O. (1994). Protein kinase C modulates the insulin secretory process by maintaining a proper function of the β-cell voltage activated channels. J. Biol. Chem. 269, 2743–2749.

Babb, E.L., Tarpley, J., Landt, M. and Easom, R.A. (1996). Muscarinic activation of Ca^{2+}/calmodulin-dependent protein kinase II in pancreatic islets. Temporal idssociation of kinase activation and insulin secretion. Biochem. J. 317, 167–172.

Barnett, D.W., Pressel, D.M., Chern, H.T., Scharp, D.W. and Misler, S. (1994). cAMP-enhancing agents "permit" stimulus-secretion coupling in canine pancreatic islet β-cells. J. Membrane Biol. 138, 113–120.

Basudev, H., Jones, P.M. and Howell, S.L. (1995). Protein phosphorylation in the regulation of insulin secretion: the use of site-directed inhibitory peptides in electrically permeabilised islets of Langerhans. Acta Diabetologica 32, 32–37.

Breen, M.A. and Ashcroft, S.J.H. (1997a). Human islets of Langerhans express multiple isoforms of calcium/calmodulin-dependent protein kinase II. Biochem. Biophys. Res. Commun. 236, 473–478.

Breen, M.A. and Ashcroft, S.J.H. (1997b). A truncated isoform of Ca^{2+}/calmodulin-dependent protein kinase II expressed in human islets of Langerhans may result from trans-splicing. FEBS Lett. 409, 375–379.

Brelje, T.C., Parsons, J.A. and Sorenson, R.L. (1994). Regulation of islet β-cell proliferation by prolactin in rat islets. Diabetes 43, 263–273.

Burns, C.J., Howell, S.L., Jones, P.M. and Persaud, S.J. (1997). Glucose-stimulated insulin secretion from rat islets of Langerhans is independent of mitogen-activated protein kinase activation. Biochem. Biophys. Res. Comm. 239, 447–450.

Calle, R., Ganesan, S., Smallwood, J.I. and Rasmussen, H. (1992). Glucose-induced phosphorylation of myristoylated alanine-rich C kinase substrate (MARCKS) in isolated rat pancreatic islets. J. Biol. Chem. 267, 18723–18727.

Cano, E. and Mahadevan, L.C. (1995). Parallel signal processing among mammalian MAPKs. Trends Biochem. Sci. 20, 117–122.

Christie, M.R. and Ashcroft, S.J.H. (1984). Cyclic AMP-dependent protein phosphorylation and insulin secretion in intact islets of Langerhans. Biochem. J. 218, 87–99.

Cohen, P. (1991) Classification of protein-serine/threonine phosphatases: Identification and quantitation in cell extracts. Methods Enzymol. 201, 389–398.

Colbran, R.J., Schworer, C.M., Hashimoto, Y., Fong, Y.-L., Rich, D.P., Smith, M.K. and Soderling, T.R. (1989). Calcium/calmodulin-dependent protein kinase II. Biochem. J. 258, 313–325.

Colca, J.R., Brooks, C.L., Landt, M. and McDaniel, M.L. (1983). Correlation of Ca^{2+} and calmodulin-dependent protein kinase activity with secretion of insulin from rat islets of Langerhans. Biochem. J. 212, 819–827.

Corbin, J.D., Keely, S.L. and Park, C.R. (1975). The distribution and dissociation of cyclic adenosine 3′:5′monophosphate-dependent protein kinases in adipose, cardiac, and other tissues. J. Biol. Chem. 250, 218–225.

Cunningham, J.M., Mabley, J.G., Delaney, C.A. and Green, I.C. (1994). The effect of nitric oxide donors on insulin secretion, cyclic GMP and cyclic AMP in rat islets of Langerhans and the insulin-secreting cell lines HIT-T15 and RINm5F. Mol. Cell. Endocrinol. 102, 23–29.

Davis, P.D., Hill, C.H., Keech, E., Lawton, G., Nixon, J.S., Sedgwick, A.D., Wadsworth, J., Westmacott, D. and Wilkinson, S.E. (1989). Potent selective inhibitors of protein kinase C. FEBS Lett. 259, 61–63.

de Wit, R.J.W., Hekstra, D., Jastorff, B., Stec, W.J., Barianak, J., Van Driel, R. and Van Haastert, P.J.M. (1984). Inhibitory action of certain cyclophosphate derivatives of cAMP on cAMP-dependent protein kinases Eur. J. Biochem. 142, 255–260.

Deeney, J.T., Cunningham, B.A., Chheda, S., Bokvist, K., Juntti-Berggren, L., Lam, K., Korchak, H.M., Corkey, B.E., Berggren, P.-O. (1996). Reversible Ca^{2+}-dependent translocation of protein kinase C and glucose-induced insulin release. J. Biol. Chem. 271,18154–18160.

Ding, W.-G. and Gromada, J. (1997). Protein kinase A-dependent stimulation of exocytosis in mouse pancreatic β-cells by glucose-dependent insulinotropic polypeptide. Diabetes 46, 615–621

Ding, W.-G., Renström, E., Rorsman, P., Buschard, K., Gromada, J. (1997). Glucagon-like peptide I and glucose-dependent insulinotropic polypeptide stimulate Ca^{2+}-induced secretion in rat α-cells by a protein kinase A-mediated mechanism. Diabetes 46, 792–800.

Dunlop, M., Clark, S. (1995). Glucose-induced phosphorylation and activation of a high molecular weight cytosolic phospholipase A_2 in neonatal rat pancreatic islets. Int. J. Biochem. Cell. Biol. 27,1191–1199.

Dunlop, M. and Larkins, R.G. (1986). Glucose-induced phospholipid-dependent protein phosphorylation in neonatal rat islets. Arch. Biochem. Biophys. 248, 562–569.

Easom, R.A., Hughes, J.H., Landt, M., Wolf, B.A., Turk, J. and McDaniel, M.L. (1989). Comparison of effects of phorbol esters and glucose on protein kinase C activation and insulin secretion in pancreatic islets. Biochem. J. 264, 27–33.

Easom, R.A., Landt, M., Colca, J.R., Hughes, J.H., Turk, J. and McDaniel, M.L. (1990). Effects of insulin secretagogues on protein kinase C-catalysed phosphorylation of an endogenous substrate in isolated pancreatic islets. J. Biol. Chem. 265, 14938–14946.

Easom, R.A., Tarpley, J.L., Filler, N.R. and Bhatt, H. (1998). Dephosphorylation and deactivation of the Ca^{2+}/calmodulin-dependent protein kinase II in βTC3 cells is mediated by Mg^{2+}- and okadaic-acid-sensitive protein phosphatases. Biochem. J. 329, 283–288.

Ebihara, K., Fukunaga, K., Matsumoto, K., Shichiri, M. and Miyamoto, E. (1996). Cyclosporin A stimulation of glucose-induced insulin secretion in MIN6 cells. Endocrinology 137, 5255–5263.

Eckert, B. Schwaninger, M. and Knepel, W. (1996). Calcium-mobilizing insulin secretagogues stimulate transcription that is directed by the cyclic adenosine-3',5'-monophosphate calcium response element in a pancreatic-islet beta cell line. Endocrinology 137, 225–233.

Fletcher, D.J. and Ways, D.K. (1991). Age-dependent expression of protein kinase C isoforms in rat islets. Diabetes 40, 1496–1503.

Frödin, M., Sekine, N., Roche, E., Filloux, C., Prentki, M., Wollheim, C.B. and Van Obberghen, E. (1995). Glucose, other secretagogues, and nerve growth factor stimulate mitogen-activated protein kinase in the insulin-secreting b-cell line, INS-1. J. Biol. Chem. 270, 7882–7889.

Gagliardino, J.J., Garcia, M.E. and Mundinam, C. (1994). Effect of glucose on islet protein phosphatase activity. Diabet. Res. Clin. Pract. 11, 290.

Gagliardino, J.J., Krinks, M.H. and Gagliardino, E.E. (1991). Identification of the calmodulin-regulated protein phosphatase, calcineurin, in rat pancreatic islets. Biochim. Biophys. Acta 1091, 370–373.

Gagliardino, J.J., Rossi, P.F.C. and Garcia, M.E. (1997). Inhibitory effect of sulfonylureas on protein phosphatase activity in rat pancreatic islets. Acta Diabetol. 34, 6–9.

Ganesan, S., Calle, R., Zawalich, K., Smallwood, J.I., Zawalich, W.S. and Rasmussen, H. (1990). Glucose-induced translocation of protein kinase C in rat pancreatic islets. Proc. Natl. Acad. Sci. USA 87, 9893–9897.

Ganesan, S., Calle, R., Zawalich, K., Zawalich, W., Schulman, G.I., Rasmussen, H. (1992). Immunocytochemical localisation of α-protein kinase C in rat pancreatic β-cells during glucose-induced insulin secretion. J. Cell Biol. 119, 313–324.

Gao, Z.Y., Gilon, P. and Henquin, J.-C. (1994). The role of protein kinase-C in signal transduction through vasopressin and acetylcholine in pancreatic β-cells from normal mouse. Endocrinology 135, 191–199.

Haby, C., Larsson, O., Islam, M.S., Aunis, D., Berggren, P.-O. and Zwiller, J. (1994). Inhibition of serine/threonine protein phosphatases promotes opening of voltage-activated L-type Ca^{2+} channels in insulin-secreting cells. Biochem. J. 298, 341–346.

Hagiwara, S., Sakurai, T., Tashiro, F., Hashimoto, Y., Matsuda, Y., Nonomura, Y. and Miyazaki, J.-I. (1995). An inhibitory role for phosphatidylinositol 3-kinase in insulin secretion from pancreatic B cell line MIN6. Biochem. Biophys. Res. Commun. 214, 51–59.

Hanks, S.K. and Hunter, T. (1995). The eukaryotic protein kinase superfamily: kinase (catalytic) domain structure and classification. FASEB J. 9, 576–596.

Harbeck, M.C., Louie, D.C., Howland, J., Wolf, B.A. and Rothenberg, P.L. (1996). Expression of insulin receptor mRNA and insulin receptor substrate 1 in pancreatic islet β-cells. Diabetes 45, 711–717.

Hardie, G. and Hanks, S. (1995). *The Protein Kinase Factsbook.* Vols. 1 and 2. Academic Press, London.

Harris, T.E., Persaud, S.J. and Jones, P.M. (1996a). Atypical isoforms of PKC and insulin secretion from pancreatic β-cells: Evidence using Gö 6976 and Ro 31-8220 as PKC inhibitors. Biochem. Biophys. Res. Commun. 227, 672–676.

Harris, T.E., Persaud, S.J. and Jones, P.M. (1997). Pseudosubstrate inhibition of cyclic AMP-dependent protein kinase in intact pancreatic islets: Effects on cyclic AMP-dependent and glucose-dependent insulin secretion. Biochem. Biophys. Res. Commun. 232, 648–651.

Harris, T.E., Persaud, S.J., Saermark, T. and Jones, P.M. (1996b). A myristoylated pseudosubstrate peptide inhibitor of protein kinase C: Effects on glucose- and carbachol-induced insulin secretion. Mol. Cell. Endocrinol. 121, 133–141.

Harrison, D.E., Ashcroft, S.J.H., Christie, M.R. and Lord, J.M. (1984). Protein phosphorylation in the pancreatic B-cell. Experientia 40, 1075–1084.

Harrison, D.E., Poje, M., Rocic, B. and Ashcroft, S.J.H. (1986). Effects of dehydrouramil on protein phosphorylation and insulin secretion in rat islets of Langerhans. Biochem. J. 237, 191–196.

Hartwig, J.H., Thelen, M., Rosen, A., Janmey, P.A., Nairn, A.C. and Aderem, A. (1992). MARCKS is an actin filament crosslinking protein regulated by protein kinase C and calcium-calmodulin. Nature 356, 618–622.

Hatfield, E.C.I., Hawkes, C.J., Payton, M.A. and Christie, M.R. (1997). Cross reactivity between IA-2 and phogrin/IA-2β in binding of autoantibodies in IDDM. Diabetologia 40, 1327–1333.

Hisatomi, M., Hayakawa, T., Hidaka, H. and Niki, I. (1997). Modulation of tyrosine kinase activity has multiple actions on insulin release from the pancreatic β-cell: Studies with lavendustin A. Jpn. J. Pharmacol. 74, 203–208.

Holz, G.G., Kuhtreiber, W.M. and Habener, J.F. (1993). Pancreatic beta-cells are rendered glucose-competent by the insulinotropic hormone glucagon-like peptide-1(7-37). Nature 361, 362–365.

Hug, H. and Sarre, T.F. (1993). Protein kinase C isoenzymes: divergence in signal transduction? Biochem. J. 291, 329–343.

Hughes, S.J. and Ashcroft, S.J.H. (1992). Cyclic AMP, protein phosphorylation and insulin secretion. In: *Nutrient Regulation of Insulin Secretion.* (Flatt, P.R., Ed.), pp. 271–288. Portland Press, London.

Hughes, S.J., Chalk, J.G. and Ashcroft, S.J.H. (1990). The role of cytosolic free Ca^{2+} and protein kinase C in acetylcholine-induced insulin release in the clonal β-cell line, HIT-T15. Biochem. J. 267,227–232.

Hughes, S.J. Smith, H. and Ashcroft, S.J.H. (1993). Characterisation of Ca^{2+}/calmodulin-dependent protein kinase in rat pancreatic islets. Biochem. J. 289, 795–800.

Hunter, T. (1991). Protein kinase classification. Methods Enzymol. 200, 3–37.

Hunter, T. (1995). Protein kinases and phosphatases: the yin and yang of protein phosphorylation and signalling. Cell 80, 225–236.

Inagaki, N., Tsuura, Y., Namba, N., Masuda, K., Gonoi, T., Horie, M., Seino, Y., Mizuta, M. and Seino, S. (1995). Cloning and functional characterization of a novel ATP-sensitive potassium channel ubiquitously expressed in rat tissues, including pancreatic islets, pituitary, skeletal muscle and heart. J. Biol. Chem. 270, 5691–5694.

Ito, A., Saito, N., Taniguchi, H., Chiba, T., Kikkawa, U., Saitoh, Y. and Tanaka, C. (1989). Localization of βII subspecies of protein kinase C in β-cells. Diabetes 38, 1005–1011.

Jonas, J.C. and Henquin, J.C. (1996). Possible involvement of a tyrosine kinase-dependent pathway in the regulation of phosphoinositide metabolism by vanadate in normal mouse islets. Biochem. J. 315, 49–55.

Jonas, J.C., Plant, T.D., Gilon, P., Detimary, P., Nenquin, M. and Henquin, J.C. (1995). Multiple effects and stimulation of insulin secretion by the tyrosine kinase inhibitor genistein in normal mouse islets. Br. J. Pharmacol. 114,872–880.

Jones, P.M., Fyles, J.M. and Howell, S.L. (1986). Regulation of insulin secretion by cAMP in islets of Langerhans permeabilised by high voltage discharge. FEBS Lett. 205, 205–209.

Jones, P.M., Persaud, S.J., Bjaaland, T., Pearson, J.D. and Howell, S.L. (1992a). Nitric oxide is not involved in the initiation of insulin secretion from rat islets of Langerhans. Diabetologia 35, 1020–1027.

Jones, P.M., Persaud, S.J. and Howell, S.L. (1992b). Insulin secretion and protein phosphorylation in PKC-depleted islets of Langerhans. Life Sciences 50, 761–767.

Jones, P.M., Salmon, D.M.W. and Howell, S.L. (1988). Protein phosphorylation in electrically permeabilised islets of Langerhans: effects of Ca^{2+}, cyclic AMP, a phorbol ester and noradrenaline. Biochem. J. 254, 397–403.

Jones, P.M., Stutchfield, J. and Howell, S.L. (1985). Effects of Ca^{2+} and a phorbol ester on insulin secretion from islets of Langerhans permeabilised by high voltage discharge. FEBS Lett. 191, 102–106.

Kanaka-Gantenbein, C., Dicou, E., Czernichow, P. and Scharfmann, R. (1995). Presence of nerve growth factor and its receptors in an in vitro model of islet cell development: Implication in normal islet morphogenesis. Endocrinology 136, 3154–3162.

Karlsson, S. and Ahrén, B. (1991). Cholecystokinin-stimulated insulin secretion and protein kinase C in rat pancreatic islets. Acta Physiol. Scand. 142, 397–403.

Kemp, B.E., Pearson, R.B. and House, C.M. (1991). Pseudosubstrate-based peptide inhibitors. Methods Enzymol. 201, 287–30.

Khoo, S. and Cobb, M.H. (1997). Activation of mitogen-activating protein kinase by glucose is not required for insulin secretion. Proc. Natl. Acad. Sci. USA 94, 5599–5604.

King, B.R., Ways, D.K. and Fletcher, D.J. (1996). Identification of protein kinase C isoforms in RINm5F and βTC-3 insulin-secreting cells. 10th International Congress of Endocrinology, P2–53.

Knutson, K.L. and Hoenig, M. (1994). Identification and subcellular characterization of protein kinase-C isoforms in insulinoma β-cells and whole islets. Endocrinology 135, 881–886.

Knutson, K.L., Hoenig, M. (1996). Regulation of distinct pools of protein kinase C δ in β-cells. J. Cell. Biochem. 60, 130–138.

Kobayashi, E., Nakano, H., Morimoto, M. and Tamaoki, T. (1989). Calphostin C (UCN-1028C), a novel microbial compound, is a highly potent and specific inhibitor of protein kinase C. Biochem. Biophys. Res. Commun. 159, 548–553.

Konrad, R.J., Dean, R.M., Young, R.A., Billings, P.C. and Wolf, B.A. (1996). Glucose-induced tyrosine phosphorylation of p125 in beta cells and pancreatic islets. A novel proximal signal in insulin secretion. J. Biol. Chem. 271, 24179–24186.

Konrad, R.J., Jolly, Y.C., Major, C. and Wolf, B.A. (1993). Fuel secretagogue stimulation of arachidonic acid accumulation in fresh and cultured pancreatic islets. Mol. Cell Endocrinol. 92, 135–140.

Kowluru, A. and MacDonald, M.J. (1984). Protein phosphorylation in pancreatic islets: Evidence for separate Ca^{2+} and cAMP-enhanced phosphorylation of two 57,000 Mr proteins. Biochem. Biophys. Res. Commun. 118, 797–804.

Kowluru, A., Seavey, S.E., Rabaglia, M.E., Nesher, R. and Metz, S.A. (1996). Carboxymethylation of the catalytic subunit of protein phosphatase 2A in insulin-secreting cells: Evidence for functional consequences on enzyme activity and insulin secretion. Endocrinology 137, 2315–2323.

Kuo, J.F. (1974). Guanosine 3′:5′-monophosphate-dependent protein kinases in mammalian tissues. Proc. Natl. Acad. Sci. USA 71, 4037–4041.

Lan, M.S., Lu, J., Goto, Y. and Notkins, A.L. (1994). Molecular-cloning and identification of a receptor-type protein-tyrosine phosphatase, IA-2, from human insulinoma. DNA Cell Biol. 13, 505–514.

Laychock, S.G. (1987). Effects of guanosine 3′,5′-monophosphate on glucose utilization in isolated islets of Langerhans. Endocrinology 120, 517–524.

Leiser, M. and Fleischer, N. (1996). cAMP-dependent phosphorylation of the cardiac-type a1 subunit of the voltage-dependent Ca^{2+} channel in a murine pancreatic β-cell line. Diabetes 45, 1412–1418.

Li, G., Hidaka, H. and Wollheim, C.B. (1992). Inhibition of voltage-gated Ca^{2+} channels and insulin secretion in HIT cells by the Ca^{2+}/calmodulin-dependent protein kinase II inhibitor KN-62: comparison with antagonists of calmodulin and L-type Ca^{2+} channels. Mol. Pharmacol. 42, 489–498.

Li, G., Regazzi, R., Ullrich, S., Pralong, W.-F. and Wollheim, C.B. (1990). Potentiation of stimulus-induced insulin secretion in protein kinase C-deficient RINm5F cells. Biochem. J. 272, 637–645.

Loweth, A.C., Williams, G.T., Scarpello, J.H.B. and Morgan, N.G. (1997). Evidence for the involvement of cGMP and protein kinase G in nitric oxide-induced apoptosis in the pancreatic β-cell line HIT-T15. FEBS Lett. 400, 285–288.

MacDonald, M.J., Kowluru, A. and Chang, C.M. (1985). Evidence for phosphorylation of pancreatic islet pyruvate kinase. Metabolism 34, 600–603.

Macfarlane, W.M., Smith, S.B., James, R.F.L., Clifton, A.D., Doza, Y.N., Cohen, P. and Docherty, K. (1997). The p38/reactivating kinase mitogen-activated protein kinase cascade mediates the activation of the transcription factor insulin upstream factor 1 and insulin gene transcription by high glucose in pancreatic β-cells. J. Biol. Chem. 272, 20936–20944.

Malaisse, W.J. and Sener, A. (1985). Inhibition by 1-(isoquinolinesulfonyl)-2-methylpiperazine (H-7) of protein kinase C activity and insulin release in pancreatic islets. IRCS Med. Sci. 13, 1239–1240.

Marshall, C.J. (1994). MAP kinase kinase kinase, MAP kinase kinase and MAP kinase. Curr. Opi. Genet. Dev. 4, 82–89.

Martiny-Baron, G., Kazanietz, M.G., Mischak, H., Blumberg, P.M., Kochs, G., Hug, H., Marmé, M. and Schächtele, C. (1993). Selective inhibition of protein kinase C isozymes by the indolocarbazole Gö 6976. J. Biol. Chem. 268, 9194–919.

Matsumoto, K., Fukunaga, K., Miyazaki, J.-I., Shichiri, M. and Miyamoto, E. (1995). Ca^{2+}/calmodulin-dependent protein kinase II and synapsin I-like protein in mouse insulinoma MIN6 cells. Endocrinology 136, 3784–3793.

Mayer, P., Jochum, C., Schatz, H. and Pfeiffer, A. (1994). Okadaic acid indicates a major function for protein phosphatases in stimulus-response coupling of RINm5F rat insulinoma cells. Exp. Clin. Endocrinol. 102, 313–319.

Metz, S.A. (1988). Is protein kinase C required for physiologic insulin release? Diabetes 37, 3–7.

Möhlig, M., Mayer, P., Horn, P.A., Schatz, H. and Pfeiffer, A. (1995). Overexpression of calcium/calmodulin-dependent protein kinase II in insulinoma cells by use of a retroviral vector. Exp. Clin. Endocrinol. 103, 71–73.

Möhlig, M., Wolter, S., Mayer, P., Lang, J., Osterhoff, M., Horn, P.A., Schatz, H. and Pfeiffer, A. (1997). Insulinoma cells contain an isoform of Ca^{2+}/calmodulin-dependent protein kinase II δ associated with insulin secretion vesicles. Endocrinology 138, 2577–2584.

Montague, W. and Howell, S.L. (1972). Mode of action of adenosine 3′:5′-cyclic monophosphate in mammalian islets of Langerhans: Preparation and properties of islet-cell protein phosphokinase. Biochem. J. 129, 551–560.

Morgan, N.G. and Montague, W. (1985). Studies on the mechanism of inhibition of glucose-stimulated insulin secretion by noradrenaline in rat islets of Langerhans. Biochem. J. 226, 571–576.

Murphy, L.I. and Jones, P.M. (1996). Phospho-serine/threonine phosphatases in rat islets of Langerhans: identification and effect on insulin secretion. J. Mol. Endocrinol. 117, 195–202.

Niki, I., Okazaki, K., Saitoh, M., Niki, A., Niki, H., Tamagawa, T., Iguchi, A., Hidaka, H. (1993). Presence and possible involvement of Ca/calmodulin-dependent protein kinases in insulin release from the rat pancreatic β cell. Biochem. Biophys. Res. Commun. 191, 255–261.

Öberg, C., Waltenberger, J., Claesson-Welsh, L. AND Welsh, M. (1994). Expression of protein tyrosine kinases in islet cells: possible role of the Flk-1 receptor for β-cell maturation from duct cells. Growth Factors 10, 115–126.

Öberg-Welsh, C. and Welsh, M. (1995). Cloning of BSK, a murine FRK homologue with a specific pattern of tissue distribution. Gene 152, 239–242.

Ohnishi, M., Tokuda, M., Masaki, T., Fujimura, T., Tai, Y., Itano, T., Matsui, H., Ishida, T., Konishi, R., Takahara, J. and Hatase, O. (1995). Involvement of annexin-I in glucose-induced insulin secretion from rat pancreatic islets. Endocrinology 136, 2421–2426.

Onoda, K., Hagiwara, M., Hachiya, T., Usuda, N., Nagata, T. and Hidaka, H. (1990). Different expression of protein kinase C isozymes in pancreatic islet cells. Endocrinology 126, 1235–1240.

Penn, E.J., Brocklehurst, K.W., Sopwith, A.M., Hales, C.N. and Hutton, J.C. (1982) Ca^{2+}-calmodulin dependent myosin light-chain phosphorylating activity in insulin-secreting cells. FEBS Lett. 139, 4–8.

Persaud, S.J., Harris, T.E., Burns, C.J., and Jones, P.M. (1999). Tyrosine kinases play a permissive role in glucose-induced insulin secretion and from adult rat islets. J. Mol. Endocrinol. 22, 19–28.

Persaud, S.J. and Jones, P.M. (1994). Antisense oligonucleotide inhibition of gene expression: Application to endocrine systems. J. Mol. Endocrinol. 12, 127–130.

Persaud, S.J. and Jones, P.M. (1995). Inhibition of glucose-stimulated insulin secretion by Ro 31-8220, a protein kinase C inhibitor Endocrine 3, 285–289.

Persaud, S.J., Jones, P.M. and Howell, S.L. (1990). Glucose-stimulated insulin secretion is not dependent on the activation of protein kinase A. Biochem. Biophys. Res. Commun. 173, 833–839.

Persaud, S.J., Jones, P.M. and Howell, S.L. (1991a). Activation of protein kinase C is essential for sustained insulin secretion in response to cholinergic stimulation. Biochim. Biophys. Acta 1091, 120–122.

Persaud, S.J., Jones, P.M. and Howell, S.L. (1991b). Activation of protein kinase C is not required for glyceraldehyde-stimulated insulin secretion from rat islets. Biochim. Biophys. Acta 1095, 183–185.

Persaud, S.J., Jones, P.M., and Howell, S.L. (1992). The role of protein kinase C in insulin secretion. In: *Nutrient Regulation of Insulin Secretion.* (Flatt, P.R., Ed.), pp. 247–269. Portland Press, London.

Persaud, S.J., Jones, P.M. and Howell, S.L. (1993a). Activation of protein kinase C alleviates noradrenaline inhibition of insulin secretion. Biochem. J. 289, 497–501.

Persaud, S.J., Jones, P.M. and Howell, S.L. (1993b). Staurosporine inhibits protein kinases activated by Ca^{2+} and cyclic AMP in addition to inhibiting protein kinase C in rat islets of Langerhans. Mol. Cell. Endocrinol. 94, 55–60.

Persaud, S.J., Jones, P.M. and Howell, S.L. (1993c). Stimulation of insulin secretion by cholecystokinin-8S: the role of protein kinase C. Pharmacol. Comm. 3, 39–47.

Persaud, S.J., Jones, P.M., Sugden, D., Howell, S.L. (1989a). The role of protein kinase C in cholinergic stimulation of insulin secretion from rat islets of Langerhans. Biochem. J. 264, 753–758.

Persaud, S.J., Jones, P.M., Sugden, D. and Howell, S.L. (1989b). Translocation of protein kinase C in rat islets of Langerhans: effects of a phorbol ester, carbachol and glucose. FEBS Lett. 245, 80–84.

Persaud, S.J., Wheeler-Jones, C.P.D., Jones, P.M. (1996). The mitogen-activated protein kinase pathway in rat islets of Langerhans. Biochem. J. 313, 119–124.

Peter-Riesch, B., Fathi, M., Schlegel, W. and Wollheim, C.B. (1988). Glucose and carbachol generate 1,2-diacylglycerols by different mechanisms in pancreatic islets. J. Clin. Invest. 81, 1154–1161.

Pipeleers, D.G., Schuit, F.C., In't Veld, P.A., Maes, E., Hooghe-Peters, E.L., van de Winkel, M., Gepts, W. (1985). Interplay of nutrients and hormones in the regulation of insulin release. Endocrinology 117, 825–833.

Ratcliff, H. and Jones, P.M. (1993). Effects of okadaic acid on insulin secretion from rat islets of Langerhans. Biochim. Biophys. Acta. 1175, 188–191.

Regazzi, R., Li, G., Deshusses, J. and Wollheim, C.B. (1990). Stimulus-response coupling in insulin-secreting HIT cells. Effects of secretagogues on cytosolic Ca^{2+}, diacylglycerol and protein kinase C activity. J. Biol. Chem. 265, 15003–15009.

Rothenberg, P.L., Willison, L.D., Simon, J. and Wolf, B.A. (1995). Glucose-induced insulin receptor tyrosine phosphorylation in insulin-secreting β-cells. Diabetes 44, 802–809.

Schmidt, H.H.H., Warner, T.D., Ishii, K., Sheng, H. and Murad, F. (1992). Insulin secretion from pancreatic B cells caused by L-arginine-derived nitrogen oxides. Science 255, 721–723.

Schubart, U.K., Erlichman, J. and Fleischer, N. (1980). The role of calmodulin in the regulation of protein phosphorylation and insulin release in hamster insulinoma cells. J. Biol. Chem. 255, 4120–4124.

Schwaninger, M., Blume, R., Kruger, M., Lux, G., Oetjen, E. and Knepel, W. (1995). Involvement of the Ca^{2+}-dependent phosphatase calcineurin in gene transcription that is stimulated by cAMP through cAMP response elements. J. Biol. Chem. 270, 8860–8866.

Schwaninger, M., Blume, R., Oetjen, E., Lux, G., Knepel, W. (1993). Inhibition of cAMP-responsive element-mediated gene transcription by cyclosporin A and FK506 after membrane depolarization. J. Biol. Chem. 268, 23111–23115.

Scott, J.D. (1991). Cyclic nucleotide-dependent protein kinases. Pharmacol. Ther. 50, 123–145.

Scott, J.D. and Carr, D.W. (1992). Subcellular localisation of type II cAMP-dependent protein kinase. News Physiol. Sci. 7, 143–148.

Seger, R. and Krebs, E.G. (1995). The MAPK signaling cascade. FASEB J. 9,726–735.

Seino, S. (1995). CACN4, the major α1 subunit isoform of voltage-dependent calcium channels in pancreatic β cells: a minireview of current progress. Diabetes Res. Clin. Pract. 28, S99–S103.

Sekine, N., Ullrich, S., Regazzi, R., Pralong, W.F. and Wollheim, C.B. (1996). Postreceptor signaling of growth hormone and prolactin and their effects in the differentiated insulin-secreting cell line, INS-1. Endocrinology 137, 1841–1850.

Selbie, L.A., Schmitz-Peiffer, C., Sheng, Y. and Biden, T.J. (1993). Molecular cloning and characterization of PKCι, an atypical isoform of protein kinase C derived from insulin-secreting cells. J. Biol. Chem. 268, 24296–24302.

Silva, A.J., Stevens, C.F., Tonegawa, S. and Wang, Y. (1992). Deficient hippocampal long-term potentiation in α-calcium-calmodulin kinase II mutant mice. Science 257, 201–206.

Sjöholm, Å., Honkanen, R.E. and Berggren, P.O. (1993). Characterization of serine/threonine protein phosphatases in RINm5F insulinoma cells. Biosci. Rep. 13, 349–358.

Sjöholm, Å., Honkanen, R.E. and Berggren, P.O. (1995). Inhibition of serine/threonine protein phosphatases by secretagogues by secretagogues in insulin-secreting cells. Endocrinology 136, 3391–3397.

Sorenson, E.R.L., Brelje, T.C and, Roth, C. (1994). Effect of tyrosine kinase inhibitors on islets of Langerhans: evidence for tyrosine kinases in the regulation of insulin secretion. Endocrinology 134, 1975–1978.

Sorenson, R.L. and Stout, L.E. (1995). Prolactin receptors and JAK2 in islets of Langerhans: an immunohistochemical analysis. Endocrinology 136, 4092–4098.

Southern, C., Schulster, D. and Green, I.C. (1990). Inhibition of insulin secretion by interleukin-1β and tumour necrosis factor-a via an L-arginine-dependent nitric oxide generating mechanism. FEBS Lett. 276, 42–44.

Stutchfield, J., Jones, P.M. and Howell, S.L. (1986). The effects of polymyxin B, a protein kinase C inhibitor, on insulin secretion from intact and permeabilized islets of Langerhans. Biochem. Biophys. Res. Commun. 136, 1001–1006.

Sugden, M.C., Ashcroft, S.J.H. and Sugden, P.H. (1979). Protein kinase activities in rat pancreatic islets of Langerhans. Biochem. J. 180, 219–229.

Suzuki, K., Imajoh, S., Emori, Y., Kawasaki, H., Minami, Y., Ohno, S. (1987). Calcium-activated neutral protease and its endogenous inhibitor. Activation at the cell membrane and biological function. FEBS Lett. 220, 271–277.

Suzuki, S., Oka, H., Yasuda, H., Ikeda, M., Cheng, P.Y. and Oda, T. (1983). Effect of glucagon and cyclic adenosine 3′,5′-monophosphate on protein phosphorylation in rat pancreatic islets. Endocrinology 112, 348–352.

Takasawa, S., Ishida, A., Nata, K., Nakagawa, K., Noguchi, N., Tohgo, A., Fujisawa, H. and Okamoto, H. (1995). Requirement of calmodulin-dependent protein kinase II in cyclic ADP-ribose-mediated intracellular Ca^{2+} mobilization. J. Biol. Chem. 270, 30257–30259.

Taylor, S.S., Buechler, J.A. and Yonemoto, W. (1990). cAMP-dependent protein kinase: Framework for a diverse family of regulatory enzymes. Ann. Rev. Biochem. 59, 971–1005.

Tazi, A., Le Bras, S., Lamghitnia, H.O., Vincent, J.D., Czernichow, P. and Scharfmann, R. (1996). Neurotrophin-3 increases intracellular calcium in a rat insulin-secreting cell line through its action on a functional TrkC receptor. J. Biol. Chem. 271, 10154–10160.

Thams, P., Capito, K. and Hedeskov, C.J. (1988). Stimulation by glucose of cyclic AMP accumulation in mouse pancreatic islets is mediated by protein kinase C. Biochem. J. 253, 229–234.

Thams, P., Capito, K., Hedeskov, C.J. and Kofod, H. (1990). Phorbol-ester-induced down-regulation of protein kinase C in mouse pancreatic islets. Potentiation of phase 1 and inhibition of phase 2 of glucose-induced insulin secretion. Biochem. J. 265, 777–787.

Thomas, T.P., Ellis, T.R. and Pek, S.B. (1989). Insulin release in RINm5F cells and glyceraldehyde activation of protein kinase C. Diabetes 38, 1371–1376.

Thorens, B., Deriaz, N., Bosco, D., DeVos, A., Pipeleers, D., Schuit, F., Meda, P. and Porret, A. (1996). Protein kinase A-dependent phosphorylation of GLUT2 in pancreatic β cells. J. Biol. Chem. 271, 8075–8081.

Tian, Y.-M., Urquidi, V. and Ashcroft, S.J.H. (1996). Protein kinase C in beta-cells: Expression of multiple isoforms and involvement in cholinergic stimulation of insulin secretion. Mol. Cell. Endocrinol. 119, 185–193.

Urquidi, V. and Ashcroft, S.J.H. (1995). A novel pancreatic β-cell isoform of calcium/calmodulin-dependent protein kinase II (β_3 isoform) contains a proline-rich tandem repeat in the association domain. FEBS Lett. 358, 23–26

Vara, E., Tamarit-Rodriguez, J. (1991). Does cyclic guanosine monophosphate mediate noradrenaline-induced inhibition of islet insulin secretion stimulated by glucose and palmitate? Biochem. J. 278, 243–248

Velloso, L.A., Carneiro, E.M., Crepaldi, S.C., Boschero, A.C. and Saad, M.J.A. (1995). Glucose- and insulin-induced phosphorylation of the insulin receptor and its primary substrates IRS-1 and IRS-2 in rat pancreatic islets. FEBS Lett. 377, 353–357.

Verspohl, E.J., Tollkuhn, B. and Kloss, H. (1995). Role of tyrosine kinase in insulin release in an insulin-secreting cell line (INS-1). Cell. Signalling 7, 505–512.

Wasmeier, C. and Hutton, J.C. (1996). Molecular cloning of phogrin, a protein-tyrosine phosphatase homologue localized to insulin secretory granule membranes. J. Biol. Chem. 271, 18161–18170.

Wenham, R.M., Landt, M. and Easom, R.A. (1994). Glucose activates the multifunctional Ca^{2+}/calmodulin-dependent protein kinase II in isolated rat pancreatic islets. J. Biol. Chem. 269, 4947–4952.

Wenham, R.M., Landt, M., Walters, S.M., Hidaka, H. and Easom, R.A. (1992). Inhibition of insulin secretion by KN-62, a specific inhibitor of the multifunctional Ca^{2+}/calmodulin-dependent protein kinase II. Biochem. Biophys. Res. Commun. 189, 128–133.

Wetsel, W.C., Khan, W.A., Merchenthaler, I., Rivera, H., Halpern, A.E., Phung, H.M., Negro-Vilar, A. and Hannun, Y.A. (1992). Tissue and cellular distribution of the extended family of protein kinase C isoenzymes. J. Cell Biol. 117, 121–133.

Wollheim, C.B., Dunne, M.J., Peter-Riesch, B., Bruzzone, R., Pozzan, T. and Petersen, O.H. (1988). Activators of protein kinase C depolarize insulin-secreting cells by closing K^+ channels. EMBO J. 7, 2443–2449.

Wollheim, C.B. and Regazzi, R. (1990). Protein kinase C in insulin releasing cells. Putative role in stimulus-secretion coupling. FEBS Lett. 268, 376–380.

Yedovitzky, M., Mochly-Rosen, D., Johnson, J.A., Gray, M.O., Ron, D. and Nesher, R. (1997). Translocation inhibitors define specificity of protein kinase C isoenzymes in pancreatic β-cells. J. Biol. Chem. 272, 1417–1420

Young, S.W., Poole, R.C., Hudson, A.T., Halestrap, A.P., Denton, R.P. and Tavare, J.M. (1993). Effects of tyrosine kinase inhibitors on protein kinase-independent systems. FEBS Lett. 316, 279–282.

Zaitsev, S.V., Efendic, S., Arkhammar, P., Bertorello, A.M. and Berggren, P.-O. (1995). Dissociation between changes in cytoplasmic free Ca^{2+} concentration and insulin secretion as evidenced from measurements in mouse single pancreatic islets. Proc. Natl. Acad. Sci. 92, 9712–9716.

Zawalich, W.S. (1996). Regulation of insulin secretion by phosphoinositide-specific phospholipase C and protein kinase C activation. Diabetes Rev. 4, 160–176.

Zawalich, W.S., Zawalich, K.C., Ganesan, S., Calle, R. and Rasmussen, H. (1991). Influence of staurosporine on glucose-mediated and glucose-conditioned insulin secretion. Biochem. J. 279, 807–813.

Zhang, A., Gao, Z.-Y., Gilon, P., Nenquin, M., Drews, G. and Henquin, J.C. (1991). Vanadate stimulation of insulin release in normal mouse islets. J. Biol. Chem. 266, 21649–21656.

Zhou, J. and Egan, J.M. (1997). SNAP-25 is phosphorylated by glucose and GLP-1 in RIN 1046-38 cells. Biochem. Biophys. Res. Comm. 238, 297–300.

SECTION IV

CONCLUSION

FUTURE CHALLENGES IN β-CELL RESEARCH

Simon L. Howell

I. HISTORY

Research into the cell biology of the pancreatic β-cell was initiated by the identification of the islets of Langerhans (by Paul Langerhans) in 1869. Identification of the pancreas with diabetes mellitus is credited to Von Mering and Minkowski in 1889, and the role of islets of Langerhans in producing the diabetes-related pancreatic product, isletin (later renamed insulin) to Laguesse in 1893. Identification of insulin as the active principle in the pancreas permitted effective therapy for insulin-dependent diabetes (IDDM or Type 1 diabetes), for the first time in 1921

Advances in Molecular and Cell Biology
Volume 29, pages 337-346.

ISBN: 0-7623-0547-9

(Banting and Best), and since that time there have been two further major discoveries relating to the chemistry of insulin: first, determination of its primary amino acid sequence by Sanger in 1955, and second, its tertiary crystal structure by Hodgkin and collaborators (Blundell et al., 1972).

From the 1960s onward, advances in the understanding of the biology of the β-cell have often been based on technical advances, commonly those in biochemistry and related subjects. Thus the development of immunoassays for peptide hormones allowed accurate measurement of insulin concentrations in body and other fluids for the first time, while isolation of islets of Langerhans from rodent pancreas permitted detailed biochemical analyses of pancreatic β-cell function from the mid 1960s onward. Since that time, our understanding of the cellular physiology of rodent pancreatic β-cells and of a number of replicating β-cell lines has increased tremendously, as has been seen from the chapters in this book. It is also clear that attention is now shifting toward comparison of rodent with human pancreatic β-cells and increasingly toward defects of pancreatic β-cell function that may be seen in noninsulin-dependent diabetes (NIDDM), or Type 2 diabetes. From the perspective of the patient with diabetes, the single most important advance, aside from Banting and Best's work, was probably the discovery in the 1950s of the hypoglycemic sulphonylurea drugs, which were found to be effective in controlling blood glucose concentrations in patients with Type 2 diabetes. While these drugs have been used for many years, it was not until the late 1980s that the mechanism of their action on the pancreatic β-cell became clear.

Increasingly, both researchers and patients with diabetes will be looking to practical benefits from the research into β-cell biology, which has been pursued with increasing intensity during the past 30 years. Many of the major advances in this period have been described in the chapters of this book. This final chapter attempts a brief overview and to summarize some of the ways forward.

II. PANCREATIC β-CELL DEVELOPMENT AND REPLICATION

As shown by Jonathan Slack in the first chapter of this volume, the basic facts about the developmental biology of the pancreatic β-cell were laid out by classical histological studies in the 1960s. Much more recently, the application of the vast range of tools available to developmental biologists has substantially advanced this subject and the mechanisms regulating this process of development are becoming well understood.

An example of the benefits of the application of modern developmental biology of the pancreatic β-cell has been the attention which is now focused on PDX-1, which is expressed in the pancreatic buds during an early stage of development. In mature β-cells PDX-1 is among the specific transcription factors regulating insulin gene expression (see the chapter by Kevin Docherty in this volume) and, interestingly, it has been identified as one of the genes that may be involved in some cases of maturity onset diabetes of the young (MODY).

In the adult, it is clear that β-cell replication occurs only slowly, and in critically important situations (e.g., in Type 1 diabetes, or in animals in streptozotocin-induced diabetes) any residual pancreatic mass cannot, by itself, replicate to replace the stock of β-cells that has been destroyed. Studies of this process are hampered by the absence of good experimental models to study the initiation and regulation of replication. This area is important since we know from studies of circulating connecting peptide levels in recently diagnosed patients with Type 1 diabetes that there is a residual population of β-cells which could, in principle, replicate to produce the adequate β-cell mass if this process could be initiated and then regulated.

β-cell division by mitosis, of course, occurs readily in a variety of replicating cell lines and these are comprehensively reviewed in Shanta Persaud's contribution. Two features of these cell lines stand out. First, none of them reflect entirely the characteristics of normal β-cells of any mammalian species. Thus, the regulation of insulin secretion, particularly with respect to glucose, is consistently lower than in normal cells, while in many cases their insulin content is also substantially lower. The second feature that is missing is the availability of replicating human β-cell lines, despite repeated attempts to create such lines by a variety of means. Nevertheless, these cell lines have provided a very valuable resource to studies of the pancreatic β-cell, particularly for investigations in which relatively large quantities of tissue are essential; isolated islets of Langerhans from rodent pancreas produce at best a few milligrams wet weight of tissue in any one isolation. Cell lines are also essential for many of the molecular biological manipulations currently required for modern cell biology. We may therefore anticipate research in this area to focus most on the creation of human pancreatic β-cell lines that could potentially be used as a source of cells for β-cell transplantation in patients with diabetes, and in producing rodent cell lines that respond better to physiological stimuli, in particular, to produce glucose stimulation in a physiological range of 5 to 8 mmol/l of glucose.

III. PANCREATIC β-CELL DEATH

This is another field which has lain relatively uninvestigated until the last 10 years when, as pointed out by Décio Eizirik and Anne Hoorens in the third chapter, there has been an explosion of work, especially on the inflammatory and metabolic agents that can produce β-cell dysfunction and death. Particularly well investigated have been the possible role of cytokines in the induction of β-cell destruction in IDDM, along with the involvement of nitric oxide and superoxide and hydroxyl radical mechanisms for producing β-cell death. It has been shown that this is produced by induction of DNA strand breaks in the target cells, and interestingly, that this process can be inhibited at least to some extent by agents that activate the enzyme poly-(ADP ribose) polymerase, such as nicotinamide. The existence of these protective agents may explain why human pancreatic islets seem more resistant than rodent islets to damage induced by β-cell toxins such as cytokines, hydrogen peroxide, streptozotocin, or alloxan. Thus, human β-cells express more heat shock protein 70 (hsp70), catalase, and superoxide dismutase compared to their rodent counterparts. Further studies of these agents is clearly important in attempts to prevent induction of Type 1 diabetes by whatever cytotoxic agents may affect the cells.

The other aspect of the death of pancreatic β-cells that has recently been investigated extensively is programmed cell death or apoptosis, a genetically controlled and metabolically active mode of cell death that is present in many cell types. Apoptosis has been demonstrated in pancreatic β-cells and in at least some pancreatic β-cell lines (see the chapter by Persaud), although less convincingly in normal β-cells or in intact islets. Thus, apoptosis may supplement the classical process of cellular necrosis as a method of maintaining pancreatic β-cell mass well adapted to insulin demand. Equally in some models of Type 1 diabetes in mice, it may be that apoptosis is accelerated as a mechanism of β-cell destruction following an autoimmune attack by cytokines or cytotoxic T cells.

As shown eloquently in Michael Christie's chapter, the β-cell is the agent of its own destruction in the autoimmune process that leads to Type 1 diabetes. Among the proteins expressed in β-cells that could lead to their destruction are:

- FAS, the cell surface receptor which, following interaction with FAS ligand, can induce cell death by apoptosis.
- Adhesion molecules, which have been shown to be expressed on vascular endothelial cells in the region of islets, at least in animal models.
- MHC Class 1 molecules, which have long been known to be a common feature of the pancreas in Type 1 diabetes.

With regard to autoantigens, antibodies to glutamic acid decarboxylase have been shown to circulate in a vast majority of patients with Type 1 diabetes. There are only two major forms of glutamic acid decarboxylase, or GAD (GAD65 and GAD67), and which form is expressed appears to be species specific. GAD is involved in the synthesis of γ-amino butyric acid (GABA), and there have been repeated suggestions of the role of GABA in the regulation of glucagon secretion which might, through paracrine effects, also influence insulin secretion. At least two other autoantigens with similar molecular weights to GAD have been identified. These are protein tyrosine phosphatase (PTP-like proteins) designated IA2 and IA2β. Antibodies to IA2 have been detected in approximately 70% of new onset patients with Type 1 diabetes, but as in other studies of autoimmunity in diabetes, the role of IA2 in the nonobese diabetic (NOD) mouse and BB rat has been hard to demonstrate. Finally, insulin itself has been shown to be a major target of autoimmunity in Type 1 diabetes, and indeed, of the autoantigens listed above, it is the only one that is solely expressed in pancreatic β-cells. As elsewhere in this complex field, the results obtained in animal models of Type 1 diabetes are not readily translatable to the human situation, and there is to date no evidence of a single primary islet antigen that triggers autoimmunity in human Type 1 diabetes. Identification of such an autoantigen would have substantial implications for predicting and preventing the onset of the disease.

IV. SYNTHESIS AND SECRETION OF INSULIN

The regulation of insulin gene expression has been reviewed by Kevin Docherty in the fifth chapter of this volume. It is clear that, while the physical relationship of the insulin gene promoter with the gene itself is relatively straightforward, much work remains to be done in obtaining a detailed understanding of the various regulatory elements and their interactions. In part, this complexity relates to apparent differences in activity of promoters in isolation and in the context of surrounding promoter elements, but also in the differences between the normal pancreatic β-cells and insulin-secreting cell lines. This theme echoes that of Persaud's chapter, in which various characteristics of β-cell lines are summarized.

The two chapters dealing with cellular mechanisms of insulin synthesis and secretion, by John Hutton and colleagues and by Romano Regazzi, present an interesting snapshot of the state of development of this field. Insulin was one of the first hormones to be shown to be synthesized by a biosynthetic precursor, proinsulin (Steiner et al., 1967) and the discovery of proinsulin and the mechanisms of it in processing have, over 25 years, put the pancreatic β-cell at the forefront of research in prohormone processing. As indicated in the chapter by Hutton's group, research has been both very intensive, using an extensive range of molecular biological approaches, and extremely productive, with the enzymes involved in prohormone processing in the pancreatic β-cell being extensively characterized in both normal rodent β-cells and in animal models of Type 2 diabetes. Recently these studies have been extended to look at defects of prohormone processing in conditions in which high levels of circulating proinsulin are found in the blood from both genetic and nongenetic causes. This transfer of our detailed knowledge of cell biology in animal models to the patient and to the clinic, however, has not been completely straightforward and it is clear that some of the in vitro findings are not immediately applicable in physiological and pathological conditions. It seems safe to conclude at present that defects in prohormone processing do not provide a major cause of Type 2 diabetes except in rare cases. The future of this field seems likely to focus increasingly on the transfer of our detailed knowledge of cell biology and cell physiology to clinical situations.

In contrast, the study of exocytosis, which is a common mechanism for extrusion of hormones and other polypeptides from cells in many different cell types, has been relatively little explored. Indeed as Romano Regazzi shows in his chapter, much of the work in exploring the detail of molecular and cell biology of this process has been obtained from nonmammalian systems and only recently has the information been applied in detail to pancreatic β-cells. At this point it seems that many of the processes identified to regulate and undertake exocytosis in eukaryotic cells are likely to be applicable to the pancreatic β-cell, although much remains to be done to elucidate how the targeting and fusion processes of the granules with their adjacent plasma membrane are regulated. In view of the relatively sparse information about the detailed processes in β-cells, it is premature to speculate how these might become defective during Type 2 diabetes, although it is notable that in this condition the β-cell complement of insulin remains high, while its regulation, and possibly the mechanism, of its secretion are clearly defective.

Another major "black hole" remains in our understanding of the mechanisms by which the newly synthesized granules, which are present in large quantities and provide a reservoir of insulin within the β-cell, are transported to the plasma membrane to begin the docking and fusion process. It has been suggested for many years that elements of the cytoskeleton are involved in these processes. Again, much information is now accumulating in other secretory systems to elucidate the exact mechanism of organelle transport within the cells (both secretory vesicles and endogenous organelles). Application of knowledge gained in these systems to the β-cell remains an obvious and important area for research in the next few years.

V. REGULATION OF INSULIN SECRETION

At the start of a detailed review of the mechanisms of signal recognition, transduction, and translation into the exocytosis process, the physiologically important nonglucose stimuli and inhibitors of insulin secretion are summarized by Bo Ahrén. He points out that the structural organization of the islet of Langerhans itself provides a complex regulatory environment since the islets are innervated by autonomic nerves secreting a variety of neuropeptides as well as neurotransmitters, norepinephrine and acetylcholine. The heterogeneous cellular structure, comprising 70% β-cells, 25% α-cells and 5% δ-cells, also raises possibilities of complex autocrine and paracrine regulation within the islet itself. It is thus not surprising that a bewildering variety of peptides are able to inhibit and stimulate insulin secretion. While these effects are easy to demonstrate in a variety of experimental systems, it is much more difficult to evaluate their physiological and pathophysiological significance. An integrative physiological approach to this problem, while experimentally difficult, seems likely in the future to be more productive than reductionist studies of individual peptides and their mechanism of action.

In the chapter on glucose metabolism in the pancreatic β-cell and how glucose sensing is achieved, Emile van Schaftingen and Franz Schuit summarize the unique situation of the pancreatic β-cell in its assembly of glucose metabolism isozymes. As a result of this configuration, once phosphorylated, glucose is rapidly burnt in what he terms the "mitochondrial furnace." The particular specializations required to achieve this are nonrate-limiting glucose transport, low affinity glucokinase, and high activity of enzymes involved in the rest of glycolysis and subsequent mytochondrial metabolism. This system is exquisitely adapted to provide ATP/ADP ratios that are altered in relation to glucose abundance. The way in which this system copes with alternative nonglucose substrates and may be defective in various forms of diabetes, including MODY, is only just beginning to be explored.

An area of tremendously rapid advance in the past few years has been a quite complete understanding of the way in which the enzymatic processes of glucose metabolism in β-cells to produce ATP are eventually translated into opening of voltage-dependent calcium channels. In her chapter with Arun Rajan, Lydia Aguilar-Bryan sets out the detailed information that is now known about the structure and function of the ATP-sensitive potassium channels, and the way in which interaction with sulphonylurea receptors and channel activity may be regulated. Demonstrations of a direct link in defects of K^+-ATP channel function and abnormal insulin secretion in infants with persistent hyperinsulinemic hypoglycemia (PHHI) naturally led to speculation that a similar defect might occur in NIDDM. For the present, however, genetics studies using candidate gene approach in patients with Type 2 diabetes have not suggested the existence of any such defect.

Issues surrounding the role of calcium and of calcium influx in regulation of insulin secretion are dealt with by Jean-Claude Henquin and colleagues. They point out that, while the generation of a trigger signal by closure of K^+-ATP channels, depolarization, and stimulation of calcium entry is the primary mechanism for β-cell stimulation by glucose, other calcium-independent actions of glucose cannot be excluded. Equally, the mobilization of calcium from intracellular stores to contribute to intracellular free calcium concentrations in β-cells has been an important mechanism of response to some stimuli. The exact details of the intracellular localization of free calcium, as indicated by fluorescent calcium-sensitive compounds and confocal microscopy, and the significance of periodic oscillations of calcium concentration within the cells in the face of a consistent stimulus, are not yet fully understood. Advances in technology should enable better spacial and temporal resolution of these changes and their significance for overall regulation of insulin secretion. Equally, detailed study of the metabolism calcium in normal human β-cells and in patients with Type 2 diabetes is a potentially fruitful line of research.

In his chapter, Stewart Metz comprehensively reviews a rapidly growing field, encompassing the role of phospholipids and phospholipase as co-factors in the regulation of insulin secretion. It is now known that pancreatic β-cells contain a range of phospholipases (A_2, C, and D), which can be stimulated by arachidonic acid. Conclusions to date in this field are that phospholipases are probably not directly involved in mediating glucose-stimulated insulin secretion, but may have an important role in insulin release induced by certain receptor agonists (e.g., cholinergic agents). A classical inhibitor approach to evaluation of different enzymatic pathways and their role in insulin secretion has been bedevilled by problems of specificity of the inhibitors, and of the interacting pathways with which they may be involved. The techniques of molecular biology, including a feasible knockout of phospholipase enzymes and transient expression of particular PLA_2 isoforms in models of phospholipase deficiency may be required before we can evaluate its real role in regulation of insulin secretion.

Mechanisms by which the second messengers involved in mediating effects of glucose and also the range of other peptide stimuli inhibitors listed in Ahrén's chapter, are addressed in Peter Jones and colleagues' discussion of the role of protein kinases in stimulus transduction toward the physical process of secretion by exocytosis. They review the evidence that protein phosphorylation by a variety of protein kinases and their subtypes provide the vital linking step between signal transduction by the various methods discussed and the activation of exocytosis, including cytoskeletal and plasma membrane–associated effects. The roles of cyclic adenosine monophosphate (cAMP)–dependent protein kinase and of protein kinase C and its subtypes have been defined for many years as modulating glucose-induced insulin secretion and providing a direct pathway of implementation of effects of a variety of other stimuli that exert their effect by generating cyclic AMP in diacyl glycerol, there is little evidence that they are directly involved in glucose stimulated insulin secretion. Equally, there is a substantial body of evidence that CAM kinase 2 is particularly important in mediating effects of all stimuli which exert their effects through increases in intracellular calcium. The exact role of the more recently identified phosphorylation enzymes, protein tyrosine kinase and mitogen-activated protein kinase, and of the phosphoprotein phosphatases, is less clear. New tools are becoming available to cell biologists to address these questions in detail.

VI. IMPACT OF THE HUMAN GENOME PROJECT

In the past few years, however, advances in technology driven through the human genome project have meant that valuable clues as to important areas of β-cell physiology and of their potential defects have come from a completely new source, other than mainstream thinking in cell biology. The clearest and best example of this to date has been the impact of the identification of genes associated with MODY, a minor subtype of Type 2 diabetes, accounting for perhaps 3% of cases. This is a monogenic dominantly inherited disease, and genes responsible can, with luck, be identified by a study of individual large pedigree families. The second such MODY gene to be identified, glucokinase, fitted well with the β-cell physiologist's perception of how this field should go. A critical role of glucokinase as a glucose sensor had been well recognized by Matschinsky, Randle, Malaisse, and others in the early 1970s, and the fact that mutations in glucokinase could cause changes in insulin secretion did not come as a surprise. During 1996 and 1997, however, the identity of four other genes identified in different MODY families had become known, and here the outcome was far from predictable. Three of the genes regulate the development of liver cells, hepatic nuclear factors 4α, 1α, and 4β. None of these have previously been known to be expressed in β-cells, yet it seems likely that mutations in their function lead to defects in insulin secretion. This has opened up a whole new area of interest among β-cell developmental biologists, and for studies of the regulation of insulin secretion, and of gene expression. The fourth MODY gene is PDX-1. PDX-1 might have been predicted as a candidate for gene studies once it was realized that some of these genes could exert their effects through regulation of gene expression, and has already been identified as a potential factor in pancreatic β-cell differentiation.

Transgenic animal studies of these MODY genes indicate that the β-cells and the animals themselves can indeed develop, but that defects of insulin secretion occur as a result of some downstream effects of their expression in a mutated form in these particular animals. Unfortunately, there are substantial technical problems in illustrating the function of these genes in β-cell systems in vitro. While interpretation of data from transgenic animals may possibly be distorted by the fact that they are also expressed in a variety of other cell types, notably the liver. Of course it would be of particular interest to identify the genes involved in susceptibility of patients to Type 2 diabetes in the common forms that affect 95% of these patients. Here progress has been painfully slow. All the candidate gene studies in which our knowledge of detailed regulation of insulin secretion has been used to predict genes that might show mutations, have proved almost uniformly negative.

In the case of Type 2 diabetes it seems likely that, as in the case of MODY, some relevant genes may belong to a family that, at least to β-cell biologists, was completely unpredictable. However, this sort of approach will produce evidence and ideas for lines of investigation that are of direct relevance to the defects in Type 2 diabetes in a way that the careful "jigsaw" approach to our understanding of β-cell biology and its defects could not achieve. Equally, understanding and identification of the non-HLA genes, which are responsible for susceptibility to Type 1 diabetes may provide valuable clues, both as to how that susceptibility is conferred and how it may be converted to protection, eliminated, or its impact avoided.

VII. CONCLUSION

The small band of dedicated biologists whose efforts are summarized in this volume, and others throughout the world, will (funding permitting) continue to study the basic mechanisms of insulin β-cell production and death, and of the regulation, synthesis, and secretion of insulin. They will increasingly use human β-cells as their preparation of choice as differences between rodent and human β-cells become clearer. They will continue to undertake high quality scientific research at the forefront of cell biology and, in particular, endocrinology. Many of them, however, hope for something more: an impact on the lives of people with diabetes. From these lines of research, two outcomes seem to be achievable.

Firstly, it seems probable that engineered cell lines or replicating human β-cells will become available that can be used in transplantation as a treatment for Type 1 diabetes. There seems little doubt that this is the best option for replacement of destroyed β-cells, despite the risk that further implants of cells might also be destroyed by autoimmune processes or rejection.

Secondly, and probably through interaction of the field of human molecular genetics and β-cell biology, it seems likely that new insights and ultimately new treatments of Type 2 diabetes may emerge from understanding of the specific biological defects that arise from specific mutations. They could, in principle, lead to the prevention of Type 2 diabetes by pharmacological intervention. In the nearer future, it should be possible to target specific therapies to patients whose subtype of diabetes is based on genetic diagnosis—an individualized therapy (or preventative strategy based on individual risk assessment) for each patient.

Such outcomes, a prevention or effective treatment by transplantation of Type 1 (IDDM) and the prevention of Type 2 (NIDDM) would have an impact on these diseases, which are rapidly increasing in incidence in all societies of the world. Such outcomes are probably achievable, but their timescale is not predictable. Either would represent an achievement comparable to those of Minkowski or Banting and Best.

REFERENCES

Banting, F.G. and Best, C.H. (1922). Pancreatic Extracts. J. Lab. Clin. Med. 7, 464–472.

Blundell, T., Dodson, G., Hodgkin, D., and Mercola, D. (1972). Insulin: The structure in the crystal and its reflection in chemistry and biology. Adv. Protein Chem. 26, 279–402.

Brown, H., Sanger, F., and Kitai, R. (1955). The structure of pig and sheep insulin. Biochem. J. 60, 356–365.

Laguesse, E.G. (1893). Sur la formation des îlots de Langerhands dans le pancréas. Comptes Rendus des Séances et Mémoires de la Société de Biologie, Paris V, 819–820.

Langerhans, P. (1869). Beiträge zur mikroskopischen anatomie der bauchspeicheldrüse zur erlangung der doctorwürde. Inaugural Dissertation, Berlin.

Steiner, D.F., Cunningham, D., Spigelman, L., and Aten, B. (1967). Insulin Biosynthesis: Evidence of a precursor. Science 157, 697–700.

von Mering, J. and Minkowski, O. (1889). Diabetes mellitus nach pankreas exstirpation. Archiv für Experimentelle Pathologie und Pharmakologie 26, 372–387.

INDEX

www.ingramcontent.com/pod-product-compliance
Ingram Content Group UK Ltd.
Pitfield, Milton Keynes, MK11 3LW, UK
UKHW020234290926
15470UKWH00002B/541

9 780762 305476